Decision Engineering

Series Editor

Professor Rajkumar Roy
Department of Enterprise Integration
School of Industrial and Manufacturing Science
Cranfield University
Cranfield
Bedford
MK43 0AL
UK

Other titles published in this series

Cost Engineering in Practice
John McIlwraith

IPA – Concepts and Applications in Engineering
Jerzy Pokojski

Strategic Decision Making
Navneet Bhushan and Kanwal Rai

Product Lifecycle Management
John Stark

From Product Description to Cost: A Practical Approach
Volume 1: The Parametric Approach
Pierre Foussier

From Product Description to Cost: A Practical Approach
Volume 2: Building a Specific Model
Pierre Foussier

Decision-Making in Engineering Design
Yotaro Hatamura

Composite Systems Decisions
Mark Sh. Levin

Intelligent Decision-making Support Systems
Jatinder N.D. Gupta, Guisseppi A. Forgionne and Manuel Mora T.

Knowledge Acquisition in Practice
N.R. Milton

Global Product: Strategy, Product Lifecycle Management and the Billion Customer Question
John Stark

Enabling a Simulation Capability in the Organisation
Andrew Greasley

Network Models and Optimization
Mitsuo Gen, Runewei Cheng and Lin Lin

Management of Uncertainty
Gudela Grote

Xinjie Yu · Mitsuo Gen

Introduction to Evolutionary Algorithms

Springer

Xinjie Yu, PhD
Department of Electrical Engineering
Tsinghua University
100084 Beijing
China
yuxj@tsinghua.edu.cn

Mitsuo Gen, PhD
Fuzzy Logic Systems Institute (FLSI)
680-41 Kawazu
820-0067 Iizuka
Japan
gen@flsi.or.jp

ISBN 978-1-4471-2569-3 ISBN 978-1-84996-129-5(eBook)
DOI 10.1007/978-1-84996-129-5
Springer London Dordrecht Heidelberg New York

British Library Cataloguing in Publication Data
A catalogue record for this book is available from the British Library

Cover design: eStudioCalamar, Figueres/Berlin

Printed on acid-free paper

Springer is part of Springer Science+Business Media (www.springer.com)

To our families, friends, and students

Preface

This is a textbook on evolutionary algorithms (EAs). In preparing the proposal and the manuscript, the following questions were always kept in our minds.

- Is this book convenient for teaching, studying, and self-study, i.e., have the contents been arranged in a pedagogically sound way?
- Does this book introduce the state of the art of EAs?
- Does this book cover the contents as comprehensively as possible?
- Does this book contain specific programming codes so that the reader can use them directly?

For the first two questions, we would like to say yes. We want this textbook to be intuitive because intuitive ideas, and not strict proofs, lead to the kind of innovation we seek. We also want to build a bridge connecting the basics and the cutting edge so that our students can reach the peak quickly yet with a solid grasp of the material. Our answer to the remaining two questions is no. This textbook is neither an encyclopedia nor a cookbook. We selected only interesting and important topics and left the reader to implement the codes for the algorithms after we have deliberately removed all understanding obstacles. The reason for the latter consideration lies in the belief contained in the following saying: "Tell me and I'll forget; show me and I may remember; let me try, and I can understand."

EAs have attracted considerable interest in recent years. To understand how "hot" the topic is, consider the number of papers indexed by the Science Citation Index (SCI) every year. If the reader has the access to the SCI, he/she is encouraged to do a search to verify the hotness of this topic.

The reason the topic is so hot is mainly because of its effects on various problems, i.e., it really works. We will introduce various application examples to show how simple ideas can be expanded to solve complex problems.

The procedure of designing or analyzing an algorithm for solving optimization or learning problems is full of challenges, i.e., problems are difficult. This is irrelevant. We will accompany and assist the reader when necessary. We will demonstrate the basic ideas behind the scary equations and try our best to make things easy to understand. Incidentally, this cumbersome procedure is also interesting. Before and during the procedure of climbing Mount Fuji, we thought it would be torture. While at the summit, we felt that all the sufferings were interesting.

The prerequisites of this book lie in two aspects. In the mathematics part, a basic understanding of linear algebra, multivariable calculus, probability, and statistics is necessary.[1] The other demand is derived from a proficiency in programming. We assume a firm grasp of at least one programming language, i.e., you can implement an algorithm. Since these requirements are in the common curriculum of junior undergraduate programs, senior undergraduate or graduate students should be able to read this textbook without knowledge obstacles.

The pedagogical approaches used in this text can be summarized as follows:

- Building the mansion from the bricks. We will always focus on the key elements of an algorithm because later they might become your tools.
- From specific to general. We will always discuss specific simple examples before formal expressions.
- From idea to implementation. We will always introduce the initial notions before discussing the concrete contents.
- Explaining the critical part but leaving questions unanswered deliberately. We will leave some obstacles deliberately in the context to activate your thoughts.

R.W. Hamming[2] gave his motto as the following statement: "The purpose of computing is insight, not numbers" in his famous book *Numerical Methods for Scientists and Engineers*. Here we would like to use it again to represent our own attitude toward EAs. As algorithm designers, we care more about the solution landscape of the problem and the corresponding search ability of the algorithms,[3] although we do seek the optimal solution to the problem. From this perspective, there will be less "how-to" in this textbook for specific instructions. On the contrary, there will be plenty of "why-to"s to explain the insights and the rationale of a given algorithm so that one can generate one's own cookbook according to these "why-to"s. We believe that's what the reader wants!

Other teaching and self-study considerations include plenty of footnotes for mentioning easily ignored yet important points, "Suggestions for Further Reading" sections in most chapters for further study, exercises with various degrees of difficulty to deepen one's understanding and even promote one's own "small-scale" research.

The pedagogical relationship among the ten chapters of this textbook is illustrated by Fig. 0.1, where the dotted circles separate the ten chapters into three categories: the basics, optimization problems, and other branches. The solid lines are the suggested pedagogical path for teaching or self-study. Chapters 1 to 3 are all necessary for subsequent chapters. After that, readers can skip around depending on personal interests. Dashed lines represent the improvements, e.g., an understanding

[1] Knowledge of operations research, e.g., linear programming, nonlinear programming, and combinatorial optimization, might contribute to one's understanding but not be necessary.

[2] Father of the Hamming distance, which will be used often in this textbook.

[3] These terms will be explained explicitly and used throughout the book. Here they can be taken as the procedure to find the optimal solution.

of Chap. 4 might contribute to, but not be a prerequisite of, understanding Chap. 6 and *vice versa*. We hope Fig. 0.1 helps the reader form a mental map for different branches and applications of EAs.

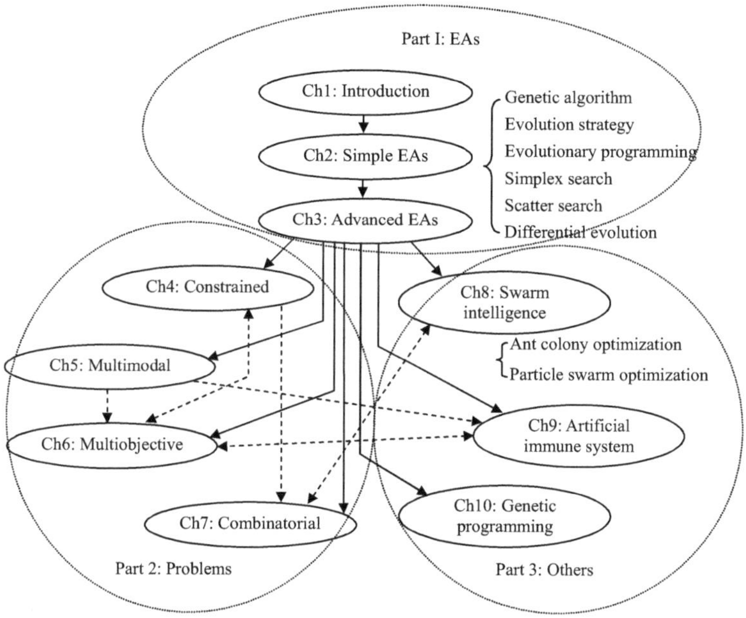

Fig. 0.1 The pedagogical relationship of the ten chapters in this textbook

This book is the result of the development of the course Evolutionary Computation and Its Applications given by the first author at Tsinghua University since 2002 and the numerous intensive courses taught by the second author around the world. Too many names require mention in the acknowledgments. But we wish to express our appreciation for our students first, especially those in the seminar Evolutionary Algorithms 2010. Without your critical comments during the seminar, we would have no appreciation of the reader's position, which is perhaps the most distinctive feature of this textbook. Many of old friends gave encourage and comments on the book, including Runwei Cheng, Baoding Liu, Kap Hwan Kim, etc. The fast response and the professional skill of Claire Protherough at Springer and Nadja Kroke at le-tex impressed us deeply. The final, but most important, acknowledgement belongs to Lin Wang and Eiko Gen because you are the backbones of our lives.

Beijing China,
Kitakyushu Japan,
May 2010

Xinjie Yu
Mitsuo Gen

Contents

Part I
Evolutionary Algorithms

Chapter 1
Introduction

Abstract Perhaps this is a required textbook for a course, perhaps you want to learn about evolutionary algorithms (EAs), or perhaps you just pick up this book occasionally. In this simple chapter, we will discuss the necessity, definition, original idea, branches, and information resources of EAs. We hope it will command your attention and stimulate you to read the other chapters.

1.1 What Are Evolutionary Algorithms Used For?

In the procedure of exploring the natural laws in science or fabricating useful products in engineering, we often face multiple problems. Two of them are the main focus of this textbook, i.e., the optimization problem and the learning problem.

For optimization, we want to find the optimal solution within the variables' definition domain. Provided that you have a basic grasp of multivariable calculus, for a differentiable scalar function $f(x,y,z)$ in a 3-D Cartesian coordinate system, its gradient is defined as $\nabla f = \left(\frac{\partial f}{\partial x}\mathbf{i}, \frac{\partial f}{\partial y}\mathbf{j}, \frac{\partial f}{\partial z}\mathbf{k}\right)$, where $\mathbf{i}$, $\mathbf{j}$, and $\mathbf{k}$ are unit vectors for three coordinates. The direction of ∇f is the fastest increasing direction of function f and the norm of ∇f is the extent of the fastness. Thus optimization problems seem easy. We can start from any initial point and use the gradient to guide the search until we find the point whose gradient is near the zero vector.

Unfortunately this is only applicable for some types of problems. For combinatorial optimization, there is no definition of differentiation. For multiobjective optimization, we face a vector function $\mathbf{f}$, and generally there is no point in the definition domain whose multiple objective values are all better than those of other points. For constrained optimization, searching along the gradient might converge to a solution that violates some constraints. Even in differentiable unconstrained optimization, gradients might not work. Let us take the Griewank function as an example [1]. The Griewank function can be scaled to n dimensions. Here we only discuss the two-variable version as follows:

$$\min f(x,y) = \frac{x^2+y^2}{4000} - \cos(x)\cos\left(\frac{y}{\sqrt{2}}\right) + 1 \tag{1.1}$$

It is quite easy to understand that the global optimal solution is $x = y = 0$.[1] We can draw the mesh graph for Eq. 1.1 with the definition domain $(-30 \leq x, y \leq 30)$, illustrated by Fig. 1.1.

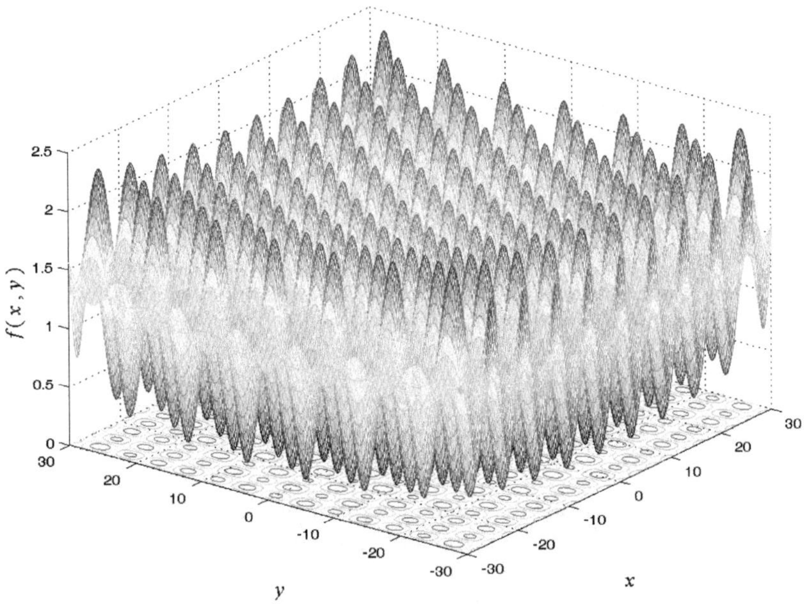

Fig. 1.1 The mesh graph of the 2-D Griewank function

This is bad! There are so many local optimal solutions (multimodal) that the gradient method with a randomly generated initial solution will converge to one of them with a large probability.

These four types of optimization problems, i.e., constrained optimization, multimodal optimization, multiobjective optimization, and combinatorial optimization, form the central core of optimization and thus the main topics in the field of *operations research* (OR). Researchers in OR have suggested many techniques for these problems. Among them, EAs have unique properties and have been attracting increasing attention.[2]

For learning, we want the computer to have human "intelligence" to some extent, i.e., *artificial intelligence*(AI). A simple example is illustrated by the scatter graph of the "chainlink dataset" in Fig. 1.2 [2].

---

[1] But computers do not know.

[2] Chapters 4 – 7 mainly discuss optimization problems.

The chainlink dataset has 1000 data and each of them belongs to one of the two groups. In Fig. 1.2, we use + to represent group 1 and ∘ to represent group 2. As human beings, we can differentiate these two groups easily even if they are all represented with the same symbol. If our algorithms can identify the two groups (with or without knowledge of the group number) only with the 3-D coordinates of each datum, i.e., elucidating which datum belongs to which group, we say the algorithm has the "intelligence" of *clustering*. This type of learning problem is called *unsupervised learning*, i.e., no teaching process in the clustering process.

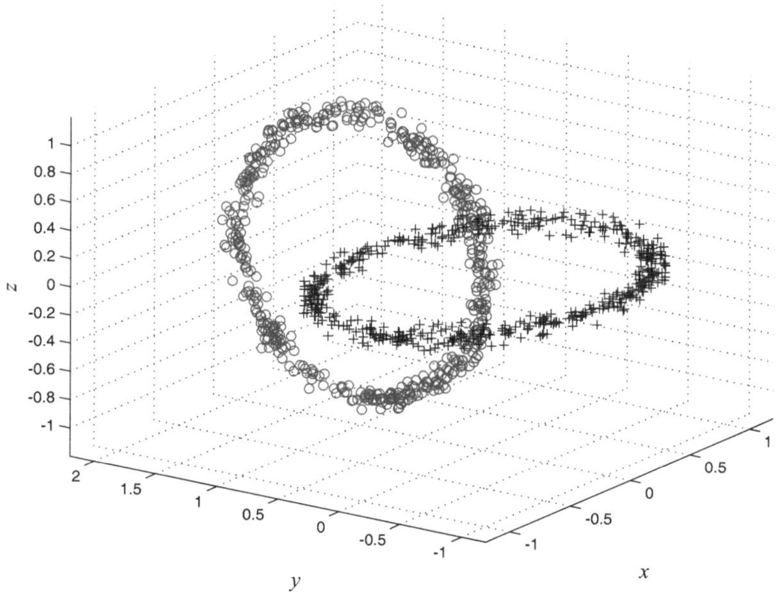

Fig. 1.2 The chainlink dataset

The chainlink dataset can also be used in another way. We randomly select 500 of them as the training set with their coordinates and group tags and use these 500 data to "train" our algorithm. After training, we use the other 500 data as the testing set with only coordinates to test whether our algorithm can classify them correctly. This type of learning problem is called *supervised learning*, i.e., a training process takes place before the testing process. The aforementioned learning problem is also called *classification* because its results are a discrete class number. If the tags of the data is not the class number but the real value, i.e., we consider the previous coordinates as the input of a system and the tags as the output of the system, this kind of leaning problem is called *regression*. Many learning algorithms have been proposed by researchers in AI. Among them are EAs.[3]

[3] Chapters 8 – 10 mainly discuss learning problems.

Optimization problems and learning problems are correlated with each other. Sometimes the function that needs to be optimized is so complex that we cannot afford to calculate the objective value for every solution. Then the learning algorithm could be used to do fitness approximation, i.e., "intelligently" generate a simpler function using previously evaluated solutions to replace the original function. On the other hand, for both clustering problems and regression problems, we could suggest a function to represent the clustering performance or the regression error. Then the goal of improving the cluster performance or decreasing the regression error would be equal to an optimization problem.

These two types of problems come from OR and AI respectively, and could be applied to many other science and engineering disciplines. The relationship between OR, AI, EAs, and other application disciplines is illustrated by Fig. 1.3

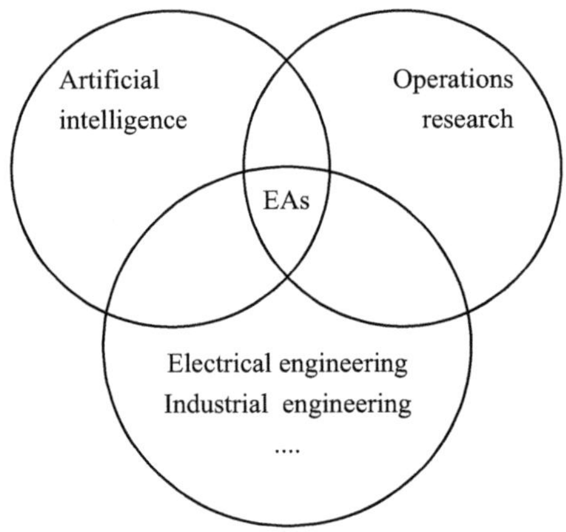

Fig. 1.3 The relationship between OR, AI, EAs, and other application disciplines

1.2 What Are Evolutionary Algorithms?

Many great inventions have been the result of bionics, i.e., the application of biological or natural principles to the study and design of human systems. We mimic bats to invent radar, fish to invent submarine, etc. The natural evolution of species could be looked at as a process of learning how to adapt to the environment and optimizing the fitness of the species. So we could mimic the viewpoint of modern genetics, i.e., "survival of the fittest" principle, in designing optimizing or learning algorithms. Enter EAs.

EAs[4] are algorithms that perform optimization or learning tasks with the ability to evolve. They have three main characteristics:

[4] EAs are also called *evolutionary computation* (EC) and *evolutionary intelligence* (EI).

- **Population-based.** EAs maintain a group of solutions, called a *population*, to optimize or learn the problem in a parallel way. The population is a basic principle of the evolutionary process.
- **Fitness-oriented.** Every solution in a population is called an *individual*. Every individual has its gene representation, called its *code*, and performance evaluation, called its *fitness value*. EAs prefer fitter individuals, which is the foundation of the optimization and convergence of the algorithms.
- **Variation-driven.** Individuals will undergo a number of variation operations to mimic genetic gene changes, which is fundamental to searching the solution space.

Since the 1960s, many algorithms with population-based, fitness-oriented, and variation-driven properties have been proposed. The timeline of the various EAs that will be introduced in this textbook is illustrated in Fig. 1.4.

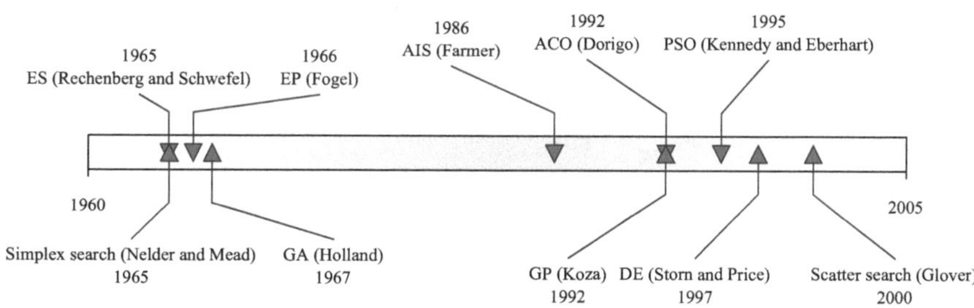

Fig. 1.4 Timeline of various EAs introduced in this textbook

Several correlated concepts need to be differentiated here.

Soft computing uses inexact solution methods for computationally hard tasks such as the solution of *NP-complete problems* [3–7].[5] It includes *expert systems*, *artificial neural networks*, *fuzzy systems*, and *EAs* conventionally, but more and more people are including *swarm intelligence* in soft computing. Soft computing puts more focus on "computing."

Computational intelligence is another, relatively new, related concept whose focus is more on "intelligence" [8–13]. Its three basic elements are also *artificial neural networks*, *fuzzy systems*, and *EAs*. Some researchers would like to include *rough set*, but others think *artificial immune system* should be included.

For hard problems, the exact algorithm that would guarantee the global optimal solution would be found within an acceptable time frame might not exist. Thus many *heuristic algorithms* have been proposed suggested by different researchers. Heuristic algorithms can generate a solution of acceptable quality relatively quickly. But no guarantee for optimality can be made and the time to generate solutions is also long in the worst case. Another notable property of heuristic algorithms is that they are problem dependent.

[5] This concept will be introduced in Sect. 7.1.2.

People use *metaheuristics* to represent those algorithms that can be widely used and generate high-quality solutions in general, i.e., metaheuristics is supposed to be at the higher level of heuristics [14–22]. Compared to the other two concepts explained above, metaheuristics focuses on "optimization." Its three elements are *EAs*, *tabu search*, and*simulated annealing*.

Figure 1.5 illustrates the number of papers indexed by the Science Citation Index (SCI) on EAs.[6] Perhaps several hundred papers per year is not impressive. But here we only use "broad" terms to do search. In the following chapters, we will see the results of specific algorithms. These data will be impressive.

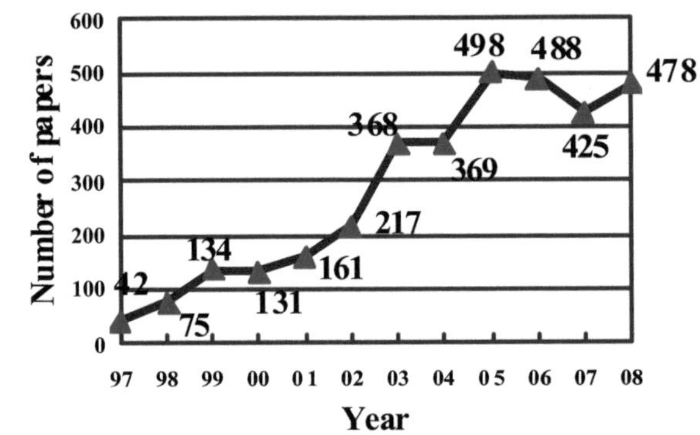

Fig. 1.5 Number of papers indexed by SCI on EAs

Suggestions for Further Reading

As has been illustrated in Sect. 1.1, EAs are interdisciplinary areas. Thus any information related to OR, AI, statistics, probability, etc. might relate to EAs. Here we only recommend several directly related conferences, journals, and books.

The two largest annual conferences on EAs are Genetic and Evolutionary Computation COnference (GECCO) and IEEE Congress on Evolutionary Computation (CEC). Apart from these conferences, the International Conference on Parallel Problem Solving from Nature (PPSN) (biannual), the European Conference on Evolutionary Computation, and various symposia organized by the IEEE Computational Intelligence Society are all conferences with considerable influence.

The top two journals on EAs are *IEEE Transactions on Evolutionary Computation* published by IEEE and *Evolutionary Computation* published by MIT Press.

---

[6] TS = ("evolutionary computation" OR "evolutionary computing" OR "evolutionary algorithms" OR "evolutionary intelligence"). The SCI index "TS" is for the search topic in the title, the keywords, and the abstract.

Other major journals are *IEEE Transactions on Systems, Man, and Cybernetics* published by IEEE; *Evolutionary Intelligence, Genetic Programming and Evolvable Machines, Journal of Heuristics, Soft Computing, Natural Computing* published by Springer; *Computers & Industrial Engineering, Computers & Operations Research,* and *European Journal of Operational Research* published by Elsevier.

Since the birth of EAs, many textbooks have been written to represent the cutting edge at that time. The notable books include those by Goldberg [23], Bäck [24], Michalewicz [25, 26], Eiben [27], Haupt [28], Burke [29], Sivanandam [30], and Sumathi [31], etc. The books by Mitchell [32], Fogel [33], and Spears [34] also provide insights from various perspectives. At the application level, Gen's three books [35–37] and those by Ashlock [38] and Yu [39] provide a comprehensive introduction.

References

1. Griewank AO (1981) Generalized descent for global optimization. J Optim Theory Appl 34(1):11–39
2. Ultsch A (2005) Clustering with SOM: U*C. In: Proceedings of the Workshop on Self-Organizing Maps, 75–82
3. Aliev RA, Aliev R (2001) Soft computing and its applications. World Scientific, Singapore
4. Tettamanzi A, Tomassini M (2001) Soft computing: integrating evolutionary, neural, and fuzzy systems. Springer, Berlin Heidelberg New York
5. Karray FO, Silva CWD (2004) Soft computing and intelligent systems design: theory, tools and applications. Addison-Wesley, Reading, MA
6. Pratihar DK (2007) Soft computing. Alpha Science, Oxford, UK
7. Maimon OZ, Rokach L (2008) Soft computing for knowledge discovery and data mining. Springer, Berlin Heidelberg New York
8. Konar A (2005) Computational intelligence: principles, techniques and applications. Springer, Berlin Heidelberg New York
9. Andina D (2007) Computational intelligence. Springer, Berlin Heidelberg New York
10. Eberhart RC, Shi Y (2007) Computational intelligence: concepts to implementations. Morgan Kaufmann, San Francisco
11. Engelbrecht AP (2007) Computational intelligence: an introduction, 2nd edn. Wiley, New York
12. John F, Jain LC (2008) Computational intelligence: a compendium. Springer, Berlin Heidelberg New York
13. Rutkowski L (2008) Computational intelligence: methods and techniques. Springer, Berlin Heidelberg New York
14. Resende M, de Sousa JP (2003) Metaheuristics: computer decision-making. Springer, Berlin Heidelberg New York
15. Glover FW, Kochenberger GA (2003) Handbook of metaheuristics. Springer
16. Gandibleux X, Sevaux M, Sörensen K *et al* (2004) Metaheuristics for multiobjective optimisation. Springer, Berlin Heidelberg New York
17. Rego C, Alidaee B (2005) Metaheuristic optimization via memory and evolution: tabu search and scatter search. Springer, Berlin Heidelberg New York
18. Dréo J, Pétrowski A, Siarry P *et al* (2005) Metaheuristics for hard optimization: methods and case studies. Springer, Berlin Heidelberg New York
19. Gonzalez TF (2007) Handbook of approximation algorithms and metaheuristics. Chapman and Hall/CRC, Boca Raton, FL

20. Siarry P, Michalewicz Z (2007) Advances in metaheuristics for hard optimization. Springer, Berlin Heidelberg New York
21. Blum C, Aguilera MJB, Roli A *et al* (2008) Hybrid metaheuristics: an emerging approach to optimization. Springer, Berlin Heidelberg New York
22. Talbi E (2009) Metaheuristics: from design to implementation. Wiley, New York
23. Goldberg DE (1989) Genetic algorithms in search, optimization, and machine learning. Addison-Wesley, Boston, MA
24. Bäck T (1996) Evolutionary algorithms in theory and practice: evolution strategies, evolutionary programming, genetic algorithms. Oxford University Press, Oxford, UK
25. Michalewicz Z (1998) Genetic algorithms + data structures = evolution programs. Springer, Berlin Heidelberg New York
26. Michalewicz Z, Fogel DB (2004) How to solve it: modern heuristics. Springer, Berlin Heidelberg New York
27. Eiben AE, Smith JE (2003) Introduction to evolutionary computing. Springer, Berlin Heidelberg New York
28. Haupt RL, Haupt SE, L R (2004) Practical genetic algorithms, 2nd edn. Wiley, New York
29. Burke EK, Kendall G (2006) Search methodologies: introductory tutorials in optimization and decision support techniques. Springer, Berlin Heidelberg New York
30. Sivanandam SN, Deepa SN (2007) Introduction to genetic algorithms. Springer, Berlin Heidelberg New York
31. Sumathi S, Hamsapriya T, Surekha P (2008) Evolutionary intelligence: an introduction to theory and applications with Matlab. Springer, Berlin Heidelberg New York
32. Mitchell M (1998) An introduction to genetic algorithms. MIT Press, Cambridge, MA
33. Fogel DB, Michalewicz Z (2001) An introduction to evolutionary computation. IEEE, Piscataway, NJ
34. Spears WM (2004) Evolutionary algorithms: the role of mutation and recombination. Springer, Berlin Heidelberg New York
35. Gen M, Cheng R (1997) Genetic algorithms and engineering design. Wiley-Interscience, New York
36. Gen M, Cheng R (1999) Genetic algorithms and engineering optimization. Wiley-Interscience, New York
37. Gen M, Cheng R, Lin L (2008) Network models and optimization: multiobjective genetic algorithm approach. Springer, Berlin Heidelberg New York
38. Ashlock D (2006) Evolutionary computation for modeling and optimization. Springer, Berlin Heidelberg New York
39. Yu T, Davis L, Baydar C *et al* (2008) Evolutionary computation in practice. Springer, Berlin Heidelberg New York

Chapter 2
Simple Evolutionary Algorithms

Abstract In Chap. 1, we drew a graph for EAs with the statement that EAs are interesting, useful, easy-to-understand, and hot research topics. Starting with Chap. 2, we will demonstrate EAs in a pedagogical way so that you can enjoy the journey around EAs with active reading. We strongly encourage readers to implement their basic EAs in this chapter in one programming environment and improve its search ability through other chapters. Footnotes, exercises, and possible research projects are of great value for an in-depth understanding of the essence of the algorithms.

2.1 Introductory Remarks

Before introducing some standard EAs,[1] we need to standardize some terms used throughout the book.

Classical EAs, including genetic algorithms (GAs), evolution strategy (ES), evolutionary programming (EP), and genetic programming (GP),[2] are all random-based solution space searching metaheuristic algorithms. So the most important thing before discussing a concrete algorithm is how to generate and manipulate random numbers in a programming environment.

We summarize the functions for handling random numbers in MATLAB®, C/C++, and Java in Table 2.1. As you can see, MATLAB® has advantages in generating random numbers in a flexible way. So we suggest using MATLAB® as the programming environment while learning EAs.

Even though MATLAB® has provided a Genetic Algorithm and Direct Search Toolbox, we strongly suggest that readers make their own EA source code from scratch if your purpose in reading this textbook is to really understand how EAs work and maybe improve some famous EAs to some degree.

---

[1] We sometimes call these standard evolutionary algorithms "simple" algorithms in the sense that they are simple compared to the improvements introduced in Chap. 3. But we need to mention that simple algorithms here do not mean weak performance.

[2] We will discuss the first three algorithms in this chapter and introduce GP in Chap. 10.

Table 2.1 The most common functions related to random numbers

	Distribution	MATLAB®	C/C++	Java
1	Uniform distribution $U(0,1)$	rand	(float)rand()/ RAND_MAX	Math.random
2	Normal distribution $N(0,1)$	randn		nextGaussian
3	Random permutation between 1 and integer n	randperm		
4	Round towards infinity	ceil	ceil	ceil
5	Round towards negative infinity	floor	floor	floor
6	Round towards nearest integer	round		

The density function of a uniform distribution random number in the range $(0,1)$, denoted as $\xi \sim U(0,1)$, is as follows:

$$p(\xi) = \begin{cases} 1 & 0 < \xi < 1 \\ 0 & \text{otherwise} \end{cases} \tag{2.1}$$

Every programming language provides a uniform distribution[3] random function. After generating $\xi \sim U(0,1)$, there are many ways to convert it into other distributions [1, 2]. The most important of these are normal distributions, denoted $\eta \sim N\left(\mu, \sigma^2\right)$, whose density function is as follows:

$$p(\eta) = \frac{1}{\sqrt{2\pi}\sigma} \mathrm{e}^{-\frac{(\eta-\mu)^2}{2\sigma^2}} \tag{2.2}$$

where μ is the expectance and $\sigma > 0$ is the standard deviation. It is easy to verify that if $\xi \sim N(0,1)$, then $\eta = (\xi \times \sigma + \mu) \sim N\left(\mu, \sigma^2\right)$. So $N(0,1)$ is very critical for generating normal distribution random numbers.

A permutation in the range $[1,n]$ is often used in evolutionary combinatorial optimization. Thus one should be familiar with how to generate permutations and random permutations.

Sometimes we need to generate an integer random number with uniform distribution in the range $[1,n]$, where n is an integer number. We can either combine the round function and $n * rand()$ or select the first cell in the random permutation in the range $[1,n]$.

In EAs, we often say that an operator (such as a crossover or mutation, discussed later) needs to be carried out with probability p. How does one implement such a simple statement in a given programming environment? We give the example in MATLAB ® as follows.[4] These are similar in other environments.

[3] Also called a Gaussian distribution.

[4] We strongly encourage readers to make it clear why this "If" statement could represent that the operator needs to be carried out with probability p.

```
%Operator A is carried out with probability p
    If rand < p
        Operator A
    End
```

where *rand* $\sim U(0,1)$.

With respect to *selection*, there are two ways to carry it out. We often pick up one solution randomly from current solutions and determine whether it could be selected. We will discuss the selection criteria in detail in later sections of this chapter and in Chap. 3. Here we would just like to determine how to handle the one being picked up. If it is put back, regardless of whether it is selected or not, and has the chance to be picked up again in the next time, we call this *selection with replacement*. If it is discarded, regardless of whether it is selected or not, and will never be picked up again, we call this *selection without replacement*.[5]

Another term that often appears in the EA literature is *norm*, which could be seen as some kind of length measure of vectors. $\|\mathbf{u}\|$ is vector **u**'s norm. Suppose **u** is a real vector with n variables. The general $p(\geq 1)$ norm for **u** is

$$\|\mathbf{u}\|_p = \left(|u_1|^p + \cdots + |u_n|^p\right)^{1/p} \tag{2.3}$$

where $|u_j|$ is the absolute value of u_j. If $p = 1$, then Eq. 2.3 is *1-norm*; it is the sum of the absolute values of all cells.

$$\|\mathbf{u}\|_1 = \left(|u_1| + \cdots + |u_n|\right) \tag{2.4}$$

If $p = 2$, *2-norm*, it is the Euclidean distance.

$$\|\mathbf{u}\|_2 = \sqrt{u_1^2 + \cdots + u_n^2} \tag{2.5}$$

If $p = \infty$, *∞-norm*.

$$\|\mathbf{u}\|_\infty = \max\left(|u_1|, \cdots, |u_n|\right) \tag{2.6}$$

The final concept is *convex function* and *concave function*. A function f is convex if any two points x_1 and x_2 in the definition domain satisfy Eq. 2.7 for any $0 \leq r \leq 1$, which is illustrated in Fig. 2.1.[6]

$$f\left(rx_1 + (1-r)x_2\right) \leq rf\left(x_1\right) + (1-r)\,f\left(x_2\right) \tag{2.7}$$

For points x_1 and x_2, $rx_1 + (1-r)x_2$ is their *convex combination*. Inverting the inequality defines a concave function.

---

[5] Sometimes we call this *sampling with replacement* or *sampling without replacement*.

[6] To call a curve convex or concave according to its geometric shape depends on where it is facing. In the definition, we look at the origin.

Fig. 2.1 A convex function

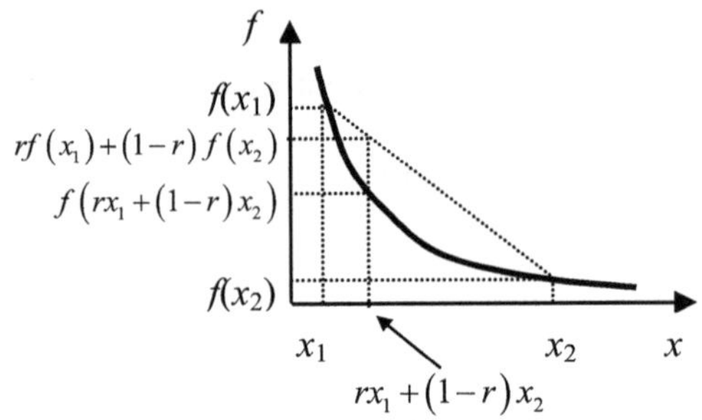

2.2 Simple Genetic Algorithm

2.2.1 An Optimization Problem

The basic ideas of GAs were introduced in Chap. 1. In this section, we'd like to introduce the details of implementing GAs with an optimization algorithm. The problem is formulated by Eq. 2.8 and illustrated by Fig. 2.2.

$$\begin{aligned} &\max f(x) = x\sin(10\pi x) + 2.0 \\ &\text{s.t.} \quad -1 \leq x \leq 2 \end{aligned} \tag{2.8}$$

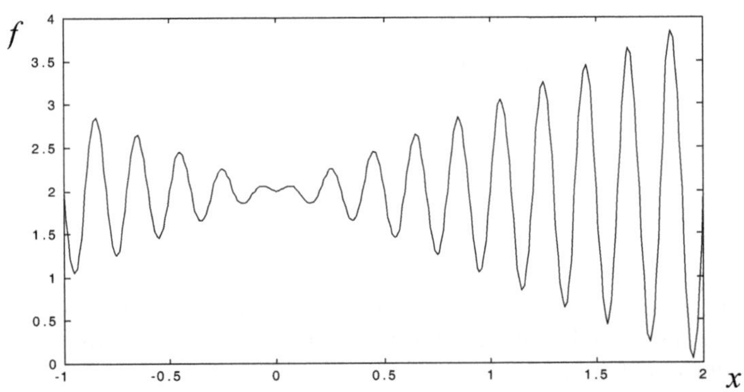

Fig. 2.2 A numerical example of an SGA

The curve of the objective function in Fig. 2.2 constitutes the *solution landscape* in which we are searching for the optimal solution. In this problem, it has many local optima, and, based on Fig. 2.2, the problem may seem difficult. We will demonstrate the strength of *simple genetic algorithm* (SGA) on this problem by the discussion in the following subsections.

In SGA, we maintain many *individuals* in a *population*. The number of individuals in the population is the *population size* (*popsize*). Each individual[7] has two properties: its location (*chromosome*[8] composed of *genes*) and its quality (*fitness value*). After obtaining the quality of all individuals, we use the selection process to generate a *mating pool*. Individuals with higher quality should have a higher probability of being selected into the mating pool so that the good ones will have more chances to breed and the bad ones will not be selected. Individuals in the mating pools are called *parents*. Generally two parents might be selected randomly from the mating pool to generate two *offspring*[9] without replacement and every offspring might undergo small changes to become a new individual. Then the newly generated population replaces the old one and another *generation* starts.

The relationship between the concepts and the operators described above and the principle of Darwinian natural selection theory is listed below.

- Selection $\Longleftrightarrow$ survival of the fittest
- Two parents generate two offspring $\Longleftrightarrow$ crossover or recombination
- Small changes in the location of the offspring $\Longleftrightarrow$ mutation

A good individual has more chances to be selected into the mating pool so that it has more chances to mate than low-quality individuals have. Then the information contained in the good individual has more chances to be preserved and passed onto next generation. Information exchange between two parents and small changes in the offspring promote the search for better individuals. Combining these two factors, the population will become more and more fit until the optimal or near optimal solution has been found, if we are lucky. In this way, GAs could gain favor over traditional gradient-based algorithms. We will discuss these operators step by step.

2.2.2 Representation and Evaluation

We can use a real number, in the range $[-1,2]$, to represent a solution in Eq. 2.8 directly. Many operators can handle real number representation. But we use the *binary code* or *binary representation* here for two reasons. GAs were originally proposed to be binary code to imitate the genetic encoding of natural organisms. On the other hand, binary code is good for pedagogy. A binary chromosome is necessary to represent a solution x in the scale $[-1,2]$. The same holds for the binary representation of real numbers in a computer.

In binary code, we cannot represent a real number completely correctly, so a tradeoff is necessary. A tolerance needs to be defined by the user, which means the errors below the tolerance are extraneous. If we divide the definition domain into

---

[7] An individual may be understood as an agent.

[8] Sometimes a chromosome is called a *genome*. These two terms have different meanings in genetics and biology, but we disregard them in EAs.

[9] Sometimes they are called *progeny* or *descendants*.

$2^1 = 2$ parts evenly and select the smallest number in the parts to represent any number in the division, we can only represent -1 and 0.5 by 0 and 1 respectively. $2^2 = 4$ divisions make the 00, 01, 10, and 11 represent -1, -0.25, 0.5, and 1.25, respectively. The larger division number we select, the less error there is in representing a real number on binary code. Suppose we use 100 binary codes to represent a real number in the range $[-1,2]$; the maximum error is $3/2^{100} \approx 2.37^{-30}$, which would be satisfactory for most users. In this way,[10] we can represent a real number with any accuracy requirements.

In this problem, we use $l = 12$ binary codes to represent one real number[11] as follows, which constitutes a chromosome to be evolved.

11	10	9	8	7	6	5	4	3	2	1	0
1	0	1	0	0	1	1	0	1	0	1	1

Fig. 2.3 Binary representation of a real number

For 12 binary codes in the chromosome, every part is called a *gene*. A gene has two properties: its value (sometime called *allele*), which is the number in the square, and its location (sometimes called *locus*), which is the number above the square. For the binary representation of a real number with l genes, its counterpart real number is

$$x = \frac{\sum_{i=0}^{l-1} a_i 2^i}{2^l} \times \left(\bar{x} - \underline{x}\right) + \underline{x} \tag{2.9}$$

where a_i is the allele of locus i, $\bar{x}$ and $\underline{x}$ are the upper and lower bounds for the real number, respectively. For the chromosome in Fig. 2.3, its counterpart real number is $x = \left(2^{11} + 2^9 + 2^6 + 2^5 + 2^3 + 2^1 + 2^0/2^{12}\right) \times (2+1) - 1 = 0.9534$. The process of finding one way to represent a solution to the problem is called *representation*. Figure 2.3 illustrates binary *encoding* and Eq. 2.9 represents binary *decoding*.

We can use both 101001101011 and 0.9534 to represent an individual (or a chromosome); the former is called a *genotype* representation of an individual and the lattera *phenotype* representation of an individual.

Every individual has two properties: its location and its quality. The location is the chromosome we described above and the quality is its objective value, which can be evaluated by Eq. 2.8. The objective value of our example individual is $f(x) = 0.9534 \times \sin(10\pi \times 0.9534) + 2.0 = 1.0520$. In GAs, we often use the *fitness value* to evaluate how much the individual fits the problem. So the function used to calculate the fitness value is called the *fitness function*, which is exactly the same as

[10] Even though this might weaken your faith in computers.

[11] What is the maximum error for this representation? Why do we only use $l = 12$? Why not $l = 300$?

the objective function in Eq. 2.8.[12] The process of obtaining a fitness value from a chromosome is called *evaluation*.

2.2.3 Initialization

SGA operates on a group of individuals. Generally the algorithm designers need to define how many individuals will be in the population, which is often represented by a variable *popsize*.[13] We need to generate *popsize* individuals to start the evolving process. This procedure is called *initialization*.

Sometimes we know which part of the definition domain contains better solutions. But in most situations, we do a blind search over the solution landscape.[14] So we could just initialize the population randomly. Another reason for random initialization is that SGA is capable of a global search, which will be illustrated later. In some complicated real-world problems, workable solutions may be in the local optimal solution's *domain of attraction*[15] so that starting from this domain may not be good for a global search. So why not depend on the global search ability of the SGA?

There are many ways to initialize *popsize* individuals randomly. The simplest way might be to generate every individual with uniform distribution. Specifically, for every gene in the chromosome, its allele is 1 with probability 0.5, and *vice versa*.[16]

Another way to generate evenly distributed individuals in the definition domain is to divide the domain into several grids. Initially, a grid is randomly selected and a solution in that grid is in turn randomly selected. We need the encoding procedure to transfer the phenotype to the genotype to get the chromosome.[17] We count the

---

[12] After reading the Sect. 2.2.4 below, consider why we could use the objective value as the fitness value directly. What is the fitness function in other situations, i.e., with a negative objective value, minimum optimization, etc.? Chap. 3 will discuss this interesting problem.

[13] Defining *popsize* without any *a priori* knowledge about the problem is a very hard job for algorithm designers. So either we need some kind of trial-and-error adjustment or we adopt some information from a current population and change *popsize* according to that information. We will discuss the latter interesting idea in Chap. 3.

[14] The results of other optimization algorithms, currently workable solutions, and uneven sampling on the definition domain according to the preference of users are examples of nonblind initialization.

[15] The domain of attraction means a subset of the definition area. For some search technologies, starting at any point in the domain of attraction will converge to the optimum in that domain even if it is a local optimum. Consider any $x \in [-0.9, -0.8]$ in Fig. 2.2; it will converge to the local optimum $f = 2.8$ with any up-hill algorithm.

[16] Readers unfamiliar with this should review Sect. 2.1.

[17] In our binary representation example, readers may check for methods of transforming a real number into its binary code in the computer basis textbook.

number of individuals generated from a grid. The larger the number for a grid is, the less opportunity the grid will have to generate new ones.[18]

After the initial chromosomes have been generated, their fitness values are calculated using a decoding process (Eq. 2.9) and the objective function (Eq. 2.8). It bears mentioning again that an individual is comprised of a chromosome and a fitness value.

For the problem illustrated by Eq. 2.8 and Fig. 2.2, we set *popsize* = 10 and obtain the ten randomly generated initial individuals, illustrated by Fig. 2.4.

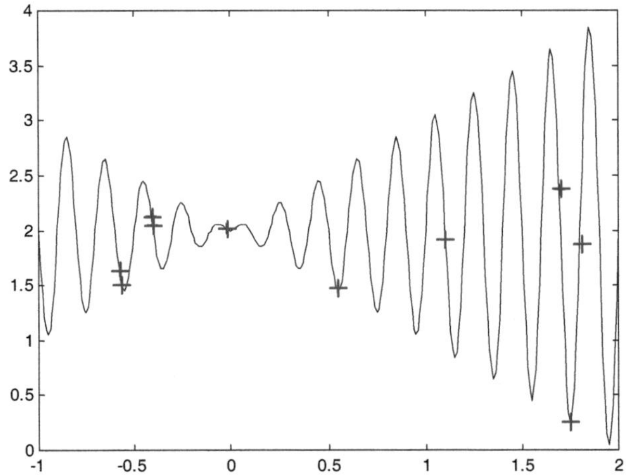

Fig. 2.4 Initial population of an SGA

The crosses in Fig. 2.4 are the initial individuals. As can be seen, the performance of the initial population is not impressive at all.

2.2.4 Selection

After initialization, the SGA enters the main loop. It starts with a *selection process*, which imitates natural selection by granting fitter individuals higher opportunity to breed, and ends with two *variation operators*,[19] crossover and mutation, which imitate natural reproduction by exchanging genes of parents to generate offspring.

In programming, we need to open another memory to reserve the individuals selected to breed. This memory is called the *mating pool*.

[18] Consider how to implement this idea in your program.

[19] These operators are also called *reproductive operators*.

There are many ways to embody the idea of natural selection. We will discuss one here and introduce others in Chap. 3. Individual i[20] in the current population has a fitness value f_i . According to natural selection, fitter individuals have more advantages in breeding. So we could define the *relative fitness value* of individual i as follows:

$$p_i = \frac{f_i}{\sum_{i=1}^{popsize} f_i} \tag{2.10}$$

It is easy to verify that $\sum_{i=1}^{popsize} p_i = 1$. Thus the relative fitness value can also be seen as the *probability of being selected* for individual i, i.e., p_i could be seen as the probability of being selected as one candidate in the mating pool for individual i.[21]

How can we implement this in a programming environment? We can do the selection as in roulette way.[22] The difference between real roulette and our selection is that the holes of the roulette wheel in our selection have different sizes. The larger p_i is, the larger hole individual i has. So the ball could drop into the hole with higher probability. The size of the hole for i is propositional to p_i, which is illustrated by Fig. 2.5. Instead of digging a hole on the wheel, we use sectors with different central angles to represent individuals. The central angle of individual i is $2\pi p_i$. The thick arrow in Fig. 2.5 represents the ball. We want the arrow to start rotating clockwise and stop randomly with a central angle $2\pi \cdot rand$, where $rand \sim U(0,1)$. If the arrow stops at one sector, the corresponding individual is selected into the mating pool. It is clear that fitter individuals have more chances to be selected.

How is the roulette wheel implemented in a programming environment? We could simulate the rotation process by an accumulated process. After obtaining *rand*, we know where the arrow will be. We memorize the individual we have already passed during the rotation process, accumulate the probabilities, and compare it to *rand*. Whenever we find an individual that is satisfied that $\sum_{i=1}^{k} p_i < rand < \sum_{i=1}^{k+1} p_i$, we know that the arrow stops in sector i[23] and select individual i into the mating pool. Then the arrow returns to the original location, just like in Fig. 2.5. We can do the above procedure *popsize* times until there are *popsize* individuals in the mating pool. This selection procedure is called *roulette wheel selection* (RWS).

In this way, some individuals in the population will be selected more than once and some will never be selected. The probability of being selected for individual i is its relative fitness. It is also necessary to mention that fitter individuals are not to be selected by RWS if they are unlucky enough never to have the arrow stop at their sector *popsize* times. We call this phenomenon *selection bias*. Many studies have

---

[20] Sometimes we also use ind_i to represent individual i.

[21] This sentence has the implicated meaning that we want to select the candidates in the mating pool in a serial way. We will introduce a parallel way in Chap. 3.

[22] Roulette is a gambling game in which a ball is dropped onto a wheel with numbered holes in it while the wheel is spinning round.

[23] How can we make such as statement?

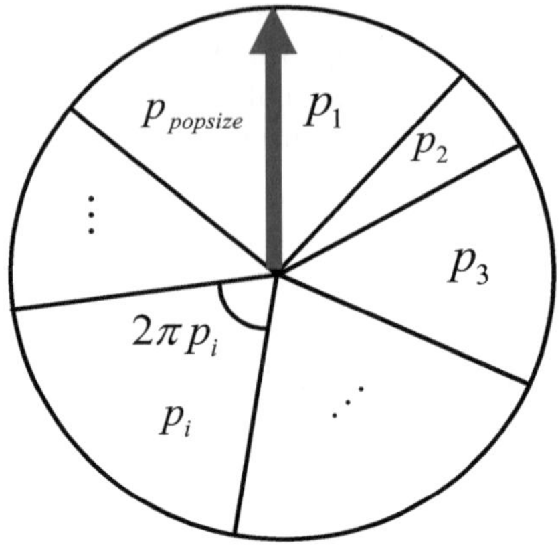

Fig. 2.5 Roulette wheel selection

been done to minimize the selection bias, and we will introduce some of them in later chapters.

Another consideration about RWS is that the problem needs to be maximum and all the objective values need to be greater than zero so that we can use objective values as fitness values and take RWS as the selection process directly.[24]

2.2.5 Variation Operators

There are many *variation operators* to change information in individuals in the mating pool. If information exchange, i.e., gene exchange, is done between two or more individuals[25], this variation operator is called *crossover* or *recombination*. If the genes of one individual changes on its own, this variation operator is called *mutation*. We will introduce single-point crossover and bit-flip mutation here.

There are two ways to select two individuals in the mating pool to determine whether or not to cross over them. One is to shuffle the mating pool randomly and assign individuals 1 and 2 without replacement to be a crossover pair, 3 and 4 without replacement to be another pair, etc. The other is to generate a random integer permutation, *per*, between $[1, popsize]$. $per(i) = j$ means the ith element in the permutation is the jth individual in the mating pool. Then we assign $ind_{per(1)}$ and $ind_{per(2)}$ without replacement as the first crossover pair, $ind_{per(3)}$ and $ind_{per(4)}$ without replacement as the second crossover pair, etc.

[24] Why do we need two such requirements for RWS to handle objective values directly?

[25] We will give examples of multiparent crossover in Chap. 3.

Generally, we will assign the probability of crossover p_c, called the *crossover rate*, to control the possibility of performing a crossover.[26]

For two individuals selected to cross over, we assign a point between 1 and $l-1$ randomly, where l is the length of the chromosome. This means generating a random integer in the range $[1, l-1]$. The genes after the point are changed between parents and the resulting chromosomes are offspring. We call this operator a *single-point crossover*. Figure 2.6 illustrates this.

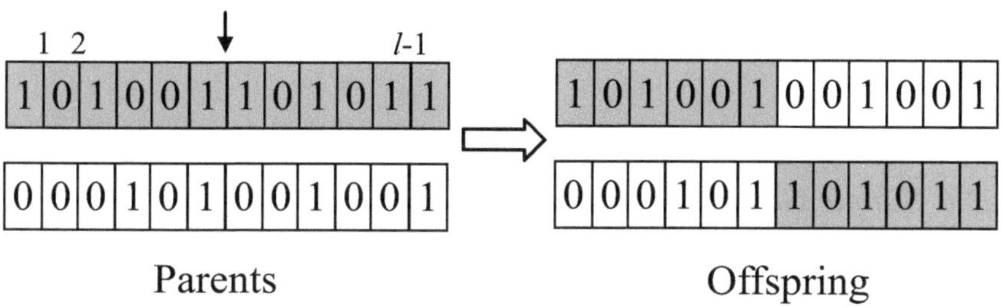

Fig. 2.6 Single-point crossover

As can be seen from Fig. 2.6, two new individuals are generated by crossover, which is generally seen as the major exploration mechanism of SGA.

If two parents do not perform a crossover according to probability p_c, their offspring are themselves.

Now we discuss mutation. There are also two ways to implement mutation. One way is to open another memory with size *popsize* to store the results of crossover, and mutation is carried out in that memory. The other way is to mutate the offspring of crossover directly. We use the latter way.

For every gene in an individual, we mutate it with probability p_m, called the *mutation rate*.[27] Provided gene j needs to be mutated, we make a bit-flip change for gene j, i.e., 1 to 0 or 0 to 1. We call this operator a *bit-flip mutation*. Figure 2.7 illustrates the bit-flip mutation. The individual after mutation is called the *mutant*.

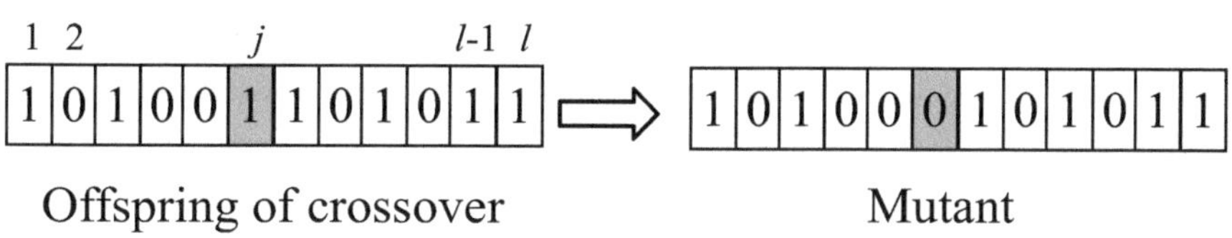

Fig. 2.7 Bit-flip mutation for gene j of the offspring

---

[26] How do we implement the statement "Individual i and individual j cross over with probability p_c"?

[27] How do we implement the statement "Gene j mutates with probability p_m"?

As can be seen from Fig. 2.7, small changes are introduced into a chromosome by bit-flip mutation,[28] which could be seen as one local search method and is generally considered a minor exploring mechanism of SGA. If every gene of an offspring does not mutate according to probability p_m, the mutant is the offspring itself.

After the variation operators, a new population with *popsize* individuals is generated. We need to evaluate the fitness for every individual and then replace the old population with the new one.[29] The process of replacing the current population with the new population is called *replacement.*

Although we call the results of crossover offspring and the results of mutation mutants, offspring is also used to represent the results after performing all variable operators.

2.2.6 Simple Genetic Algorithm Infrastructure

The selection, crossover, mutation, and replacement discussed above constitute one *generation* or *iteration* of an SGA. Then the evolving process continues to the next generation. As mentioned above, selection grants fitter individuals greater odds opportunity of propagating their high-quality genes, and crossover and mutation explore the solution landscape. We could have a rational expectation that the population will become better and better. But when do we stop? Here we just assign the maximum generation number *maxgen*.[30] If the generation number exceeds *maxgen*, the SGA stops and the individuals in the final population are the results.

So the infrastructure, *solution process*, of an SGA can be illustrated as follows:

Solution Process of SGA

Phase 1: Initialization.

Step 1.1: Assign the parameters for SGA, such as p_c, p_m, *popsize*, *maxgen*, etc.

Step 1.2: Generate *popsize* uniformly distributed individuals randomly to form the initial population and evaluate their fitness values. $gen = 0$.

Phase 2: Main loop. Repeat the following steps until $gen > maxgen$.

Step 2.1: Select *popsize* individuals from current population using RWS to generate the mating pool.

---

[28] How large a perturbation will the bit-flip mutation introduce in a chromosome? Does something not seem right? We will discuss this question in Chap. 3.

[29] It seems cruel and unreasonable because perhaps we want to keep the good ones in the current population. How do we solve this problem? We will discuss it in Chap. 3. Sect. 2.3 also gives some hints.

[30] We will discuss other more flexible techniques to stop EAs in Chap. 3.

Step 2.2: Repeat the following operations until a new population with *popsize* individuals has been generated. Select two individuals from the mating pool randomly without replacement to perform single-point crossover with probability p_c, and perform bit-flip mutation for every gene of the offspring with probability p_m. Then insert the mutant into the new population.

Step 2.3: Evaluate the fitness value for every new individual in the new population.

Step 2.4: Replace the current population with the new population. $gen = gen + 1$.

Phase 3: Submitting the final *popsize* individuals as the results of the SGA.

For the problem described by Eq. 2.8 and Fig. 2.2, we use an SGA with $popsize = 10$, $maxgen = 10$, $p_c = 0.8$, and $p_m = 0.01$[31] and obtain the results illustrated in Fig. 2.8. The cross in Fig. 2.8 represents the final individuals.

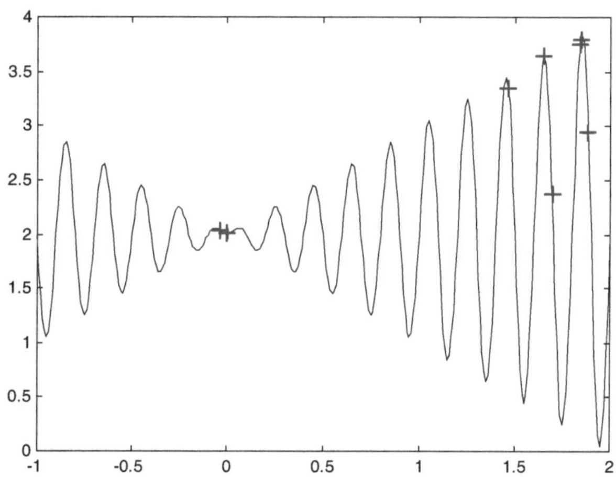

Fig. 2.8 Final population of SGA

As can be seen from comparing Figs. 2.2 and 2.8, an SGA can find a point very close to the global optimal solution in $10 \times 10 = 100$ samples over a solution landscape without any requirement of gradient information for such a not-so-easy problem, which could illustrate that SGA, with the help of selection, crossover, and mutation, is an effective search method.

To demonstrate the evolving process of SGA, we can draw a graph with the horizontal axis representing generation, *gen*, and the vertical axis representing the best fitness values in one generation, f_{best}. For our implementation, the graph is

[31] Why do we assign such a small p_m?

Fig. 2.9. The global optimal solution is $f(1.85) = 3.85$ and the SGA finds 3.8 at generation 9.

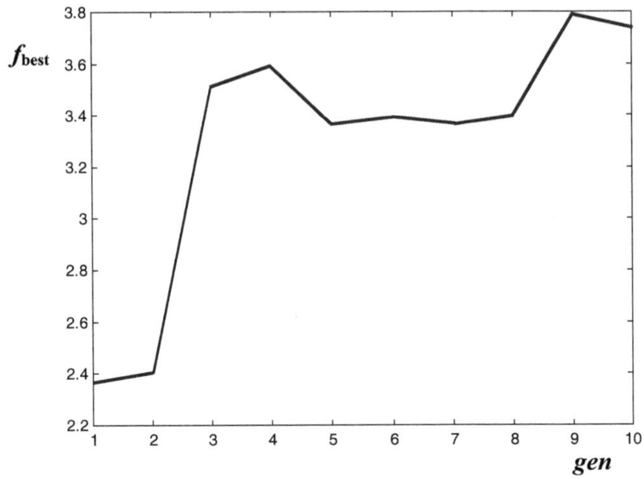

Fig. 2.9 The evolving process of the best fitness value

As can be seen from Fig. 2.9, the randomly generated initial best fitness value is close to 2.4 and it evolves to about 3.8 with only nine generations, which could demonstrate the global search and local fine-tuning ability of the SGA. It is necessary to mention two considerations for Fig. 2.9. The first thing is that connecting the best values of different generations with direct lines is meaningless because the points on the lines do not have any meaning. But we'd like to use this form to emphasize that EAs are evolutionary processes, even though they are implemented in a discrete environment. The second thing is that after implementing and running an SGA one time, drawing the results as in Fig. 2.8 and the evolving process as in Fig. 2.9 is far from thoroughly evaluating the global and local search ability of an algorithm. We will discuss this important issue, statistical performance evaluation, in Chap. 3.

Figure 2.10 illustrates the number of papers indexed by the SCI on GAs).[32]

From Fig. 2.10 we can say that GA is becoming more and popular in recent days.

---

[32] TS = ("genetic algorithm" OR "genetic algorithms"). The SCI index "TS" is for the search topic in the title, the keywords, and the abstract.

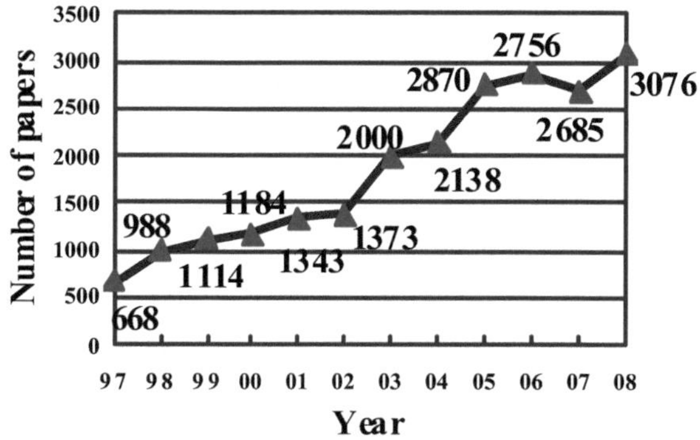

Fig. 2.10 Number of papers on GAs indexed by SCI

2.3 Evolution Strategy and Evolutionary Programming

2.3.1 Evolution Strategy

In designing an ES, Rechenberg and Schwefel use real numbers to represent alleles of genes and make normal distribution mutation the most important exploration technique over a solution landscape.

Let us take the simple example described by Eq. 2.8 and Fig. 2.2 again to illustrate the procedure of ES.

Suppose we have μ individuals in the current population. For every individual,[33] its chromosome is x, which is a real number in the range $[-1,2]$. We first randomly select two of them, x_1 and x_2, to do the crossover and generate one offspring x as

$$x = \frac{x_1 + x_2}{2} \tag{2.11}$$

This crossover operator is called an *intermediate crossover*.[34] There are other options for crossover operators in ES. Because crossover is considered a minor exploring tool in ES by Rechenberg and Schwefel, we do not discuss it here. Real code crossover operators will be discussed in Chap. 3.

We complete the crossover λ times to generate λ offspring (λ is often larger than μ[35]). For every offspring, we want to give a normal distribution disturbance on every variable (the example has only one variable).

For a normal distribution $N\left(\xi,\sigma^2\right)$ with mean ξ and standard deviation σ, it can be generated by $N\left(\xi,\sigma^2\right) = \xi + \sigma N(0,1)$, where $N(0,1)$ is a standard normally distributed random number with mean 0 and standard deviation 1. In MATLAB®,

[33] Without any preference for fitter individuals.

[34] Why?

[35] So in the previous crossover, we need to sample parents with replacement.

the *randn* function can generate $N(0,1)$ directly. For other programming environments, readers should refer to textbooks on probability for generating normally distributed random numbers from uniformly distributed random numbers [1].

In ES, we often want to make changes for every variable based on its current value. So we only use σ to represent the scale of mutation in ES.[36] For the ith variable x_i of one offspring, execute

$$x_i^{'} = x_i + N_i\left(0, \sigma^2\right) = x_i + \sigma N_i(0,1) \tag{2.12}$$

where $x_i^{'}$ is the mutant of x_i.[37] By using Eq. 2.12 for every gene of every offspring, we can get new individuals. This mutation operator is called a *normal mutation*. Then their fitness values can be calculated using the objective function. The standard deviation σ might be the same for all variables and it also might be different for every variable, depending on the algorithm's designer.

Then we combine μ current individuals and λ new individuals and then pick the μ best ones according to their fitness values to form new population. So the solution process of ES can be illustrated as follows:

Solution Process of $(\mu+\lambda)$*-ES*

Phase 1: Initialization.

Step 1.1: Assign the parameters for ES, such as λ, μ, and σ.

Step 1.2: Generate μ uniformly distributed individuals randomly to form the initial population and evaluate their fitness values. $gen = 0$.

Phase 2: Main loop. Repeat the following steps until $gen > maxgen$.

Step 2.1: Repeat the following operations until a new population with λ individuals has been generated. Randomly select two individuals with replacement to perform crossover, such as intermediate crossover (Eq. 2.11). Then perform a mutation (Eq. 2.12) for every gene of the offspring. Then insert the mutant into the new population.

Step 2.2: Calculate the fitness value for every new individual in the new population.

Step 2.3: Combine μ current and λ new individuals and pick the μ best ones to form a new population. $gen = gen + 1$.

Phase 3: Submitting the final μ individuals as the results of ES.

ES differs considerably from SGA and we will leave it to the reader to discover the differences. Here we need to point out that difference in the replacement procedure. A new population will certainly replace the old one in SGA, but the new population

[36] What is the logical relationship between this sentence and the previous one?

[37] Why do Rechenberg and Schwefel choose a normal distribution instead of a uniform distribution? What is their initial design idea behind $N(0,1)$?

is selected from the union of λ new individuals and μ current individuals in ES. Obviously, the replacement mechanism in ES conserves the best individual. Apart from replacement, the coolest part of ES is perhaps its self-adaptive control of standard deviation σ , i.e., coding σ into the chromosomes. This will be discussed in Chap. 3.

In the above solution process, μ current individuals and λ new individuals are combined and the μ best ones form a new population. Thus we call it $(\mu+\lambda)$-ES. In another type, (μ, λ)-ES, μ current individuals are used to generate λ new individuals and the μ best ones among the λ new individuals form the new population.[38] There are also other types like $(1+1)$-ES, $(1+\lambda)$-ES, and $(1, \lambda)$-ES.

2.3.2 Evolutionary Programming

Fogel used EP to solve the learning problem and used a finite state machine to represent the chromosome, which causes some difficulties for solving optimization problems with EP. In the 1990s, many researchers developed EP into an optimization field and formed many types of EP. The most cited EP is listed as the following solution process, where real numbers are used to represent variables.

Solution Process of One Type of EP Implementation

Phase 1: Initialization.

Step 1.1: Assign the parameters for EP, such as λ, μ, and σ.

Step 1.2: Generate μ uniformly distributed individuals randomly to form the initial population and evaluate their fitness values. $gen = 0$.

Phase 2: Main loop. Repeat the following steps until $gen > maxgen$.

Step 2.1: Repeat the following operations until a new population with μ individuals has been generated. Perform a mutation (Eq. 2.12) for every gene of the individual to generate a new one.

Step 2.2: Calculate the fitness value for every new individual.

Step 2.3: Combine μ current and μ new individuals and pick the μ best ones to form a new population. $gen = gen + 1$.

Phase 3: Submitting the final μ individuals as the results of EP.

Thus the above listed implementation of EP can be regarded as a $(\mu+\mu)$-ES without crossover.

---

[38] SGA can be regarded as a $(popsize, popsize)$-ES if we only consider the replacement procedure. Here we would like to raise an interesting but important question: Is $(\mu+\lambda)$-ES always better than (μ, λ)-ES in optimization? The answer is no! In Chap. 3, we will introduce the dilemma of exploration and exploitation and discuss the famous No Free Lunch Theorem. This question may be revisited after reading these materials.

Figure 2.11 illustrates the number of papers indexed by the SCI on ES and EP.[39]

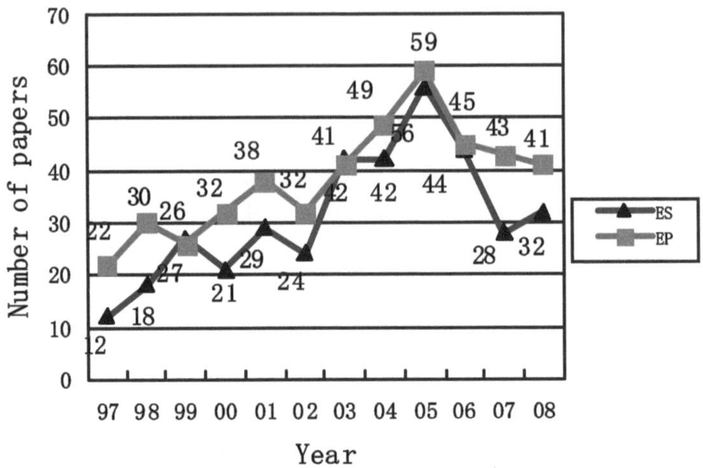

Fig. 2.11 Number of papers indexed by SCI on ES and EP

From Fig. 2.11 we can say that ES and EP attract a similar level of attention.

2.4 Direction-based Search

All the aforesaid algorithms are random-based, i.e., generating initial points, exploring new points, selecting better points, etc. In this section, we will first introduce a deterministic search method, simplex search, that can explore and exploit the solution space without the requirement for gradient information. Then two stochastic direction-based search methods, scatter search and differential evolution, are introduced. All of these algorithms use a specific direction, unlike SGA or ES discussed above, for generating new solutions. Thus we call them *direction-based search* methods.

2.4.1 Deterministic Direction-based Search

Methods that do not require gradient information to perform a search and sequentially explore the solution space are called *direct search methods*. There are many effective direct search methods, such as *simplex search*, *pattern search*, etc. All

[39] TS = ("evolution strategies") and TS = ("evolutionary programming"). The SCI index "TS" is for the search topic in the title, the keywords, and the abstract.

of them are based on the following philosophy. The methods maintain a group of points. They utilize some sort of deterministic exploration methods to search the objective space and almost always utilize a greedy method to update the maintained points.

Some of them use direction information, which might be the improvement directions, to search the objective space. Thus it might be very useful to embed these directions into one's evolutionary algorithm as either a local search method or an exploration operator.

2.4.1.1 Simplex Search

Nelder and Mead introduced the most famous deterministic direction-based search method, the *simplex search*, in 1965 [3]. Thus sometimes the simplex search is referred as the Nelder–Mead method. Do not confuse it with the simplex methods used in linear programming. But these algorithms use the same concept of simplex, which is a polytope with $n+1$ vertices in n dimensions: a line segment in one dimension, a triangle in 2-D, a tetrahedron in 3-D space, and so forth.[40]

In multidimensional spaces, the subtraction of two vectors means a new vector starting at one vector and ending at the other, like $(\mathbf{x}_2 - \mathbf{x}_1)$ in Fig. 2.12. Vectors in the space could be moved with their length and direction freely. So we often refer to the subtraction of two vectors as a direction.

The addition of two vectors can be implemented in a triangular way, moving the start of one vector to the end of the other to form an addition vector, like $\mathbf{x}_3 + (\mathbf{x}_2 - \mathbf{x}_1)$ in Fig. 2.12. We often refer to the addition of two vectors as a point.

The expression $\mathbf{x}_3 + (\mathbf{x}_2 - \mathbf{x}_1)$ can be regarded as the destination of a moving point that starts at $\mathbf{x}_3$ and has a length and direction of $(\mathbf{x}_2 - \mathbf{x}_1)$.

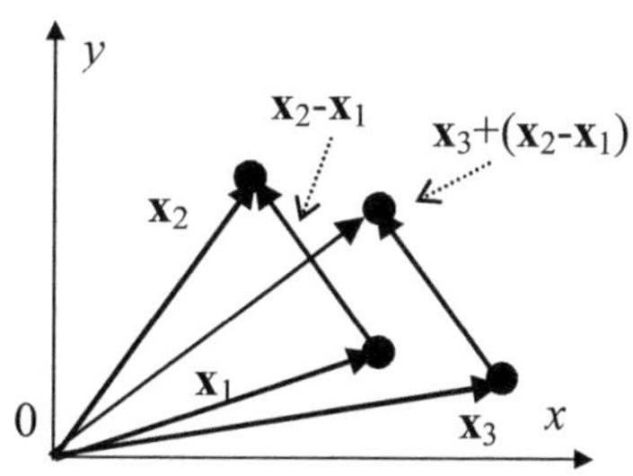

Fig. 2.12 Graphical meaning of the subtraction between two vectors

In a direct search, generally we cannot obtain the gradient information so that the idea of a step in the negative gradient direction for a minimum problem is impractical. But if the objectives of a group of solutions are available, we then know which one is the best one (suppose it is **B**). For any other solution (suppose it is

[40] A simplex can be considered the simplest set of points to make an effective search.

C), its improvement direction is unambiguously **C** → **B**.[41] This intuitive idea is the foundation of many powerful direction-based search methods and EAs.

Now let us discuss the simplex search in the 2-D minimum optimization situation. There are three maintained points that are the vertices of a triangle, each with an objective value. Let us rank and name them by Eq. 2.13, which means the first point (B) will *always* be the best one, the second point (G) will *always* be the middle one, and the third point (W) will *always* be the worst one.

$$f(\mathbf{x}_B) < f(\mathbf{x}_G) < f(\mathbf{x}_W) \tag{2.13}$$

Point B is the best one in the simplex and point W is the worst one, which means moving from W to B is a good search direction. Also moving from W to G is a good search direction. So why not combine these two considerations and move from W toward the centroid, gravity center, of B and G for a further step?

The gravity center of B and G is M. We think the direction from W to M is the optimizing direction.

$$\mathbf{x}_M = \frac{\mathbf{x}_G + \mathbf{x}_B}{2} \tag{2.14}$$

We start from point W and go in the good search direction (W→M) and extend a further step to point R, which satisfies

$$\mathbf{x}_R = \mathbf{x}_M + (\mathbf{x}_M - \mathbf{x}_W) \tag{2.15}$$

The search discussed above is called a *reflection* procedure. W is reflected with respect to M, which means a moving point starts at M and has a length and direction of $(\mathbf{x}_M - \mathbf{x}_W)$. Figure 2.13 illustrates the situation. Now B, G, and R constitute the new simplex.

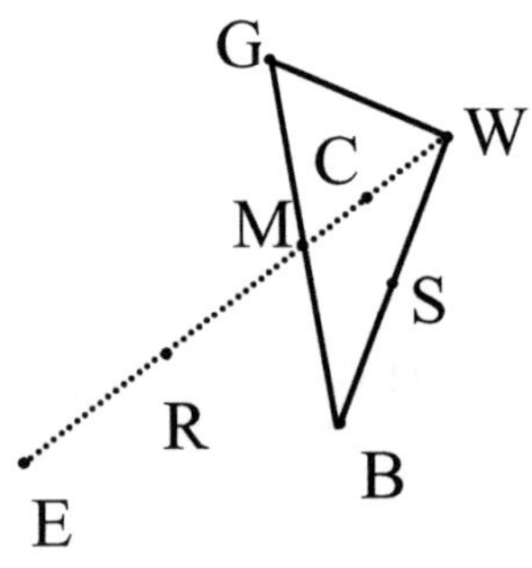

Fig. 2.13 Simplex search

If $f(\mathbf{x}_R) < f(\mathbf{x}_B)$, then the reflection improves the best points thus far and proves our guess that the direction (W→M) is good. So why not extend it more? This is

[41] Perhaps this direction is not the best or the fastest direction, but it is a workable one because **B** is better than **C**.

the *expansion* procedure, which means that we want to expand the simplex B-G-R to B-G-E. That is, we search point E by

$$\mathbf{x}_E = \mathbf{x}_M + 2(\mathbf{x}_M - \mathbf{x}_W) \quad (2.16)$$

If $f(\mathbf{x}_E) < f(\mathbf{x}_B)$, then we are fully satisfied and make E, B, and G the current simplex. If not, then the expansion procedure is not good enough, but we still have good point R. So R, B, and G constitute the current simplex.

If $f(\mathbf{x}_B) < f(\mathbf{x}_R) < f(\mathbf{x}_G)$, which means the reflection is acceptable. So B, R, G constitutes the current simplex.

If $f(\mathbf{x}_G) < f(\mathbf{x}_R)$, then the reflection does not find good points. But we still believe that the direction (W→M) is correct. The too-large step is the cause of the failure. So we shorten the step and make a *contraction* procedure, which means we want to contract the simplex B-G-R to B-G-C. That is, we search point C by

$$\mathbf{x}_C = \mathbf{x}_W + 0.5(\mathbf{x}_M - \mathbf{x}_W) \quad (2.17)$$

If $f(\mathbf{x}_C) < f(\mathbf{x}_W)$, then we accept the results and make B, G, and C the current simplex.

If $f(\mathbf{x}_W) < f(\mathbf{x}_C)$, this weakens our idea that the direction (W→M) is correct. We reject it and do a *shrink* procedure, which means we want to shrink the simplex based on point B. Then B, M, and S constitute the new simplex.

$$\mathbf{x}_S = \frac{\mathbf{x}_W + \mathbf{x}_B}{2} \quad (2.18)$$

For every new simplex, we need to assign B, G, W according to their objective values. Then the simplex search repeats reflection, expansion, contraction, and shrink again and again in a very efficient and deterministic way. Vertices of the simplex will move toward the optimal point (perhaps the local optimal solution) and the simplex will become smaller and smaller. Stop criteria could be made based on the time of function evaluation, the length of the edge, the improving rate of B, etc.

Dennis and Torczon modified the standard simplex search by a different reflection procedure [4]. Based on that, MATLAB® contains a direct search toolbox [5].

The simplex search is a group-based deterministic search method capable of exploring the objective space very fast, but sometimes becoming trapped in the local optimal point. Thus many EAs use simplex as a local search method after mutation, which can combine the global search ability of EAs and the local search ability of the simplex search. In Chap. 3, we will discuss it again as part of memetic algorithms.

2.4.2 Random Direction-based Search

2.4.2.1 Scatter Search

Although the simplex search is effective, it is more like a technique than an algorithm when facing real-world complex problems. Glover, Laguna, and Marti include the elitism mechanism in the simplex search and suggest an ES-like algorithm: the *scatter search* [6, 7].

The basic idea of the scatter search is the same as that of the simplex search. Given a group of points, the algorithm somehow finds new points, accepts the better ones, and discards the worse ones.

The scatter search has four main steps, illustrated by Fig. 2.14. The *reference set* (RS) contains b "best" solutions, b_1 of which are good with respect to their objective value ($RefSet_1$) and b_2 of which are good with respect to diversity (far away from $RefSet_1$ points) ($RefSet_2$) ($b = b_1 + b_2$).

The initialization procedure of a scatter search randomly generates solutions in such a way that the more individuals are generated in one area, the less opportunity this area will have to generate new ones.[42] In this way, the initial solutions of the scatter search can maintain maximum diversity. After the initialization procedure, the scatter search makes use of the improvement procedure, the simplex search, to make the initial solutions better. After that, $RefSet_1$ is selected from the improvement results according to the objective quality, and $RefSet_2$ is selected from the improvement results according to the smallest distance to $RefSet_1$ of the remaining improved individuals (the larger the better). Then the main loop starts. We use RS to generate subsets. The solutions in the subsets are combined in various ways to get *Psize* new solutions. Then the solutions are improved by some local search approaches (such as simplex search) to become better solutions. Finally, the improved solutions will replace some solutions of RS if they are good with respect to objective quality or diversity. The main loop is illustrated in Fig. 2.14.

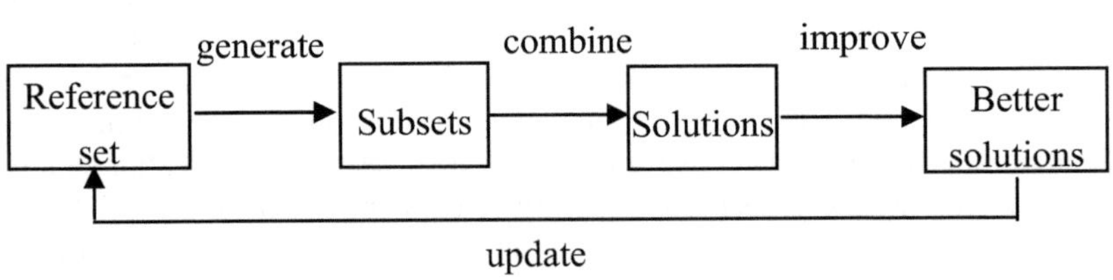

Fig. 2.14 Scatter search

There are four types of subsets to be generated in a scatter search:

1. All two-element subsets containing all pairwise combinations of the b reference set solutions.

---

[42] Readers are encouraged to reread the second method for the initialization of SGA.

2. Three-element subsets derived from two-element subsets by adding the best solution not in this subset (measured by objective value).
3. Four-element subsets derived from three-element subsets by adding the best solution not in this subset (measured by objective value).
4. Subsets containing the best i elements (measured by objective value), for $i = 5$ to b.

In this way, subsets regard the objective value as the most important factor but still contain the diversity factor.

There are many types of combinations for generating new solutions from subsets. Let us give an example for a two-element subset: $\mathbf{x}_1$ and $\mathbf{x}_2$. We can first find a vector starting at $\mathbf{x}_1$ and pointing to $\mathbf{x}_2$ as $\mathbf{d} = \frac{\mathbf{x}_2 - \mathbf{x}_1}{2}$(the length of the vector is half the distance between $\mathbf{x}_1$ and $\mathbf{x}_2$). Glover, Laguna, and Marti suggested three types of re-combination: (1) $\mathbf{x}_{\text{new}} = \mathbf{x}_1 - r\mathbf{d}$, (2) $\mathbf{x}_{\text{new}} = \mathbf{x}_1 + r\mathbf{d}$, and (3) $\mathbf{x}_{\text{new}} = \mathbf{x}_2 + r\mathbf{d}$,[43] where $r \sim U(0,1)$. Every subset can generate several new solutions according to the composition of the subset.[44]

- Both $\mathbf{x}_1$ and $\mathbf{x}_2$ belong to $RefSet_1$, which means that they are all good solutions. Four new solutions are generated by types 1 and 3 once and type 2 twice.
- Only one of $\mathbf{x}_1$ and $\mathbf{x}_2$ belongs to $RefSet_1$. Three new solutions are generated by types 1, 2, and 3 once.
- Neither $\mathbf{x}_1$ nor $\mathbf{x}_2$ belongs to $RefSet_1$, which means that they are all uncrowded solutions. Two new solutions are generated by type 2 once and type 1 or 3 once.

As has been stated, a simplex search is used by Glover, Laguna, and Marti to improve the new solutions.

If an improved solution's objective value is better than that of the worst one in $RefSet_1$, it will replace the worst one without replacement (delete it from the set of improved solutions). If an improved solution's distance to the closest RS solutions is larger than that of most crowded solutions in $RefSet_2$, it will replace the most crowded one without replacement.

If RS does not change in the updating procedure and the stop criterion has not been satisfied, then the initialization procedure will be started to construct a new $RefSet_2$.

Glover, Laguna, and Marti suggested that $Psize = \max(100, 5 * b)$. We can consider the scatter search as a special kind of $(b + Psize)$-ES. But in the updating (replacement) phase, the objective value is not the only criterion.

2.4.2.2 Differential Evolution

Differential evolution (DE) is also a kind of direction-based search, suggested by Storn and Price in 1997 [8, 9]. DE also maintains a population with Np individ-

[43] Readers may find out the geometric explanation for these formulae using Fig. 2.12.

[44] What is the intuitive idea of Glover, Laguna, and Marti for generating new solutions? This is the origin of the name "scatter."

uals and has mutation, crossover operators, and a selection process. So DE could be regarded as a real parameter coded version of GA.[45] We suppose there are n dimensions in the decision space so that individual i of DE can be expressed as

$$\mathbf{x}_i = (x_{i,1}, x_{i,2}, \cdots, x_{i,n}) \tag{2.19}$$

where $x_{i,j} \in R,\ i = 1,2,\cdots,Np,\ j = 1,2,\cdots,n$.

The main difference between DE, SGA, and ES lies in the mutation operator, where the direction is used to make the perturbation from the current individual.

Let us start with individual i. Unlike the random step size of mutation along each dimension of ES, DE uses the directional information from the current population. The standard mutation operator of DE needs three randomly selected different individuals from the current population ($\mathbf{r}_1 \neq \mathbf{r}_2 \neq \mathbf{r}_3 \neq \mathbf{x}_i$) for each individual to form a simplex like triangle. For $\mathbf{x}_i$, its mutants $\mathbf{v}_i$ is[46]

$$\mathbf{v}_i = \mathbf{r}_1 + F(\mathbf{r}_2 - \mathbf{r}_3) \tag{2.20}$$

where F is a positive real number (seldom larger than one) that controls the strength of the direction. Differential $(\mathbf{r}_2 - \mathbf{r}_3)$ forms a direction that is the origin of DE and the reason it is a direction-based search. Equation 2.20 means the mutant of $\mathbf{x}_i$ starts at $\mathbf{r}_1$ and has direction and length of $F(\mathbf{r}_2 - \mathbf{r}_3)$.

After mutation, DE utilizes a uniform crossover operator[47] to combine the information of the parents $\mathbf{x}_i$ and $\mathbf{v}_i$ into the offspring $\mathbf{u}_i$. We need to have at least one variable of $\mathbf{v}_i$, so a random integer number j_{rand} in the range $[1,n]$ should be generated before crossover. The offspring $\mathbf{u}_i$ is generated by Eq. 2.21:

$$u_{i,j} = \begin{cases} v_{i,j} & rand \leq Cr \text{ or } j = j_{\text{rand}} \\ x_{i,j} & \text{otherwise} \end{cases} \tag{2.21}$$

where $Cr \in [0,1]$ is a user-defined parameter controlling the effect of crossover, $rand \sim U(0,1)$. Equation 2.21 means that the offspring $\mathbf{u}_i$ has the possibility of Cr to select variables from mutant $\mathbf{v}_i$ and ensures at least the j_{rand}th variable will be picked from $\mathbf{v}_i$.

Then $\mathbf{u}_i$, called a *trial vector*, competes with $\mathbf{x}_i$, called a *target vector*, for survival in the next generation in a steady way,[48] which means

$$\mathbf{x}_i(t+1) = \begin{cases} \mathbf{u}_i(t) & f(\mathbf{u}_i(t)) \leq f(\mathbf{x}_i(t)) \\ \mathbf{x}_i(t) & \text{otherwise} \end{cases} \tag{2.22}$$

Mutation, crossover, and selection will be carried out for Np individuals in one generation. The initialization and the stop criteria might be the same with SGA.

---

[45] So we generally do not differentiate "individuals" between "points," "solutions," and "vectors."

[46] What is the geometric explanation of Eq. 2.20?

[47] Uniform crossover will be discussed in detail in Chap. 3.

[48] DE treats every individual with mutation, crossover, and a replacement procedure, which is a steady state EA, unlike the generational approach in SGA. These two terms will be discussed in Chap. 3.

Even though standard DE utilizes a random direction to mutate individuals, the direction may point to the promising area during the evolving process. Apart from that, more effectively assigning $\mathbf{r}_1 \neq \mathbf{r}_2 \neq \mathbf{r}_3 \neq \mathbf{x}_i$ may promote evolution toward the designated area, which will be discussed in later chapters. DE is a kind of simple but powerful EA that has been attracting increasing research interests. Figure 2.15 illustrates the number of papers indexed by the SCI on DE.[49] Mathematica® has already added DE to its numerical optimizer package.

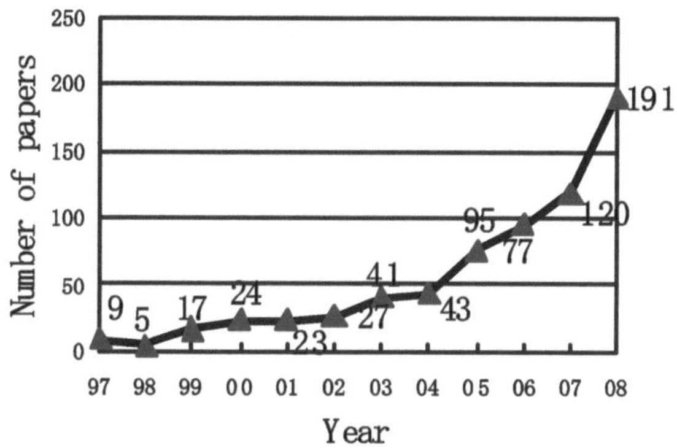

Fig. 2.15 Number of papers indexed by SCI on DE

All the algorithms introduced in this section may be regarded as EAs because they all maintain a group of individuals and have some kind of selection scheme and variation operators.

2.5 Summary

After providing the necessary mathematical and programming backgrounds in Sect. 2.1, we introduced SGA in detail using a not-so-easy problem to demonstrate the strength of EAs, and we explained the implementation key points for programming EAs. After that, five algorithms, including ES, EP, simplex search, scatter search, and DE, were introduced. We do not want readers to remember every step of these algorithms. But the intuitive ideas of these algorithms are of utmost importance because they might be the effective search techniques of your algorithms.

Observing Fig. 2.9 carefully, you might notice that SGA can improve the quality of the best fitness value very fast, and there is a saturation like character thereafter. It is almost always correct with respect to every EA that about 80% of its initial improvements from randomly generated initial individuals are done in about 20%

---

49 TS = ("differential evolution"). The SCI index "TS" is for the search topic in the title, the keywords, and the abstract.

of its evolving process and about 20% of its final improvements are made in about 80% of its evolving process.[50] In Chap. 3, we will introduce many ways to improve the final search efficiency.

Good metaheuristics, such as GA, ES, EP, scatter search, DE, etc., need at least three mechanisms to accomplish the search requirement satisfactorily.

- Global search mechanism. Good algorithms need a global search mechanism to find the domain of attraction of global optimum.
- Convergence mechanism. Good algorithms need a convergence mechanism to promote the evolution of individuals toward the best ones.
- Up-hill mechanism for minimum optimization problems.[51] Good algorithms need an up-hill mechanism to accept no-so-good individuals so that the population can escape the domain of attraction of the local optimum where the current best individual resides.

For SGA, its global search mechanism is randomly generated initial individuals and crossover and mutation operators; its convergence mechanism is RWS; its up-hill mechanism is also RWS and maintaining a group of individuals in the population. Readers are strongly suggested to find out these mechanisms for every EA we introduce hereafter.

Before concluding this chapter, we need to mention two viewpoints. The prerequisite for designing an effective algorithm for real-world problems is to grasp the innovative elements of standard EAs. The only way to grasp, instead of memorizing them, is to read actively. Apart from that, if an powerful algorithm takes advantage of various effective search techniques introduced in this chapter and the later ones, it is sometimes hard to call it a GA, ES, EP, DE, or other specific algorithm.

You need to grasp the programming details of SGA and understand the intuitive ideas behind other introduced EAs.

Suggestions for Further Reading

This chapter deals with the basics of EAs, so the suggested reading list only includes chapters of the textbooks.

For SGA, ES, and EP, we encourage interested readers to read Chaps. 2–5 of Eiben and Smith's textbook [10] and Chaps. 8–10 of Bäck *et al.*'s textbook [11]. Haupt and Haupt also give good introductions for binary GA in Chap. 2 of their second version of the textbook published in 2004 [12].

For DE, we recommend Chap. 2 of Price *et al.*'s textbook [9] and Chap. 1 of Feoktistov's textbook [13].

We introduced several EAs in this chapter. Readers interested in their taxonomy are encouraged to read paper by Calégari *et al.* [14].

---

50 This is an example of the 80/20 rule.

51 Or down-hill mechanism for maximum optimization problem.

Exercises and Potential Research Projects

2.1. Implement an SGA from scratch in any programming environment to repeat the solution process of the problem illustrated by Eq. 2.8. We insist on encouraging readers to implement their SGA without the help of any source codes to promote an in-depth understanding of EAs. MATLAB® is suggested for implementing the SGA. Comparisons between different parameter settings are encouraged.

2.2. In RWS, is the selection with replacement or without replacement?

2.3. Why do we need to define the crossover rate p_c and the mutation rate p_m? What will happen if $p_c = p_c = 1$ in an SGA?

2.4. Summarize the difference between SGA, ES, and EP in a table.

2.5. What are the meanings of $(1+1)$-ES, $(1+\lambda)$-ES, and $(1,\lambda)$-ES?

2.6. Find out the global search mechanism, the convergence mechanism, and the up-hill mechanism of ES.

2.7. Find out the global search mechanism, the convergence mechanism, and the up-hill mechanism of scatter search.

2.8. Why do we use the centroid of B and G as the reflection center in simplex search? Is there any other method? For example, if we think B should have twice as much influence on the reflection center as G, then what is the expression for M?

2.9. Read [4] and summarize its simplex search on a single sheet of paper.

References

1. Papoulis A, Pillai S (2002) Probability, random variables and stochastic processes, 4th edn. McGraw Hill Higher Education
2. Ross S (2009) A first course in probability, 8th edn. Prentice Hall
3. Nelder J, Mead R (1965) A simplex method for function minimization. Comput J 7(4):308–313
4. Dennis JE, Jr, Torczon V (1991) Direct search methods on parallel machines. SIAM J Optim 1:448–474
5. MathWorks T (2009) Genetic algorithm and direct search toolbox 2 user's guide. Tech. rep., The MathWorks, Natick, MA
6. Glover F, Laguna M, Marti R (2003) Advances in evolutionary computation: theory and applications, chap. Scatter search, 519–537. Springer, Berlin Heidelberg New York
7. Laguna M, Marti R (2003) Scatter search: methodology and implementations in C. Springer, Berlin Heidelberg New York
8. Storn R, Price K (1997) Differential evolution: A simple and efficient heuristic for global optimization over continuous spaces. J Glob Optim 11(4):341–359
9. Price KV, Storn RM, Lampinen JA (2005) Differential evolution: a practical approach to global optimization. Springer, Berlin Heidelberg New York

10. Eiben AE, Smith JE (2003) Introduction to evolutionary computing. Springer, Berlin Heidelberg New York
11. Bäck T, Fogel D, Michalewicz Z (2000) Evolutionary computation 1: basic algorithms and operators. Taylor & Francis, London, UK
12. Haupt RL, Haupt SE (2004) Practical genetic algorithms, 2nd edn. Wiley, New York
13. Feoktistov V (2006) Differential evolution: in search of solutions. Springer, Berlin Heidelberg New York
14. Calégari P, Coray G, Hertz A *et al* (1999) A taxonomy of evolutionary algorithms in combinatorial optimization. J Heurist 5(2):145–158

Chapter 3
Advanced Evolutionary Algorithms

Abstract The numerical example in Chap. 2 demonstrates that EAs are powerful yet simple search methodologies. In this chapter, several critical topics will be discussed in a clear way through which you can really claim that you know EAs and have the courage to design your own EA for the problem you face. We hope it's a productive journey reading this chapter.

3.1 Problems We Face

In Chap. 2, we showed that randomly generated initial individuals efficiently evolve with the help of a selection process and variation operators toward the global optimal solution. But we need to ponder the simple EAs before utilizing them in real-world problems. These problems might include:

- **Why do we need binary code?** The numerical example discussed in Chap. 2 is only with one variable. What would happen if we needed to optimize a problem with *50* variables, each of them needing to be discretized into *20* binary genes? The overall length of one chromosome would be *1000*! Sure, a human being has more than ten thousand genes. But it takes more than one hundred thousand years for us to evolve. It could be anticipated that the time for EAs to find the global optimal solution will be extended with longer chromosomes due to the larger solution space where EAs conduct their search. So the first problem we face is how to represent the solution of the problem so that EAs' process and operators can handle the representation easily without the expense of too large a search space.
- **Why RWS?** As was discussed in Chap. 2, RWS can grant better individuals more chances to breed. But an alert reader might have noticed that the numerical example was designed elaborately so that the objective value for every individual is larger than zero, and the larger the objective value is, the fitter the individual is. So the second problem we face is how to handle the problem with a minimum

requirement. Another situation needs to be considered is a maximum problem whose objective values might fall below zero.

- **How does one determine the key EA elements that will greatly influence the result?** There are many decision making problems in implementing an EA. So the third problem we face is how to determine the proper *popsize*, termination criteria, crossover probability p_c, mutation probability p_m, etc.
- **Which algorithm is better?** We have introduced GA, ES, EP, DE, simplex search, and scatter search and will introduce many other EAs and their variations. Which one is most fit for your problem? So the fourth problem we face is how to evaluate the performance of an EA fairly.

These four problems constitute the main part of Chap. 3 and the main concerns of the EA community. The basic problem behind the first three problems listed above is how to balance a global search with a local search so precisely that we can find the global *attraction basin*[1] of the solution landscape quickly in the early stages and focus on fine tuning in this basin efficiently in the late stages. In the EA community, we often call this balance the tradeoff between *exploration* and *exploitation*.[2]

Another way to understand the balance could be the tradeoff between *population diversity* and *selective pressure*. In the same way, we want to distribute the population evenly on the solution landscape, i.e., keeping the population diversity, in the early stage to promote exploration and converge the population to the global optimal solution, i.e., adding selective pressure, in the late stage to promote exploitation. This relationship could also be called the tradeoff between *diversification* and *intensification*.

We have mentioned so many problems to make decisions. You might feel unsure about EAs. Cheer up! We will show you in the following sections that the EA community has considered these problems carefully. Even though we cannot say these problems have been solved satisfactorily, many conclusions have been drawn to guide the design of EAs to at least alleviate these problems. Besides, these considerations are quite helpful for understanding EAs in a more profound way. Thus we urge you to actively read this chapter.

3.2 Encoding and Operators

This section answers the first question mentioned in Sect. 3.1. Before the main part, there are two points we need to discuss.

Using EAs to solve optimization problems or learning problems might need three maps (Fig. 3.1). If EAs cannot handle the variables of the optimization or learning problems directly, we need to design some kind of codes that could be handled by

[1] An attraction basin is the set in the definition domain. Points in this set will finally converge to one point, perhaps a global or local optimal solution, with a specific search technique.

[2] These two terms are also known as *diversification* and *intensification*, respectively. Blum and Roli gave intensive discussions on these terms in 2003 [1].

EAs' operators to represent the real solution. So we need a *decoding* operator to map the codes[3] in the *operation domain* onto the real solutions[4] in the *definition domain*. Sometimes the inverse map, the *encoding* operator, is necessary. After getting the solution, we could use the objective function, which is given by the problem, to map it onto the objective value. If the problem is not maximum or the objective values are not all greater than zero, we need the third map, fitness transferral, to enable RWS for the selection process.[5]

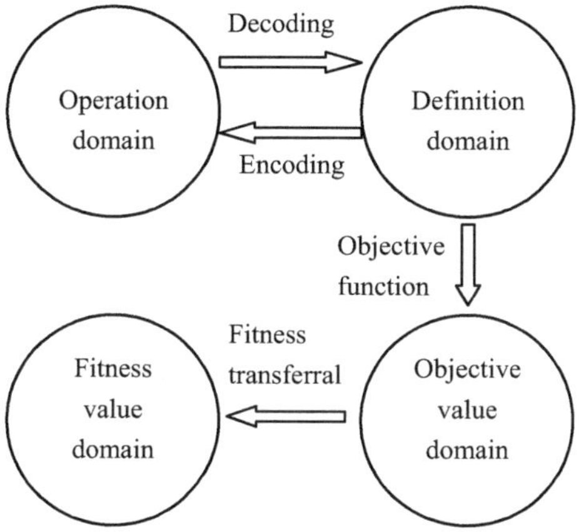

Fig. 3.1 Three maps between operation domain, definition domain, objective value domain, and fitness value domain

The map from set A onto set B means that every point $a \in A$ could be mapped onto one image $b \in B$. There are three types of maps.

- **Injective.** Different points in A will be mapped onto different images in B.
- **Surjective.** For every point in $b \in B$, there is at least one original image $a \in A$ that is mapped onto b.
- **Bijective.** Points in A and B have a one-to-one map. The sufficient and necessary conditions for a bijective map are it is both injective map and surjective map.

If we need the encoding/decoding map, we want this map to be bijective because it will neither enlarge the search space nor lose the possible global optimal solution. But for a specific problem, sometimes it is extremely hard to design a bijective map.

If the map is injective but not surjective, we need to ensure that the optimal solution in the definition domain has an original image. If not, there is a system error. For the numerical experiment discussed in Chap. 2, we divide the definition into $2^{12} \approx 4000$ discrete points. So the maximum system error might be $3/4000/2 \approx 0.00375$. Whether or not this system error is acceptable depends on the requirement of the users. We can decrease the system error of binary code for a real

[3] Genotypes.

[4] Phenotypes.

[5] The second map is given by the problem and the third map will be discussed in Sect. 3.3.

number by enlarging the length of the chromosome. But here comes the tradeoff between the solution precision and the search ability of EAs. In numerical experiments, the SGA searches the operation domain $10 \times 10 = 100$ times and gets rather good results. So we can roughly say that the SGA could do an efficient search on the problem with about $(100/4000) \times 100 = 2.5\%$ samplings on the operation domain. If we expand the operation domain too much, such as to use 100 binary genes to represent a solution, the SGA should not achieve the same result quality with the same parameter, i.e., $popsize = 10$ and $maxgen = 10$.

On the other hand, if the map is surjective but not injective, that means the map is multi-to-one, i.e., multiple chromosomes might corresponding to the same solution. Then the genetic search on these chromosomes is a waste of time. In this situation, we need a powerful search ability to explore the enlarged operation domain.

The second point we need to mention here is that the variation operators of EAs depend on the codes. Different representations have different types of chromosomes. Then different techniques are required to handle the different chromosomes. So we discuss codes together with their related operators.

3.2.1 Binary Code and Related Operators

Consider a binary chromosome with length l. It is easy to see that the operation domain contains 2^l points.

If we use a binary code to illustrate a real number, some drawbacks need to be considered. The first one is the *Hamming cliff*. Let us consider a simple problem. How do we mutate the binary code from (0 1 1 1) to (1 0 0 0)? On the one hand, (0 1 1 1) in binary means the integer 7; and (1 0 0 0) in binary means the integer 8. They are neighbors in the discrete integer domain. On the other hand, we need to be lucky enough to perform four bit-flip mutations successively to mutate (0 1 1 1) to (1 0 0 0). Generally, p_{m} is rather small. So the probability of 7 becoming 8 is rather small. It seems that there is a cliff between 7 and 8 for bit-flip mutation to climb. The *Hamming distance* can illustrate the situation. The Hamming distance is the number of positions for which the corresponding genes are different. The Hamming distance between (0 1 1 1) and (1 0 0 0) is 4, which is the largest Hamming distance in the field of 4 binary code.

Gray code is a way to solve the problem of the Hamming cliff. The binary codes of integers from 0 to 7 are 000, 001, 010, 011, 100, 101, 110, 111, but their Gray codes are 000, 001, 011, 010, 110, 111, 101, 100.[6] Thus the Hamming distance between every two neighbors is just one. So in this way, there is no Hamming cliff!

The second drawback is that the neighborhood of binary code is not the same as that of real value. In binary code, we can define the neighborhood as the Hamming distance. If the neighborhood between two binary chains is no greater than a predefined threshold, such as 1, these two chains are neighbors. The Hamming distance

[6] Interested readers may find out the mapping rule between binary code and Gray code in the computer basis textbook.

between chromosome (0 0 0) and chromosome (1 0 0) is one, which means that (0 0 0) and (1 0 0) are neighbors. Only one bit-flip mutation can change from (0 0 0) to (1 0 0). But the phenotypes of these two individuals are 0 and 4, respectively, which is rather far away in the decision space.[7] So it is hard to control the mutation strength.

The third drawback for binary code is the contradiction between the system error and the search ability of EAs, which was discussed at the beginning of this section.

Considering all these drawbacks, we may come to the conclusion that binary code for a real value variable is not a good choice. But there exist some problems that are suitable for binary code.[8] So we'd like to introduce the variation operators for binary codes.

Multiple-point Crossover

Single-point crossover, which was discussed in Chap. 2, is a way to exchange information between two parents. We can just expand single-point crossover to *multiple-point crossover*. Let us take a *two-point crossover* as an example. For two individuals selected to perform a two-point crossover, we assign two different points between 1 and $l-1$ randomly, where l is the length of the chromosome. The genes between these two points are exchanged between parents and the resulting chromosomes are offspring. Figure 3.2 illustrates the two-point crossover.

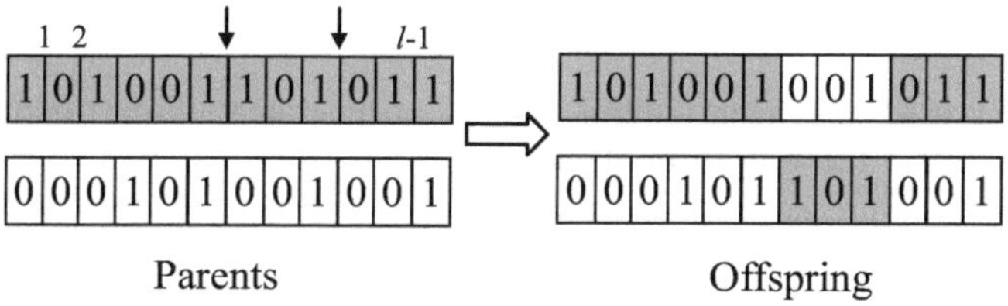

Fig. 3.2 Two-point crossover

When considering multi-variable problems, we use one binary chain to represent each variable, the length of which need not be the same, and connect these chains directly to generate a genotypic chromosome for the individual. When two individuals are selected to perform crossover, we need to make decisions on whether the operator is over the whole binary chromosome or over the separate chains and express it clearly. If we do the former single-point crossover, we should express it as a *single-point crossover over the chromosome*. If we do the former multiple-point crossover, we should express it as a *multiple-point crossover over the chromosome*. If we do the single-point crossover and multiple-point crossover separately on the

[7] This will also happen in Gray code.

[8] We will discuss one example of this problem in Chap. 7.

binary chains of variables, we should express them as a *single-point crossover over the variables* and a *multiple-point crossover over the variables*, respectively.

Uniform Crossover

A *uniform crossover* exchanges information between parents in a different way. We suppose two individuals, i.e., A and B, are selected to perform uniform crossover. Each gene has a probability of 0.5 of inheriting the gene from A, otherwise it inherits the gene from B.[9] If we want to generate two offspring from two parents, then every gene of the second offspring could be selected inversely to the corresponding gene of the first offspring, i.e., if the gene of the first offspring inherits the gene from A, then the corresponding gene of the second offspring inherits the gene from B and *vice versa*. Suppose we generate a random vector (0.55 0.03 0.67 0.77 0.30 0.25 0.99 0.78 0.10 0.58 0.87 0.73). Figure 3.3 illustrates the uniform crossover.

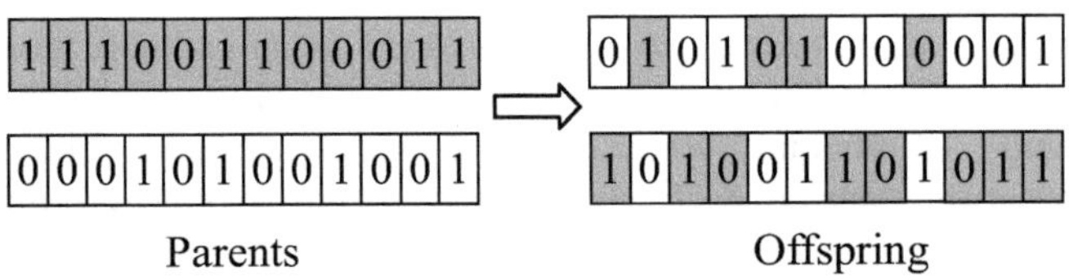

Fig. 3.3 Uniform crossover

Uniform crossover can use multiple parents to generate multiple offspring. Readers may speculate the way to implement it. Apart from that, in two-parents-one-offspring uniform crossover, we can define a probability p for the offspring to inherit genes from the first parent, and $1 - p$ for the other offspring to inherit genes from the other parent. This type of uniform crossover is called *parameterized uniform crossover*. If p is determined by the relative weight between two parents, i.e., $p = \frac{f_{p_1}}{f_{p_1}+f_{p_2}}$ for maximum problems, where f_{p_1} and f_{p_2} are the fitness values of two parents, this operator is called the *fusion operator*. The classical method discussed in the above paragraph means $p = 0.5$, i.e., 0.5 uniform crossover, which is also called *discrete crossover*.

The most frequently used binary code mutation operator is a bit-flip mutation, which was discussed in Chap. 2.

[9] How does one implement this statement in a programming environment?

3.2.2 Real Code and Related Operators

As was discussed in Sect. 3.2.1, binary code for a real-value variable is not a good choice. For the optimization problem $\min f(\mathbf{x}) = f(x_1, x_2, \cdots, x_n)$, where $x_i \in \mathbb{R}$, it is natural to express the chromosome as $(x_1, x_2, \cdots, x_n)$. EAs based on real code are called *real-code EAs* (RCEAs).

Consider a real number chromosome with length n. It is easy to see that the operation domain contains ∞ points. For a point $\mathbf{x}$ in $\mathbb{R}^n$ space, its neighborhood contains all the points whose Euclidean distance to $\mathbf{x}$ is no more than ε, where ε is the predefined threshold.

Although the single-point crossover, discussed in Sect. 2.2.5, and multiple-point crossover and uniform crossover, discussed in Sect. 3.2.1, could also be used directly, we do not encourage readers to do this. The reason is left as an exercise.

Many researchers have discussed the rationale for designing a good real code crossover operator. Kita *et al.* suggested two useful guidelines [2] and Deb and Beyer expressed theirs in a similar way [3].

1. **Preservation of statistics.** The offspring generated by good crossover operators should preserve the statistics of their parents. Those statistics might include mean, variance, and covariance.
2. **Diversity of offspring.** The offspring generated by good crossover operators should have as much diversity as possible under the constraint of guideline 1.

The crossover operators for real code could be roughly classified into two categories according to whether the operator has preference or not.

3.2.2.1 Real Code Crossover Without Preference

Let us first discuss an optimization problem with two variables to illustrate the concept of *nonseparable variables* and *separable variables*.

Suppose our search method is in a greedy way. From any point in 2-D space, we search four directions in the order $+x$, $-x$, $+y$, and $-y$ with the same small step. Any improvement of the fitness value along a direction moves the current point along that direction to the next point. The greedy search method continues until it cannot improve in all directions or it finds a solution whose fitness value is within the predefined threshold.

The first problem is $\min f = x^2 + 2y^2$, and the definition domain for both x and y is $[-3.5, +3.5]$.[10] The function is an ellipsoid function whose principal axes are the x-axis and the y-axis. The second problem is $\min f = (3x^2 + 3y^2 - 2xy)/2$, and the definition domain for both x and y is also $[-3.5, +3.5]$.[11] The function is the same

[10] The objection function could be written as $f = f_1(x) + f_2(y)$. We then say that this optimization problem is with separable variables, or it is a *decomposable* problem [4].

[11] The objection function could not be written as $f = f_1(x) + f_2(y)$. We then say that this optimization problem is with nonseparable variables.

ellipsoid as the first one, but we rotate it 45° counterclockwise so that its principal axes are $x = \pm y$. In this way, the two variables are correlated. The trajectories of the search points started from $(3, -3)$ in both cases are illustrated on the contour of the function in Fig. 3.4a and b. The step size is 0.002, and the stop criterion is $f(\mathbf{x}) < 10^{-9}$. In Fig. 3.4a, the greedy method needs 6000 steps to stop, while in Fig. 3.4b it needs 7500 steps to stop.

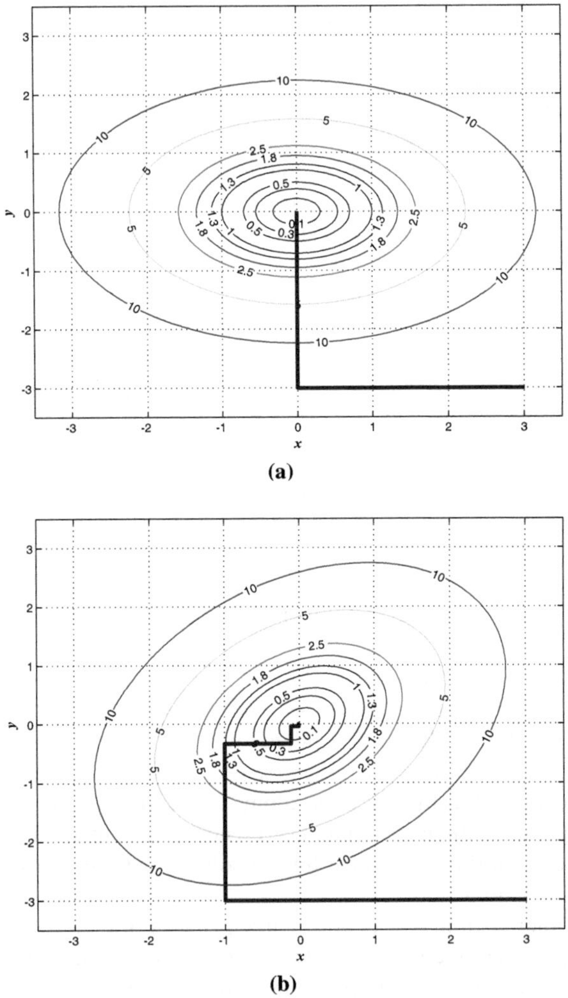

Fig. 3.4 Problems with separable and nonseparable variables: (a) Searching the separable variables, and (b) Searching the nonseparable variables

For our greedy method, problems with nonseparable variables are harder and have a longer search time with the same initial point, step size, and stop criterion.

This conclusion could be extended to other algorithms, i.e., generally problems with nonseparable variables are harder.

To solve problems with nonseparable variables effectively, we need to design powerful variation operators so that the correlation between variables could be handled. Several operators discussed below have the power to deal with such problems.

Arithmetic Crossover

For two individuals $\mathbf{x}^1 = (x_1^1, x_2^1, \cdots, x_n^1)$ and $\mathbf{x}^2 = (x_1^2, x_2^2, \cdots, x_n^2)$, their *arithmetic crossover* result, i.e., offspring $\mathbf{y} = (y_1, y_2, \cdots y_n)$, could be expressed as

$$\mathbf{y} = \alpha \mathbf{x}^1 + (1-\alpha)\mathbf{x}^2 \tag{3.1}$$

where $\alpha \sim U(0,1)$. If we want to generate two offspring from two parents, simply assign $\beta = (1-\alpha)$ and use β as the random number in Eq. 3.1.

Equation 3.1 is also called a *whole arithmetic crossover* because it uses one uniformly distributed random number for all variables. Apart from that, you can generate n independent uniformly distributed random numbers in $(0,1)$, $\boldsymbol{\alpha} = (\alpha_1, \cdots, \alpha_n)$ and do the following local arithmetic crossover:

$$\mathbf{y} = \boldsymbol{\alpha} \cdot \mathbf{x}^1 + (\mathbf{1} - \boldsymbol{\alpha}) \cdot \mathbf{x}^2 \tag{3.2}$$

where $\cdot$ is for the dot product of two vectors.

The intermediate crossover discussed in Sect. 2.3 could also be regarded as a special deterministic arithmetic crossover.

Blend Crossover

The standard arithmetic crossover is actually a linear interpolation of two individuals, so it could only generate offspring on the line connecting two parents. So the second guideline discussed above is hard to satisfy. Eshelman and Schaffer suggested *blend crossover* (BLX) to expand the range of arithmetic crossover [5].

BLX is performed on the gene level. For gene x_i^1 and x_i^2, suppose $x_i^1 < x_i^2$ without loss of generality; their offspring gene is

$$y_i = rand((x_i^1 - \alpha(x_i^2 - x_i^1)), (x_i^2 + \alpha(x_i^2 - x_i^1))) \tag{3.3}$$

where $rand(a,b)$ is a function to generate a uniformly distributed random number in the range (a,b),[12] α is a user-defined parameter that controls the extent of the expansion. So we often use BLX-α to make things clear. Eshelman and Schaffer reported that $\alpha = 0.5$ is a good choice for most situations. So the most frequently used BLX is BLX-0.5.[13]

---

[12] How does one implement it in a programming environment?

[13] What is the geometric interpretation of BLX-0.5?

Simplex Crossover

The *simplex crossover* (SPX), proposed by Tsutsui *et al.* in 1999 [6], uniformly select offspring in the simplex generated by parents. For an n-dimensional space, its simplex is $n+1$ points in the space.[14]

For a 2-D space, its simplex is a triangle, its vertices are $\mathbf{x}^1$, $\mathbf{x}^2$, and $\mathbf{x}^3$ (Fig. 3.5). Their centroid is $\mathbf{k} = (\mathbf{x}^1 + \mathbf{x}^2 + \mathbf{x}^3)/3$. We then expand the simplex along the directions $\mathbf{x}^1 - \mathbf{k}$, $\mathbf{x}^2 - \mathbf{k}$, and $\mathbf{x}^3 - \mathbf{k}$, respectively, with a step size of ε to simulate the extrapolation of BLX. So the new simplex is constituted by $\mathbf{y}^1 = \mathbf{x}^1 + \varepsilon(\mathbf{x}^1 - \mathbf{k})$, $\mathbf{y}^2 = \mathbf{x}^2 + \varepsilon(\mathbf{x}^2 - \mathbf{k})$, and $\mathbf{y}^3 = \mathbf{x}^3 + \varepsilon(\mathbf{x}^3 - \mathbf{k})$.[15] The simplex and its expansion results are illustrated in Fig. 3.5. Then we need to sample uniformly in the triangle formed by the new simplex to generate offspring.[16] If more offspring are required, we can execute the sampling procedure in the simplex again and again.

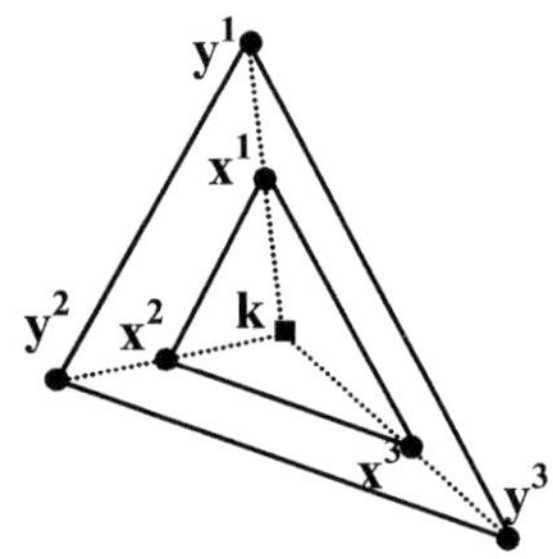

Fig. 3.5 Simplex crossover

3.2.2.2 Real Code Crossover with Preference

The preference depends on the designer's consideration. Some examples are preference for points near the parents or a preference for points near the centroid of the parents.

Simulated Binary Crossover

Deb and Agrawal researched the single-point crossover for binary number code and found that the offspring of the single-point crossover have the same centroid as that of the parents. They then suggested *simulated binary crossover* (SBX) to simulate this property in a real code crossover [3, 7].

They defined a term called a *spread factor* β_i for real code gene i as follows:

[14] We introduced the concept of simplex in Sect 2.4.

[15] Readers unfamiliar with this are referred to read Sect. 2.4.

[16] How to implement the sampling is left as an exercise.

$$\beta_i = \left| \frac{c_i^1 - c_i^2}{p_i^1 - p_i^2} \right| \tag{3.4}$$

where c_i^1 and c_i^2 are gene i of two offspring, and p_i^1 and p_i^2 are gene i of two parents. If $\beta_i < 1$, the operator is called a *contracting crossover*. If $\beta_i > 1$, the operator is called an *expanding crossover*. If $\beta_i = 1$, the operator is called a *stationary crossover*.

To simulate the sharing centroid property of single-point crossover for binary number code, Deb and Agrawal want the offspring to be symmetric to the centroid of the two parents.

They also want the real code crossover to have a high probability of being a stationary crossover and a low probability of being an contracting crossover and an expanding crossover for every variable. To implement this, they regarded β_i as a random number and suggested a probability density function for β_i as follows:

$$p(\beta_i) = \begin{cases} 0.5(n+1)\beta_i^n, & \beta_i \leqslant 1 \\ 0.5(n+1)/\beta_i^{n+2}, & \beta_i > 1 \end{cases} \tag{3.5}$$

where n is a control parameter. Figure 3.6 illustrate two conditions where $n = 2$ and $n = 5$. According to Fig. 3.6 β_i has a high probability of being 1. Larger n promotes this preference further.

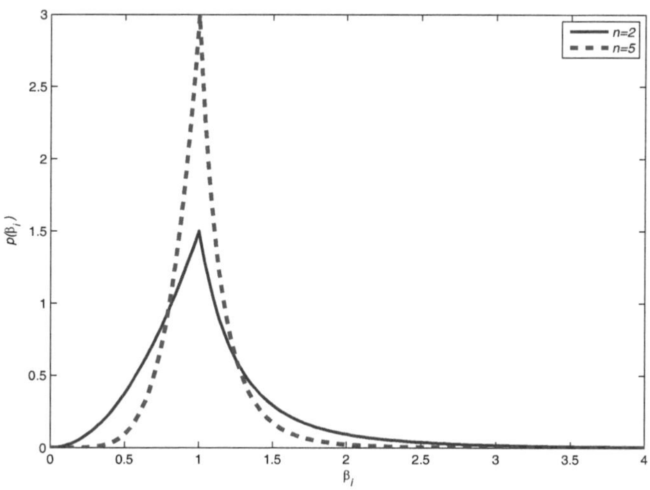

Fig. 3.6 Probability density function for SBX

But how do we generate a random number with the probability density function illustrated by Eq. 3.5? Interested readers are referred to relevant textbooks such as [8] and [9]. Here we just use the inverse transformation to generate a random number β_i with density function Eq. 3.5 from uniformly distributed random number $u_i \sim U(0,1)$.

The distribution function of β_i is the integration of Eq. 3.5, i.e.,

$$F(\beta_i) = \begin{cases} \int_0^{\beta_i} 0.5(n+1)x^n \mathrm{d}x = 0.5\beta_i^{n+1}, & \beta_i \leqslant 1 \\ 0.5 + \int_1^{\beta_i} 0.5(n+1)x^{-(n+2)} \mathrm{d}x = 1 - 0.5\beta_i^{-(n+1)}, & \beta_i > 1 \end{cases} \tag{3.6}$$

Let $u_i = F^{-1}(\beta_i)$ and solve β_i as

$$\beta_i = \begin{cases} (2u_i)^{\frac{1}{n+1}}, & u_i \leqslant 0.5 \\ (2(1-u_i))^{-\frac{1}{n+1}}, & u_i > 0.5 \end{cases} \tag{3.7}$$

Then for every $u_i \sim U(0,1)$, β_i, which satisfies Eq. 3.5, could be generated with Eq. 3.7.

According to Eq. 3.7, smaller u_i[17] means smaller spread factor β_i, which makes the crossover a contracting crossover, i.e., the offspring are close to the centroid; larger u_i[18] means a larger spread factor β_i, which makes the crossover an expanding crossover, i.e., the offspring are far from the centroid. But according to Fig. 3.6, SBX is likely to be stationary crossover with a high probability.

Larger β_i means farther distance from the centroid and *vice versa*. So Eq. 3.8 could be carried out on every gene of the offspring.

$$\begin{aligned} c_i^1 &= 0.5\left(p_i^1 + p_i^2\right) + 0.5\beta_i\left(p_i^1 - p_i^2\right) \\ c_i^2 &= 0.5\left(p_i^1 + p_i^2\right) + 0.5\beta_i\left(p_i^2 - p_i^1\right) \end{aligned} \tag{3.8}$$

After substituting Eq. 3.8 into Eq. 3.4 we can see that Eq. 3.8 is just the implementation method of Eq. 3.4 after we get β_i from Eq. 3.7. Because β_i has a high probability of being close to 1 according to Fig 3.6, the offspring of SBX have a high probability of being close to the parents according to Eq. 3.8, which guarantees that the offspring have the same centroid as that of the parents.

If the parents are far from each other, the offspring of SBX are far from each other with a high probability because of the high probability of $\beta_i = 1$, which could promote the exploration in the early stage of EAs. If the parents are close, the offspring of SBX are close with a high probability, which could promote exploitation in the late stage of EAs.

SBX is based on axes. So its search ability is limited when dealing with problems with nonseparable variables. Deb *et al.* suggested *parent-centric crossover* (PCX) with the property of multi-parent crossover to improve SBX [10].

---

[17] Close to zero.

[18] Close to one.

Unimodal Normal Distribution Crossover

Ono *et al.* proposed unimodal normal distribution crossover with both the nonseparable variable searching ability and the preference for centroid [11–13]. They want the offspring to be close to the centroid of the parents, which can also satisfy the previously discussed guidelines. Apart from that, they introduced the variance direction orthogonal to the direction of the two parents. Let us discuss a 2-D example, illustrated in Fig. 3.7.

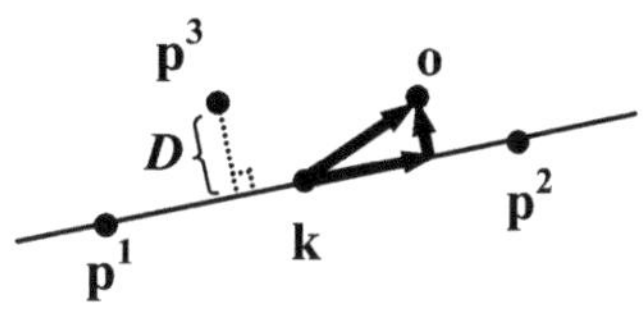

Fig. 3.7 Unimodal normal distribution crossover

We need to pick up three parents, i.e., $\mathbf{p}^1$, $\mathbf{p}^2$, and $\mathbf{p}^3$, to generate one offspring $\mathbf{o}$. First, the centroid of $\mathbf{p}^1$ and $\mathbf{p}^2$, i.e., $\mathbf{k}$, is calculated with $\mathbf{k} = (\mathbf{p}^1 + \mathbf{p}^2)/2$. Then $\mathbf{d} = \mathbf{p}^2 - \mathbf{p}^1$ is the vector and starts at $\mathbf{p}^1$ and ends at $\mathbf{p}^2$. D is used to illustrate the distance from the third parent to vector $\mathbf{d}$.[19] Using $\mathbf{d}$ and any other $n-1$ independent vectors in $\mathbb{R}^n$ space, we can generate n unit orthogonal bases using Gram–Schmidt orthonormalization.[20] We use $\mathbf{e}^1, \cdots, \mathbf{e}^{n-1}$ to illustrate other $n-1$ unit orthogonal bases besides $\mathbf{d}/|\mathbf{d}|$. Then we can generate the offspring in the following way.

$$\mathbf{o} = \mathbf{k} + \xi \mathbf{d} + D \sum_{i=1}^{n-1} \eta_i \mathbf{e}_i \tag{3.9}$$

where $\xi \sim N(0, \sigma_\xi^2)$ and $\eta_i \sim N_i(0, \sigma_\eta^2)$. From Eq. 3.9 we can see that the offspring starts from the centroid of two parents $\mathbf{p}^1$ and $\mathbf{p}^2$, i.e., $\mathbf{k}$, and extends in the direction $\mathbf{d}$ and length ξ, and then extends in other directions orthogonal to $\mathbf{d}$. Because ξ and η are all zero mean normally distributed random numbers, the offspring is close to the centroid of $\mathbf{p}^1$ and $\mathbf{p}^2$ with high probability. Thus the guidelines are satisfied.

The first two parts in Eq. 3.9 are the primary search component, and the third part is the secondary search component. Generally this will generate offspring within the ellipsoid whose principal axis is $\mathbf{d}$. Thus Kita *et al.* called it a *unimodal normal distribution crossover* (UNDX). They also suggested that $\sigma_\xi = 0.5$ and $\sigma_\eta = 0.35/\sqrt{n}$.

Apart from the Gram–Schmidt orthonormalization for orthogonal bases, Kita also suggested that independent random vectors orthogonal to $\mathbf{d}$ are also acceptable. Let us discuss a simple example in 2-D space. Suppose $\mathbf{p}^1 = (4,2)$ and $\mathbf{p}^2 = (8,4)$, then $\mathbf{d} = (4,2)$. We just need to generate a random number $u_1 \sim U(0,1)$, for example 0.35, and find another number u_2 that can satisfy $(u_1, u_2) \cdot \mathbf{d} = 0$, where $\cdot$ is for the dot product. In our example, $u_2 = -0.7$. Then we just need to normalize (u_1, u_2) to be $\mathbf{e}_1 = (0.447, -0.894)$ and use Eq. 3.9 to do UNDX in 2-D space.

[19] Consider how to calculate D in $\mathbb{R}^n$ space.

[20] Any textbook on linear algebra will have a discussion of Gram–Schmidt orthonormalization.

UNDX could search other directions apart from **d**, so it can handle the problem with nonseparable variables effectively. Kita *et al.* suggested a multiparental extension of UNDX, i.e., UNDX-m, to further improve the search ability [2].

3.2.2.3 Real Code Mutation Without Explicit Direction

As mentioned in Chap. 2, mutation is the secondary search operator for GAs but a primary search operator for ES and EP. So here we'd like to introduce the real code mutation operator in detail.[21]

Uniform Mutation

For individual $\mathbf{x}^i = (x^i_1, x^i_2, \cdots x^i_n)$, let us suppose the definition domain for variable j is (L_j, U_j). Then for gene j, *uniform mutation* is carried out using Eq. 3.10 with probability p_{m}:

$$x^i_j = rand(L_j, U_j) \tag{3.10}$$

where $rand() \sim U(0,1)$. Uniform mutation is a rather coarse operator because it will randomly sample the definition domain at any time, which is not good for convergence in the late stages of EAs. Many researchers have suggested different ways to limit the mutation extent.

Boundary Mutation

Similar to uniform mutation, for gene j of individual $\mathbf{x}^i = (x^i_1, x^i_2, \cdots x^i_n)$, *boundary mutation* is defined by Eq. 3.11 with probability p_{m}:

$$x^i_j = \begin{cases} L_j & rand \leq 0.5 \\ U_j & rand > 0.5 \end{cases} \tag{3.11}$$

where $rand() \sim U(0,1)$ and (L_j, U_j) is the definition domain for variable j. Boundary mutation is useful in constrained optimization when the feasible optimal solution is on the boundary of the feasible domain.

Nonuniform Mutation

Considering the drawbacks of uniform mutation, we might want to mutate the individual in a nonuniform way. Then two principles are considered:

---

[21] Most of the techniques discussed in Chap. 3 could be used in your EA, no matter what you would like to call it.

1. The mutant should be sampled randomly.
2. The extent of mutation should decrease with the evolving process of EAs.

When talking about decreasing the search scale, we need to mention the very famous *simulated annealing* (SA) suggested by Kirkpatrick [14]. SA is a global search algorithm based on a local search method. The reason for its global search ability is that it can accept the worse result of the local search with specific probability.

SA uses local search to explore the solution landscape from current solution i. The local search procedure might be different for various problems and encoding methods. The local search will suggest a new solution j. The acceptance probability of j taking the place of i as the current solution for a minimum problem is determined by

$$p_t(i \rightarrow j) = \begin{cases} 1, & f(j) \leqslant f(i) \\ \exp\left(\frac{f(i)-f(j)}{t}\right) & \text{, otherwise} \end{cases} \tag{3.12}$$

where t is a control parameter called *temperature*. If j is no worse than i, j takes the place of i with probability 1. Otherwise, the probability depends on how bad j is and what the temperature is. We suppose $\Delta = f(i) - f(j) < 0$ and draw the acceptance probability of two temperatures (Fig. 3.8).

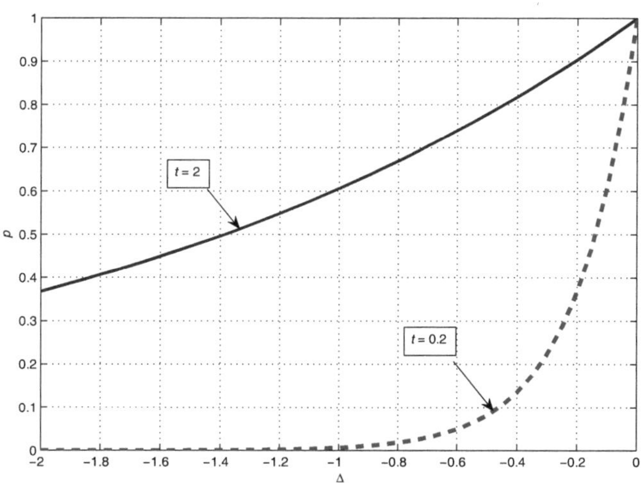

Fig. 3.8 Acceptance probability of SA

It is easy to see that the worse j is, the lower the acceptance probability it has, and the lower the temperature is, the lower the acceptance probability with the same extent of badness.

The temperature of SA will be relatively high at the beginning and decrease according to some *annealing* rules. The simplest way to anneal is $t' = \alpha t$, where $0 < \alpha < 1$ controls the speed of annealing.[22]

We can adopt the concept of annealing in many aspects of EAs. If we want to limit the extent of mutation with the evolving process of EA, we can regard generation g as temperature t and design the following *nonuniform mutation*:

$$x_j' = \begin{cases} x_j + \Delta(g, U_j - x_j), & rand \geq 0.5 \\ x_j - \Delta(g, x_j - L_j), & rand < 0.5 \end{cases} \tag{3.13}$$

where x_j' is the mutant of x_j, L_j and U_j are its lower and upper bounds, respectively, and $\Delta(g,y)$ is the mutation extension function whose variables are generation g and a possible extension length for variable j. Variable j has a probability of 0.5 of changing toward its upper limit and *vice versa*. The definition of $\Delta(g,y)$ is as follows:

$$\Delta(g,y) = y\left(1 - rand^{\left(1-g/maxgen\right)^b}\right) \tag{3.14}$$

where $rand \sim U(0,1)$, b is a parameter to control the annealing speed, g is the generation number, and *maxgen* is the maximum generation number. Suppose $rand = 0.5$ and $y = 1$, we can draw the curve of $\Delta(g)$ as Fig. 3.9.[23]

As can be seen from Fig. 3.9, a larger b means quicker annealing. Also, a larger g, i.e., later stage of EAs, means smaller perturbation.

Normal Mutation

We discussed in Sect. 2.3 *normal mutation* and we would like to rewrite it here for convenience:

$$x_j' = x_j + \sigma N_j(0,1) \tag{3.15}$$

where σ is the standard deviation of a normal distribution. Normal mutations do not have an annealing mechanism, but they can limit the extent of mutation by normal distribution, i.e., the possibility of the mutant being in the range $x_j^i - 3\sigma, x_j^i + 3\sigma$ is more than 0.99.[24]

---

[22] A real implementation of SA will consider a *reheating* process to increase the temperature under some conditions in order to help SA escape from the current local optimal solution. In EA, we could simulate the reheating process by the *restart* part of the population randomly to add population diversity [15].

[23] Here we fixed the random number at 0.5 to simplify the problem and represent the overall annealing convergence effect of nonuniform mutation. While in real implementation, $\Delta(g,y)$ might not be a monotonically decreasing function.

[24] It is called the 3σ rule. Any probability textbook on probability will illustrate this point.

Equation 3.15 only has one standard deviation for all variables, i.e., all the variables change with the same strength. The geometrical interpretation for Eq. 3.15 in a 2-D situation is that its equal probability density contour lines are circles and the radius is defined by σ. So we call this type of mutation a *uncorrelated mutation with one step size*. If we want to implement the hyperellipsoid effect for the mutation, different standard deviations for various genes are necessary, i.e., $\boldsymbol{\sigma} = (\sigma_1, \sigma_2, \cdots, \sigma_n)$. Then the *uncorrelated mutation with n step sizes* could be defined as

$$x'_j = x_j + \sigma_j N_j(0,1) \tag{3.16}$$

The geometrical interpretation of Eq. 3.16 in a 2-D situation is that equal probability density contour lines are ellipses whose principal directions are defined by x and y and the corresponding equatorial radii are defined by σ_x and σ_y, respectively.

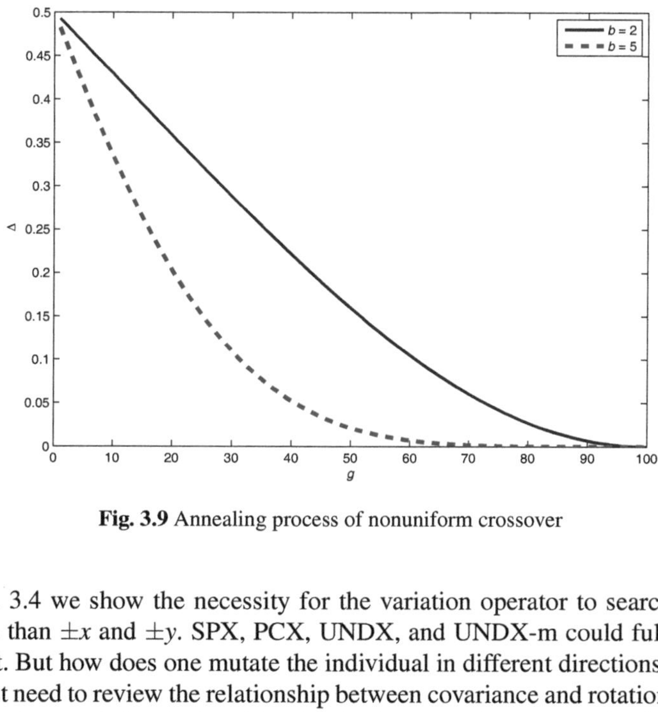

Fig. 3.9 Annealing process of nonuniform crossover

In Fig. 3.4 we show the necessity for the variation operator to search in more directions than $\pm x$ and $\pm y$. SPX, PCX, UNDX, and UNDX-m could fulfil this requirement. But how does one mutate the individual in different directions?

We first need to review the relationship between covariance and rotation in multidimensional normal distribution. The density function of a normal distribution random vector could be illustrated as follows:

$$p(\mathbf{x}) = \frac{1}{(2\pi)^{n/2}(\det \mathbf{B})^{1/2}} \exp\left\{-\frac{1}{2}(\mathbf{x}-\mathbf{a})^{\mathrm{T}}\mathbf{B}^{-1}(\mathbf{x}-\mathbf{a})\right\} \tag{3.17}$$

where $\mathbf{a}$ is the expectance vector and $\mathbf{B}$ is the covariance matrix.

We just take the 2-D case as an example. If we generate two independent normally distributed random numbers x_1 and x_2 with standard variations of σ_1 and σ_2, respectively, and zero mean, then **B** in Eq. 3.17 is $\begin{bmatrix} \sigma_1^2 & 0 \\ 0 & \sigma_2^2 \end{bmatrix}$. The covariance between two numbers is zero because they are independent. We need to introduce a theorem before continuing the discussion.

Theorem 3.1. *Suppose the linear transform of random vector* ξ *is* $\eta = \mathbf{C}\xi$ *and* $\xi \sim N(\mathbf{a}, \mathbf{B})$, *then* $\eta \sim N(\mathbf{Ca}, \mathbf{CBC}^T)$.

This theorem is a very important fact of multidimensional normal distribution. Interested readers are referred to the relevant probability textbooks.

Now we consider the rotation transform, which is a linear transform. For a vector ξ to rotate counterclockwise at angle φ, the rotation transform can be written as

$$\begin{pmatrix} \eta_1 \\ \eta_2 \end{pmatrix} = \begin{pmatrix} \cos\varphi & -\sin\varphi \\ \sin\varphi & \cos\varphi \end{pmatrix} \begin{pmatrix} \xi_1 \\ \xi_2 \end{pmatrix} \tag{3.18}$$

Then the $\mathbf{CBC}^{\mathrm{T}}$ can be illustrated as

$$\mathbf{CBC}^{\mathrm{T}} = \begin{pmatrix} \cos^2\varphi\sigma_1^2 + \sin^2\varphi\sigma_2^2 & \sin\varphi\cos\varphi\left(\sigma_1^2 - \sigma_2^2\right) \\ \sin\varphi\cos\varphi\left(\sigma_1^2 - \sigma_2^2\right) & \sin^2\varphi\sigma_1^2 + \cos^2\varphi\sigma_2^2 \end{pmatrix} \tag{3.19}$$

We take the standard deviation of the new random vector as $\sigma_1^{'}$ and $\sigma_2^{'}$, respectively, and solve σ_1 and σ_2 from Eq. 3.19 as

$$\sigma_1^2 = \frac{-\sigma_1^{'2}\cos^2\varphi + \sigma_2^{'2}\sin^2\varphi}{\sin^2\varphi - \cos^2\varphi}, \quad \sigma_2^2 = \frac{\sigma_1^{'2}\sin^2\varphi - \sigma_2^{'2}\cos^2\varphi}{\sin^2\varphi - \cos^2\varphi} \tag{3.20}$$

Substituting σ_1^2 and σ_2^2 into Eq. 3.19, we can solve the covariance between two new random numbers as

$$c_{12} = c_{21} = \frac{1}{2}\tan(2\varphi)\left(\sigma_1^{'2} - \sigma_2^{'2}\right) \tag{3.21}$$

From Eq. 3.21 we arrive at the conclusion that rotating a 2-D zero-mean uncorrelated normal distribution vector could generate a covariance between the two numbers, and the relationship between the rotating angle and the covariance is given by Eq. 3.21. The largest covariance happens with the rotation angle 45°.

The geometrical interpretation for Eq. 3.17, whose covariance is not zero in a 2-D situation, is that its equal probability density contour lines are ellipses whose principal directions are defined by the eigenvectors of the covariance matrix **B**, the rotation angle is determined by Eq. 3.21, and the corresponding equatorial radii are defined by the inverse of the square root of the eigenvalues, i.e., $\sigma_1^{'}$ and $\sigma_2^{'}$, respectively.

Up to now, we know that covariance in normal distribution means rotation operation. But how do we generate a 2-D normal distribution random vector whose covariance is not zero?

Again, we need to use Theorem 3.1. Suppose we are given a positive definite covariance matrix **B** from which to generate a normal distribution number. We can first do the Cholesky decomposition[25] to generate a lower triangular matrix $\mathbf{L}$ that satisfies $\mathbf{L}\mathbf{L}^{\mathrm{T}} = \mathbf{B}$. Then we can generate n independent standard normal distribution random numbers $\xi = (\xi_1 \quad \cdots \quad \xi_n)^{\mathrm{T}}$, $\xi_i \sim N(0,1)$. The covariance matrix of this vector is unit matrix $\mathbf{E}$. The linear transformation $\eta = \mathbf{L}\xi$ generates normal distribution vector η whose covariance matrix is $\mathbf{L}\mathbf{E}\mathbf{L}^{\mathrm{T}} = \mathbf{B}$, which is exactly what we want.

We can do the following numerical experiment to verify the above procedure. In MATLAB®, `randn(m,n)` is used to generate n normal distribution random vectors, each of which contains m uncorrelated standard random numbers. `cov(x)` is used to calculate the covariance matrix for many random vectors. Each row in matrix **x** is a random vector. So we can generate 10000 two-dimension standard normal distribution random vectors and determine their covariance matrix as follows:

```
old_matrix=randn(2,10000);
covariance_old=cov(old_matrix');
```

MATLAB® gets $\begin{bmatrix} 1.0223 & -0.0067 \\ -0.0067 & 0.9962 \end{bmatrix}$, which shows that these 10000 random vectors are uncorrelated standard normal distribution vectors.

Suppose we want to generate the 2-D random vector with covariance matrix as $\begin{bmatrix} 5 & 2 \\ 2 & 1 \end{bmatrix}$; the following commands could do the Cholesky decomposition:

```
B=[5 2; 2 1];
cho=chol(B);
```

MATLAB® gets $\begin{bmatrix} 2.2361 & 0.8944 \\ 0 & 0.4472 \end{bmatrix}$. It is an upper triangular matrix. So we just need to do the transposition and time it with `old_matrix`:

```
new_matrix=cho'*old_matrix;
covariance_new=cov(new_matrix');
```

MATLAB® gets $\begin{bmatrix} 5.1366 & 2.0823 \\ 2.0823 & 1.0445 \end{bmatrix}$. It is obvious that the Cholesky decomposition-based method could generate the normal distribution with a given covariance matrix.

Sometimes, It is hard to ensure that the user-generated covariance matrix, which is used to mutate an individual, is a positive definite matrix. So we can use the rotation transform to generate the positive definite covariance matrix easily. After given the standard deviation for each gene, we can generate the covariance between every two genes using Eq. 3.21 if we know the degree of rotation between every two axes. In this way, we can generate a normal distribution deviation from the current point with any angle and any strength. So it is called a *correlated mutation*.

We can illustrate the power of a correlated mutation with Fig. 3.10.

Suppose that in this 2-D situation, we only use an uncorrelated mutation with standard deviation $\sigma_x = 1$ and $\sigma_y = 1/3$. According to the 3σ rule, the main part

[25] Cholesky decomposition is discussed in all linear algebra textbooks.

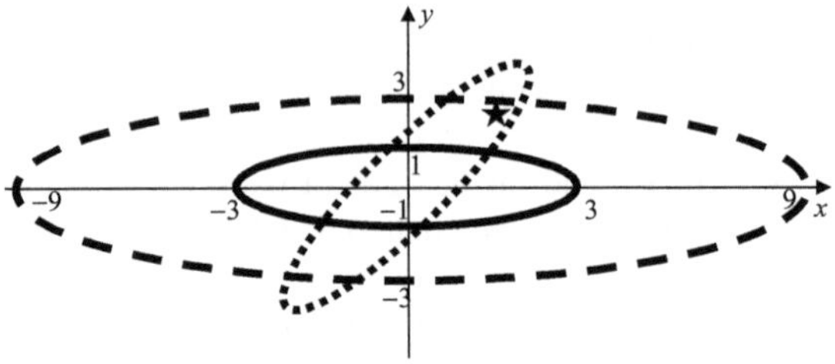

Fig. 3.10 The power of correlated mutation

of the possible sampling region is in the ellipse with a solid line in Fig. 3.10. If the global optimal solution is the star in Fig. 3.10, then the above mutation has little change to find it because of the 3σ rule. To achieve the goal, we need to enlarge the standard deviation, e.g., $\sigma_x = 3$ and $\sigma_y = 1$. Then the main part of the possible sampling region is in the ellipse with a dashed line in Fig. 3.10. But the ellipse with the dashed line is rather large so that the possibility of sampling the star is still rather slight. But if we could generate the covariance matrix considering covariance by Eq. 3.21, the normal distribution, whose main part of the possible sampling region is in the ellipse with a dotted line in Fig. 3.10, might find the star with high probability. It is obvious that considering covariance might improve the search significantly.

The correlated mutation with n variables could be implemented with the following steps.

1. Assign standard deviation, σ_j, for each variable j and rotation angles, φ_{ij}, for every two variables i and j.[26]
2. Use Eq. 3.21 to calculate the covariance between every two variables i and j. Thus the positive covariance matrix **B** is generated.
3. Use Cholesky decomposition to get the lower triangular matrix **L** from **B**.
4. Generate random vector $\xi = (\xi_1 \quad \cdots \quad \xi_n)^{\mathrm{T}}$ with n independent standard normal distribution random numbers.
5. Generate the multivariate normal distribution random vectors with covariance matrix **B** by $\mathbf{L}\xi$.
6. Generate the mutant of individual **x** as follows:

$$\mathbf{x}' = \mathbf{x} + \mathbf{L}\xi \tag{3.22}$$

Even through correlated mutation is powerful, it requires a lot of parameters predefined by users[27] and thus consumes much computation cost. So whether or not one uses correlated mutation depends on the tradeoff between performance requirements and computation ability.

[26] ES and EP use a self-adaptive control method to encode these control parameters into chromosomes, which will be discussed in Sect. 3.5.1. Here we just assume they are user defined control parameters.

[27] This feature promotes the self-adaptive control on them, which will be discussed in Sect. 3.5.1, and the covariance matrix adaptation, which will be introduced in Sect. 3.5.2.6.

Cauchy Mutation

Apart from normal distribution, Yao *et al.* used a Cauchy distribution to implement the wider mutation scale in EP [16]. The density function of a Cauchy distribution is as follows:

$$p(\xi) = \frac{1}{\pi} \frac{t}{t^2 + \xi^2} \tag{3.23}$$

where $t > 0$ is a scale parameter. The density functions of the standard normal distribution and Cauchy distribution ($t = 1$) are illustrated as follows:

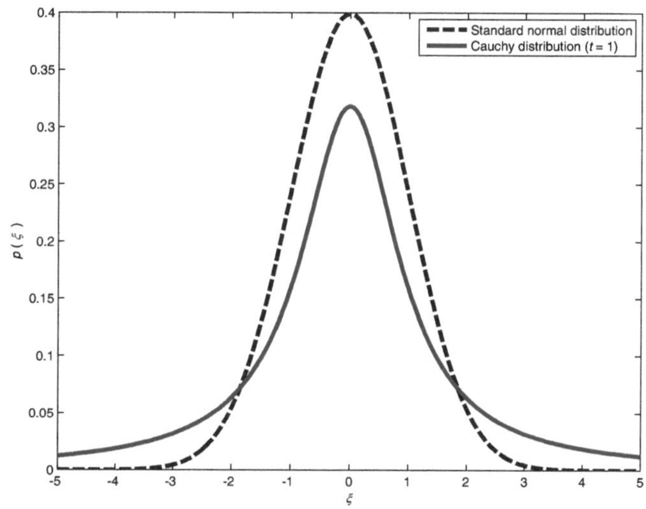

Fig. 3.11 The density functions of the standard normal distribution and Cauchy distribution ($t = 1$)

A Cauchy mutation differs little from a normal mutation. Here we use uncorrelated mutation with one step size as an example; other ways are all straightforward.

$$x_j' = x_j + \sigma C_j \tag{3.24}$$

where C_j is a Cauchy distribution random number generated by Eq. 3.23.

As can be seen from Fig. 3.11, the Cauchy mutation is very good at searching in a large neighborhood. So *Yao et al.* called it a "fast EP." To further promote the global search ability of mutation in EP, Lee and Yao suggested a mutation based on Lévy distribution [17].

Polynomial Mutation

Deb and Goyal suggested the *polynomial mutation* based on the polynomial distribution as follows [18]:

$$p(\xi) = 0.5\,(n+1)\,(1-|\xi|)^n \tag{3.25}$$

where n is a control parameter and $-1 \leq \xi \leq +1$. Figure 3.12 illustrates two conditions where $n=2$ and $n=5$. According to Fig. 3.12, ξ has a high probability of being 0. Larger n promotes this preference further.

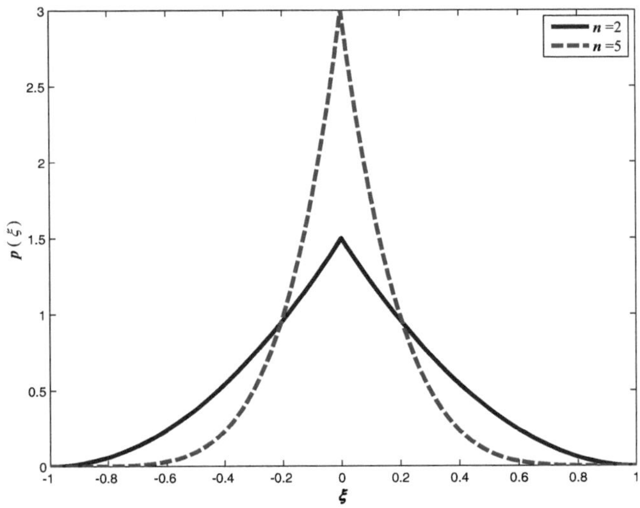

Fig. 3.12 Probability density function of polynomial distribution

As discussed in Sect. 3.2.2.2, we can generate the polynomial distribution number ξ using uniform distribution $u \sim U(0,1)$ as follows:

$$\xi = \begin{cases} (2u)^{\frac{1}{n+1}} - 1, & u < 0.5 \\ 1 - [2\,(1-u)]^{\frac{1}{n+1}}, & u \geq 0.5 \end{cases} \tag{3.26}$$

The mutant of i's variable is as follows:

$$x_j' = x_j + \Delta_{\max} \times \xi_j \tag{3.27}$$

where ξ_j is a polynomial distribution random number generated by Eq. 3.26, and $\Delta_{\max}$ is the maximum permissible perturbation of x_j.

3.2.2.4 Directional Mutation

We introduced several direction-based search methods in Sect 2.4. Here we will discuss some other directional mutation methods for DE apart from Eq. 2.18. For the sake of completeness, we rewrite it here. The mutant $\mathbf{v}_i$ of individual $\mathbf{x}_i$ is as follows:

$$\mathbf{v}_i = \mathbf{r}_1 + F(\mathbf{r}_2 - \mathbf{r}_3) \tag{3.28}$$

where $\mathbf{r}_1 \neq \mathbf{r}_2 \neq \mathbf{r}_3 \neq \mathbf{x}_i$ are randomly selected individuals from the current population and generally $0 \leq F \leq 1$ controls the strength of the direction. Equation 3.28 uses one random start point and one random direction to generate the mutant, so it would belong to the *rand/1* category.

Other categories include:

best/1

$$\mathbf{v}_i = \mathbf{x}_{\text{best}} + F(\mathbf{r}_1 - \mathbf{r}_2) \tag{3.29}$$

current to best/1

$$\mathbf{v}_i = \mathbf{x}_i + K(\mathbf{x}_{\text{best}} - \mathbf{x}_i) + F(\mathbf{r}_1 - \mathbf{r}_2) \tag{3.30}$$

best/2

$$\mathbf{v}_i = \mathbf{x}_{\text{best}} + K(\mathbf{r}_1 - \mathbf{r}_2) + F(\mathbf{r}_3 - \mathbf{r}_4) \tag{3.31}$$

rand/2

$$\mathbf{v}_i = \mathbf{r}_1 + K(\mathbf{r}_2 - \mathbf{r}_3) + F(\mathbf{r}_4 - \mathbf{r}_5) \tag{3.32}$$

current/2

$$\mathbf{v}_i = \mathbf{x}_i + K(\mathbf{r}_3 - \mathbf{x}_i) + F(\mathbf{r}_1 - \mathbf{r}_2) \tag{3.33}$$

where F and K are positive real numbers that control the strength of the direction, $\mathbf{r}_1, \mathbf{r}_2, \mathbf{r}_3, \mathbf{r}_4, \mathbf{r}_5$ are different randomly selected individuals from the current population, and $\mathbf{x}_{\text{best}}$ is the current best individual. Equations 3.28–3.33 are different with respect to how they learn from other individuals, so these ideas are called *learning strategies*.

DE's authors, Storn and Price, consider Eqs. 3.28–Eq. 3.33 mutation operators, but they actually use several parents to generate an offspring. So many GA participants also consider them crossover operators.

3.2.3 Other Topics on Code and Operators

Whenever a new code and its related operators are defined, we need to know the following things.

- How large is the operation domain? This determines the search space for EAs.
- Is the encoding and decoding map injective, surjective, and bijective? This determines whether there are *illegal* solutions and the size of the search space of EAs.[28]
- How is the distance on the operation domain defined? It will be used by many EA operators.
- What is the definition of neighborhood in the operation domain? It will be used in the possible local search.
- Is there any effective local search technique available for the operation domain? This will accelerate the convergence of EA.

There are many other codes for different problems and different EAs. Permutation code and its related operators will be discussed in Chap. 7. Tree code and its related operators will be discussed in Chap. 10.

Another very interesting and quite useful topic in code is variable length. Sometimes variable length code is a direct and efficient way to express the solution. But a special decoding process and variation operators need to be designed elaborately. A messy GA might be the first variable length code, which was suggested by Goldberg *et al.* [19]. Then the SAGA, suggested by Harvey *et al.* [20], and the SVLC, suggested by Hutt and Warwick [21], are all famous implement of variable length codes.

Here we would like to discuss the effects of crossover and mutation for exploration and exploitation in different stages of EAs. In the early stage, we need to explore the solution landscape, so the crossover is mainly responsible for exploration because the initial individuals are generated randomly and crossing them over may allow for a large-scale search for new solutions. In the late stage, the EA approaches convergence. So individuals are similar. Crossover then changes into a fine-tuning technique to exploit the attraction basin of an optimal solution.

On mutation, we first assume that the strength of mutation remains unchanged. Mutation has the role of exploring all through the evolving process if its strength remains large, or always exploiting the current point if its strength remains small, or changing from exploitation to exploration compared to crossover if its strength remains moderate.[29] If there is some kind of annealing technique used in mutation that shrinks the search scale with the evolving process, mutation has an effect similar to that of crossover.

---

[28] If some points in the operation domain do not have images in the definition domain, these points are illegal.

[29] Under this condition, we say that crossover and mutation are complement.

In this section, we have introduced many crossover operators that can generate one offspring with many parents. Then the parent selection process should be with replacement to ensure generating *popsize* offspring.

Some variation operators, such as SBX, PCX, UNDX, UNDX-m, and normal mutation, might generate illegal solutions if there is a definition region for every variable. In this case, special techniques should be adopted to ensure the legality of offspring and mutant. The most simple way is to just assign the value as the lower bound if it is smaller than the lower bound and *vice versa*. Another way is to reflect it back in the following way:

$$x_i = \begin{cases} 2L_i - x_i, & x_i < L_i \\ 2U_i - x_i, & x_i > U_i \\ x_i, & \text{otherwise} \end{cases} \tag{3.34}$$

where x_i is variable i of individual x in real value code and L_i and U_i are lower and upper bounds for that variable, respectively. Other advanced techniques will be discussed in Chap. 4.

It is quite obvious that operators performing multidirection search, such as PCX, UNDX, UNDX-m, and normal mutation, have advantages in optimizing problems with nonseparable variables. But the time complexity of these operators is higher than that of those basic operators and they are relatively hard to implement. So algorithm designers need to make tradeoff between search ability and search speed.

Another topic on the selection of parents is also helpful. SGA has no preference on this issue, i.e., any pair could cross over with probability p_c, which is known as *panmictic mating* or *random mating*. Actually, we can adjust population diversity using a *mating restriction*. If two similar individuals (genotype or phenotype) have more chances to breed, this type is called *positive assortative mating*. By contrast, *negative assortative mating* promotes crossover between dissimilar parents. Negative assortative mating could be implemented by comparing the distance between two individuals with a predefined threshold. If the distance is larger than the threshold, crossover can happen and *vice versa*. It is quite obvious that negative assortative mating promotes exploration. Thus population diversity could be maintained in this way. On the other hand, positive assortative mating promotes exploitation and pushes the population toward convergence. We will discuss an adaptive way to control the intensity of negative assortative mating in Sect. 3.5 and a positive assortative mating to find different optimal solutions simultaneously in Chap. 5.

3.3 Selection Methods

3.3.1 Dilemmas for Selection Methods

As was discussed in Chap. 2, we want the selection process to grant the better individuals more chances to breed. So selection is the main driving force for EAs to converge toward a global optimal solution.

But for real optimization or learning problems, designing an ideal selection process is not a trivial matter. We need to first emphasize something that is detrimental to the ideal selection and then discuss the solution for those dilemmas.

Genetic Drift

We consider a very simple algorithm that does not have any variation operators. The algorithm has *popsize* individuals in the population, each of which has only one binary gene, i.e., the allele is 0 or 1. The only operation in one generation is selection and we do not consider the performance of every individual, i.e., everyone has the same chance of being selected into the next generation. This is a trivial selection without any selective pressure. We can use a uniformly distributed random integer j in the range $[1, popsize]$ to select the current individual j as one individual in the next generation. This operation is repeated until *popsize* individuals have been selected. Remember, this selection is a sampling with replacement. The initialization is done in a random way, i.e., all individuals have a probability of 0.5 of being 1 and *vice versa*.

We define $number\_of\_1$ as a variable to illustrate how many individuals are 1. Of course, $number\_of\_0 = popsize - number\_of\_1$. So $number\_of\_1$ is a random integer, which is defined between 0 and *popsize*. In the evolving process, $number\_of\_1$ might change with generation number *gen*, which makes the evolving process a discrete time process, and $number\_of\_1(gen+1)$ is only determined by $number\_of\_1(gen)$. So it is a Markov chain. There are $popsize+1$ states for $number\_of\_1$, i.e., $\{0, 1, \cdots, popsize\}$. For $number\_of\_1(gen)$, we can get the probability changing from the current state to any state in the next generation. In this way, we can constitute $(popsize+1) \times (popsize+1)$ state transition matrix[30] $\mathbf{C}$ as follows:

$$\mathbf{C} = \begin{bmatrix} 1 & 0 & \cdots & 0 \\ p_{1,0} & p_{1,1} & \cdots & p_{1,popsize} \\ \vdots & \vdots & \vdots & \vdots \\ p_{popsize-1,0} & p_{popsize-1,1} & \cdots & p_{popsize-1,popsize} \\ 0 & \cdots & 0 & 1 \end{bmatrix} \tag{3.35}$$

[30] Here we do not deduce explicitly how to generate a stage transition matrix for this selection process. Interested readers should consult the relevant textbooks, such as [8] or [22].

where $p_{i,j}$ means the probability changing from state i to state j, which must satisfy $\sum_{j=0}^{popsize} p_{i,j} = 1$.[31] It is quite obvious that $p_{0,0} = 1$ and $p_{(popsize,popsize)} = 1$ because of our selection rule, i.e., state 0 and *popsize* are absorbing states. So for any initial state, it always has a probability to enter state 0 or *popsize* during this trivial selection. Whenever it enters 0 or *popsize*, it will never come out. Using the conclusions from the Markov chain, we can calculate the expectant absorbing generation, i.e., *takeover time*. But here we just use numerical simulation to illustrate this phenomenon.

We run the above algorithm until the population converges, i.e., individuals are all 1 or 0, count the generation number for the population to converge, and record which state it converges to. We run the algorithm with different *popsize* and run 1000 times for each *popsize*. Then we average the convergence generation and calculate the probability of converging to state *popsize*[32] and draw Fig. 3.13.

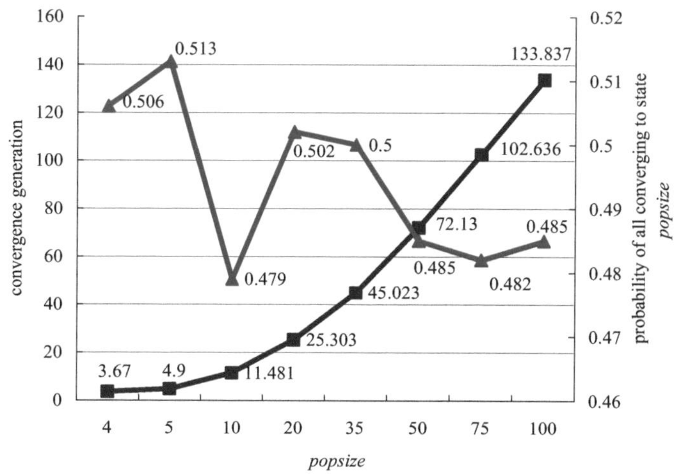

Fig. 3.13 Genetic drift. The *left vertical axis* is the average convergence generation for different *popsize*, the *squares* and the *links* between them correspond to this axis. The *right vertical axis* is the probability of converging to state *popsize*, the *triangles* and the *links* between them correspond to this axis

It is quite obvious from Fig. 3.13 that even without any selective pressure, the algorithm will converge to one absorbing state with almost the same probability after al relatively short time. This phenomenon is called *genetic drift*. If the genes have finite states and the population has finite *popsize*, genetic drift will happen and cause the algorithm to converge toward the wrong solution. The binary number code, the integer number code, and the permutation code all suffer from genetic

[31] What is the meaning of this constraint?

[32] You can get the probability of converging to state 0 in this way.

drift. So we need to design a good selection process to push the population in the right direction.

Premature

We need to push the population to converge to a global optimal solution with the help of a selection process. But sometimes a "super individual" that is far better than other individuals in the current population, appears in the early stages of EAs. It is actually a local optimal solution. Suppose the fitness value of "super individual" a is 200, the global optimal value is 205, and the mean fitness value of nine other normal individuals is 30. Then RWS will generate $10 \times 200/(30 \times 9 + 200) = 4.25 \approx 4$ copies of a if there is no selective bias,[33] which means a will soon dominate the whole population with the help of RWS. This phenomenon is called *premature*. To prevent premature, we want the selection process to preserve the population diversity as much as possible while giving good individuals more chances to breed.

Random Walk

The counterpart of premature is *random walk*. In the late stage of EAs, the population approaches the global optimal solution, if we are lucky enough. All the individuals are of high quality, i.e., their fitness values are high and similar. Their probabilities of being selected are similar. So in this situation, EAs sample the population randomly, which is like a random walk in the optimal basin. The random walk weakens the fine-tuning ability of EAs in the late stage. So we want the selection process to exaggerate the difference in the late stage of EAs.

High Expense of Calculating the Fitness Value

The fitness function of the numerical experiment in Chap. 2 is very simple. In a real-world problem, sometimes it is hard to get a specific number for every individual to evaluate its fitness. If we want to compose music using EAs, we need to evaluate the chromosome, which is a composition of music. Generally, human beings are used to evaluating music subjectively. It is easy for a listener to tell which one he or she prefers when comparing two pieces of music, but very hard to assign an exact number to evaluate the music. This EA type is often referred to as an *interactive EA*. Giving a specific fitness value for every individual is useful for most algorithms. But it is hard in some conditions, e.g., the above mentioned music evaluation. Another way to evaluate individuals is to rank them, which requires less information. The least requirement for evaluating an individual is to indicate a preference in the comparison process, i.e., the individual is better than, worse than, or the same as another

---

[33] We will discuss selective bias later.

individual, which often happens in multiobjective optimization and interactive EAs. The selection process should be able to deal with these requirements.

Fitness Transferral

This situation was discussed at the beginning of this chapter. We need to change the objective value into a maximum problem and require all the fitness values to be positive to enable RWS. Or we need some other selection processes that can deal with minimum and negative objective values directly.

3.3.2 Proportional Selection

If one selection process can fulfill the requirement that the expected number of an individual be proportional to its relative fitness value, i.e., probability of being selected, we call it a *proportional selection.* RWS is a typical proportional selection.

We rewrite the probability of being selected p_i for individual i, Eq. 2.10, as follows:

$$p_i = \frac{f_i}{\sum_{i=1}^{popsize} f_i} \tag{3.36}$$

where f_i is the fitness value of individual i. The expected number of an individual in proportional selection is as follows:

$$n_i = p_i \times popsize = \left\lfloor popsize \times \frac{f_i}{\sum_{i=1}^{popsize} f_i} \right\rfloor \tag{3.37}$$

where f_i is the fitness value of individual i, *popsize* is the population size, and $\lfloor\ \rfloor$ is the round function, which might be toward negative infinity or toward the nearest integer. We define the *selective error* as the difference between n_i and the real number of individual i after the selection.

In RWS, all the results are randomly determined, which means every individual might be lost whatever its relative fitness value is. So RWS has selection errors. Then we could say that RWS suffers from *selective bias.*

How does one decrease the selective bias or even eliminate the selective error?

Many methods have been proposed, such as remainder stochastic sampling with (or without) replacement. But the most often used one is *stochastic universal sampling* (SUS) suggested by Baker [23].

Let us discuss SUS in a simple four-individual population, illustrated in Fig. 3.14.

RWS has one arrow. Each time the arrow rotates $rand \times 2\pi$ angles, where $rand \sim U(0,1)$. If the arrow stops at area j, we say that individual j is selected. So RWS is carried out in a serial way.

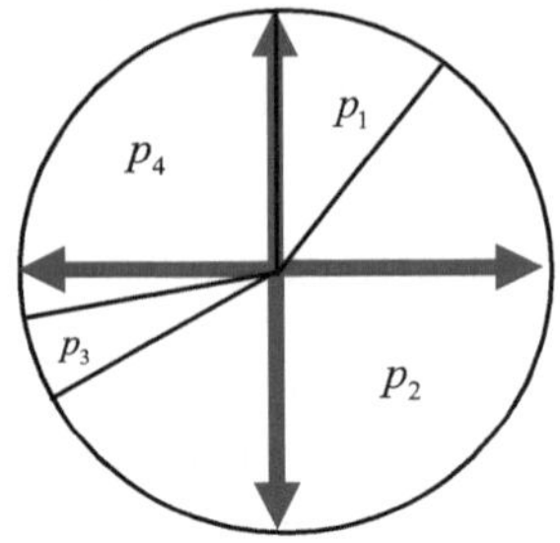

Fig. 3.14 Stochastic universal sampling

Unlike RWS, SUS has *popsize* evenly distributed arrows. The angle between them is $2\pi/popsize$. Each time the arrows rotate $rand \times (2\pi/popsize)$ degrees, where $rand \sim U(0,1)$. If any arrow i stops at area j, we say that individual j is selected by arrow i. So SUS is carried out in a parallel way.[34]

It is quite obvious from Fig. 3.14 that if the expected number of an individual is greater than one but less than two, such as individual 4 in Fig. 3.14, it will be selected at least once and perhaps twice. Individual 2 in Fig. 3.14 will be selected at least twice and perhaps three times. In this way, SUS could do a proportional selection without bias. The unbiased characteristic makes SUS the most popular proportional selection.

3.3.3 Fitness Scaling and Transferral

In this subsection, we will first try to solve the contradiction between premature and random walk and then discuss the fitness transferral.

Linear Scaling

Scaling is a very useful way to change the shape of a curve. Let us suppose that the curve represents the alteration of temperature changing with time. If the scale for the vertical axis is between −10 and 150°C but the real temperature changes between 15 and 25°C, the curve then looks more like a flat line. We can change the scale of the vertical axis, a procedure called *scaling*, to between 10 and 30°C to clear the temperature change. Mathematically speaking, the above scaling procedure is $y' = 0.125 \times y + 11.25$, where y is the original temperature graduation and y' is the scaled graduation.[35] Even though it is an affine transformation, not a strict linear transformation,[36] we often call it *linear scaling*.

[34] The implementation of SUS in a programming environment is left as an exercise.

[35] You can substitute numbers into the function to verify it.

[36] Translational move is not a linear transformation.

Before discussing linear scaling used in EAs, we need to point out that scaling could be used in both the vertical and horizontal axis and many other nonlinear scaling methods are very useful in engineering, such as Bode plot in control theory and signal process.

As was introduced above, linear scaling change the fitness value of individuals in the following way:

$$f_i^{'} = af_i + b \tag{3.38}$$

where f_i is the fitness value of individual i before scaling, $f_i^{'}$ is the fitness value of i after scaling, and a and b are control parameters embodying the designer's consideration.

There might be many implementation methods to determine a and b in Eq. 3.38. One of them might consider the following two things:

1. The original average fitness value does not change its value after scaling.
2. The original maximum fitness value becomes twice as large as the original average fitness value.

Equation 3.39 could satisfy the above consideration.[37]

$$f_i^{'} = \frac{f_{\text{avg}}}{f_{\max} - f_{\text{avg}}} f_i + \frac{f_{\text{avg}}\left(f_{\max} - 2f_{\text{avg}}\right)}{f_{\max} - f_{\text{avg}}} \tag{3.39}$$

where f_{avg} is the original average fitness value and $f_{\max}$ is the original maximum fitness value. Then the proportional selection discussed in Sect. 3.3.2 could be used to do the selection.

We can say that the average fitness value does not change because

$$f_{\text{avg}}^{'} = \sum_{i=1}^{popsize} f_i^{'}/popsize = a\left(\sum_{i=1}^{popsize} f/popsize\right) + b = af_{\text{avg}} + b = f_{\text{avg}}$$

If there is a "super individual" in the early stage, linear scaling in Eq. 3.39 could force the expected number of the individual with the maximum fitness value to be *two*, which diminishes its influence and promotes exploration in the early stage.

If individuals hold similar fitness values in the late stage, linear scaling in Eq. 3.39 could also force the expected number of the one with the maximum fitness value to be *two*, which enhances its influence and promotes exploitation in the late stage.

But Eq. 3.39 might cause $f_{\min}^{'} < 0$, which is not acceptable for proportional selection. If this happens, we just need to modify our considerations as follows:

1. The original average fitness value does not change its value after scaling.
2. The original minimum fitness value becomes zero after scaling.

---

[37] Readers are encouraged to deduce the result. At least you should verify whether Eq. 3.39 satisfies the above considerations or not.

Equation 3.40 could satisfy the above considerations. We need to point out here that the individual with the minimum fitness value has no chance to survive in the following proportional selection:

$$f_i' = \frac{f_{\text{avg}}}{f_{\text{avg}} - f_{\text{min}}} f_i - \frac{f_{\text{avg}} f_{\text{min}}}{f_{\text{avg}} - f_{\text{min}}} \tag{3.40}$$

Fitness Transferral

If the problem has a minimum requirement, i.e., "$\min f(\mathbf{x})$," and we want to use proportional selection, the simplest way to deal with it is to time -1 with the function and change the optimization to be "$\max -f(\mathbf{x})$."

But this simple method cannot guarantee that all $-f(\mathbf{x}) \geq 0$, which is required by proportional selection. We can add a constant number c to all fitness values to ensure that $c - f(\mathbf{x}) \geq 0$.

The problem is that it's hard to assign a proper c if we are doing a blind search over the solution landscape. If c is too small, it cannot ensure that $c - f(\mathbf{x}) \geq 0$; if c is too large, the proportional selection will like a random walk again.

The pragmatic way to solve this problem is to set $c = f_{\text{max}}$, where f_{max} is the maximum fitness value in the current population. In this way, the minimum optimization problem "$\min f(\mathbf{x})$" could be changed to "$\max(f_{\text{max}} - f(\mathbf{x}))$" so that proportional selection is usable. This process is called *fitness transferral*.[38] In the following part of this subsection, we assume that fitness transferral has been done, i.e., we are facing maximum fitness values that are greater than zero.

The alert reader might have already noticed that the above fitness transferral process could be regarded as one type of linear scaling method.[39]

Sigma Truncation

Sigma truncation is a linear scaling method considering the standard deviation of the current population. It can be illustrated as follows:

$$f_i' = f_i - \left(f_{\text{avg}} - c \times \sigma\right) \tag{3.41}$$

where σ is the standard deviation of the current population, c is an integer control parameter in the range $[1,5]$,[40] and f_{avg} is the average fitness value of the current population.

If the result of sigma truncation is less than zero, simply set it to zero.

---

38 Readers are encouraged to verify the correctness of this transferral by letting $f(\mathbf{x}) = f_{\text{max}}$ and $f(\mathbf{x}) = f_{\text{min}}$. This point verification method is very useful in understanding equations in papers.

39 If the objective value for a minimum optimization $\min f(\mathbf{x})$ satisfies $f(\mathbf{x}) \in [0, \infty)$, we can use another nonlinear scaling to change it into a maximum problem, i.e., $\max \frac{1}{1+f(\mathbf{x})}$.

40 The effect of c on the selective pressure is left as an exercise.

In the early stage, f_{avg} is relatively small and σ is relatively large, which makes the $f_{avg} - c \times \sigma$ relatively small, sometimes even smaller than zero. This is equivalent to subtracting a small positive number or even adding a positive number to the current fitness value, which decreases the selective pressure and promotes exploration.

In the late stage, f_{avg} is relatively large and σ is relatively small, which makes the $f_{avg} - c \times \sigma$ relatively large. This is equivalent to subtracting a large positive number to the current fitness value, which increases the selective pressure and promotes exploitation.

Power Law Scaling

Power law scaling uses the power law to implement scaling. It can be illustrated as follows:

$$f_i' = (f_i)^{\alpha} \tag{3.42}$$

where α is a control parameter. The larger the α is, the larger selective pressure is.

Boltzmann Scaling

Boltzmann scaling uses exponentiation to scale the fitness value and considers the annealing effect.[41] It can be illustrated as follows:

$$f_i' = \exp\left(\frac{f_i}{t}\right) \tag{3.43}$$

where t is the temperature. In the early stage, the temperature is relatively high. This means the results of the Boltzmann scaling are similar, which represents a low selective pressure and promotes exploration. In the late stage, the temperature is relatively low. This means the results of the Boltzmann scaling are significantly different, which represents a high selective pressure and promotes exploitation.

Suppose we have a population. The fitness values varies from 1 to 10. The mean f_{avg} and the standard deviation σ are 5 and 3, respectively. c in sigma truncation is 1, α in power scaling is 2, and temperature t in Boltzmann scaling is 1. We use Eqs. 3.40–3.43 to scale the fitness value and draw the transfer function in Fig. 3.15.

Readers are encouraged to analyze Fig. 3.15 to summarize further conclusions.

[41] Turn to page 53 if you are not familiar with annealing.

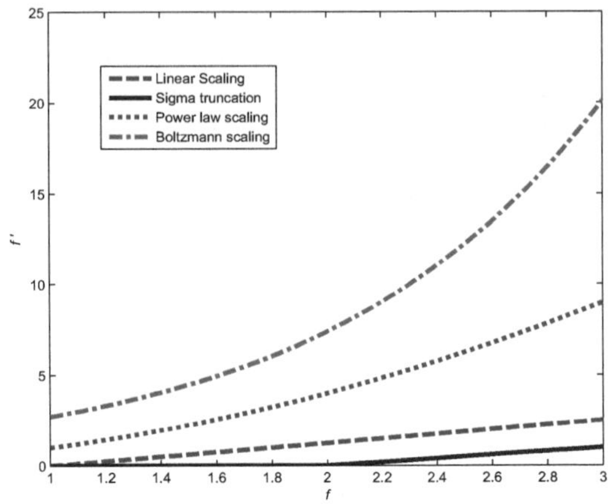

Fig. 3.15 Fitness scaling. For the reason of clearness, only fitness values in the range $[1,3]$ is presented

3.3.4 *Ranking*

We might need ranking in two conditions. The first is that sometimes we cannot get the exact fitness value but we can get the rank. So we cannot use proportional selection directly. The second reason is that we can use ranking to adjust the selective pressure effectively, which will be illustrated later. *Ranking* is to rank the individuals first and use these ranks to grant the probability of being selected. Then proportional selection can be used.

Linear Ranking

There are several methods of linear ranking. We will discuss two of them.

The first method is to rank the individual in the following way. The best one has rank 0 and the worst one has rank $popsize - 1$. Then we can assign the probability of being selected p_i for individual i as follows:

$$p_i = \frac{\alpha + \frac{rank_i}{popsize-1}(\beta - \alpha)}{popsize} \tag{3.44}$$

where α and β are parameters to control selective pressure and $rank_i$ is the rank of individual i.

The rank of the best individual is 0, its expected number in the following proportional selection is α. The rank of the worst individual is $popsize - 1$, its expected number in the following proportional selection is β.

We need to ensure that all the probabilities of being selected add up to one. So we can deduce that α and β must satisfy $\alpha + \beta = 2$.[42] The largest number for α is 2, which means the worst individual will not be selected. And we can also deduce that the expected number of an individual with average fitness is 1. So the expected number of the best individual is no more than twice that of the average one.

The second method is to rank individuals in a similar way. The best one has rank 1 and the worst one has rank *popsize*. Then we can assign the probability of being selected p_i for individual i as follows:

$$p_i = q - (rank_i - 1)\, r \tag{3.45}$$

where q is the parameter to control selective pressure, r is the parameter to ensure the sum of the probability of being selected is one, and $rank_i$ is the rank of individual i.

We can deduce that q and r must satisfy $q = \frac{r(popsize-1)}{2} + \frac{1}{popsize}$.

If $r = 0$, then $q = 1/popsize$. According to Eq. 3.45, this means that every individual has the same probability of being selected $1/popsize$, which reflects the minimum selective pressure.

If $r = \frac{2}{popsize(popsize-1)}$, then $q = 2/popsize$. This means that the worst individual will not be selected, which reflects the maximum selective pressure.

It is not difficult to get the result that the probability of being selected of the individuals in the current population decreases from q to $(\frac{2}{popsize} - q)$ linearly [43].

Nonlinear Ranking

Unlike linear ranking, the probability of being selected in nonlinear ranking is the nonlinear function of the rank. There are also many implementations for nonlinear ranking. We just discuss one example, which uses the density function of geometric distribution to construct the probability of being selected:

$$p_i = \alpha\,(1-\alpha)^{popsize-rank_i} \tag{3.46}$$

where $0 < \alpha < 1$ is the parameter to control selective pressure. If *popsize* is very large, the sum of the probability is approximately one. The best individual has rank *popsize*, which makes its probability of being selected α. The worst individual has rank 1, which makes its probability of being selected $\alpha(1-\alpha)^{popsize-1} \approx 0$.

Generally, nonlinear ranking has stronger selective pressure than linear ranking, which is like the relationship between linear scaling and nonlinear scaling.

---

42 Readers are encouraged to do the deduction.

43 Why?

3.3.5 Tournament Selection

The above discussed selection methods all consider global information, i.e., the relative fitness value or the rank. Sometimes only local information, i.e., which is the best one in a small group, is available. Then *tournament selection* is quite useful.

To implement tournament selection, we just need to pick up k individuals randomly with replacement and compare the fitness values of these k individuals, which is the tournament. The best one wins the tournament and is selected into the mating pool. Repeat the above process until *popsize* individuals have been selected.

The k is called the *tournament size*, which controls the selective pressure. If $k = 1$, then this indicates a random sample on the population without any selective pressure. But if $k = popsize$, then the mating pool will only contain the best individual in the current population. The most frequently used tournament selection is *binary tournament selection*, in which $k = 2$.

The main characteristics of tournament selection could be summarized as follows. These properties make tournament selection quite useful in some situations, such as multiobjective optimization.

- Tournament selection only uses local information.
- Tournament selection is very easy to implement and its time complexity is small.
- Tournament selection can be easily implemented in a parallel environment.

But tournament selection also suffers from selection bias, which means that the best one will not be selected if it is very unlucky and *vice versa*. To diminish the selective error, Sokolov and Whitley suggested *unbiased tournament selection* [24] and made improvements to it subsequently [25].

The idea of unbiased tournament selection is very simple. Let us discuss it with a population with five individuals. We suppose it is a minimum problem. Their fitness values are $\{4.1, 3.4, 0.5, 9.8, 7.7\}$, and the sequence represents the ID of each individual. If we want to implement the tournament selection with $k = 3$, so we just need to generate three random permutations in the range $[1, 5]$. Let us suppose they are (13452), (54123), and (32541), respectively. Then we can do the tournament selection as illustrated in Fig. 3.16.

For tournament size k, every individual will be compared with others k times in unbiased tournament selection, which ensures that the best individual will have k copies in the mating pool and the worst one will not be selected.

Goldberg added probability into tournament selection and suggested *Boltzmann tournament selection* [26]. Suppose the larger fitness value means fitter. Then in binary Boltzmann tournament selection, we just pick up randomly two individuals i and j with replacement. The probability of i winning the tournament is illustrated by Eq. 3.47.

$$p_i = \frac{1}{1 + \exp\left(\frac{f_j - f_i}{t}\right)} \tag{3.47}$$

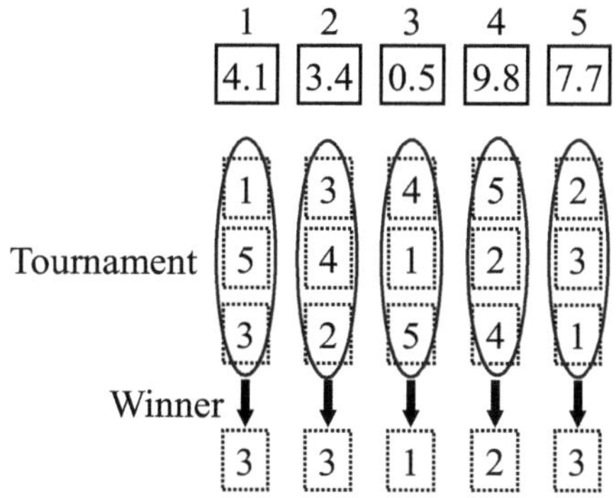

Fig. 3.16 Unbiased tournament selection

where t is for temperature, which will be decreased by an annealing process, and f_i and f_j are fitness values of individual i and j, respectively.

Let us first fix the temperature. If $f_i = f_j$, then $p_i = 0.5$, which reflects a random sampling. If $f_i > f_j$, which means i is better than j, then $p_i > 0.5$, which reflects the fact that i has a greater chance of winning the tournament. If $f_i < f_j$, which means i is worse than j, then $p_i < 0.5$, which reflects the fact that i has less chance of winning the tournament.

With the evolving process, temperature t decreases. If the temperature is very high, the differences in fitness values are negligible. The selection is just like a random walk to promote exploration in the early stage. Lower temperature will magnify the differences between individuals so that very small t will give the fitter individual probability 1 of being selected.

We have introduced many techniques in Sects. 3.3.2–3.3.5 to cope with the dilemmas discussed in Sect. 3.3.1. These methods can be summarized in Fig 3.17.

3.4 Replacement and Stop Criteria

3.4.1 Replacement

In Sect. 3.2, we introduced many variation operators in which the number of the offspring might be different from that of the parents. We could discipline ourselves to ensure that the number of the total offspring, λ, is always the same as that of the parents, μ. We can also suggest some rules to determine which offspring can go to the next generation under the condition that $\lambda > \mu$, or the rules are to determine which parents will be replaced by the offspring under the condition that $\lambda < \mu$. We often call the former circumstance *survivor selection* [44] and the latter circumstance *replacement*. In this book, we do not differentiate these two terms unless necessary.

[44] Then the selection process for generating the mating pool would be called *parental selection* accordingly for differentiation purposes.

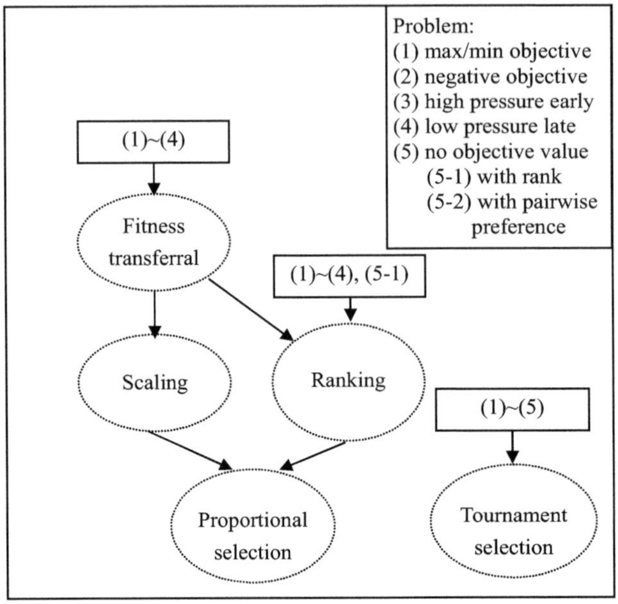

Fig. 3.17 Techniques to deal with dilemmas in Sect. 3.3.1 and adjust selective pressure

Another reason to consider replacement arises by comparing $(\mu+\lambda)$-ES with SGA, where $(\mu+\lambda)$-ES has the mechanism to maintain the best solutions found up till now.[45] From the discussions below you will know that many different replacement rules have dramatically different impacts on selective pressure. Some of them will maintain population diversity effectively. So in a dynamic environment, where the objective function changes with time, these rules take precedence [27]. We will discuss this in Sect. 3.7.

The replacement mechanism of SGA is called *nonoverlap*, which means that the new population and the old one will not overlap if parents and offspring are different.[46] We can define the extent of overlap in replacement by the term *generation gap* (*G*), which means that $G\mu$ individuals in the old population will be replaced by new individuals. $G = 1$ means nonoverlap replacement and is equivalent to SGA if $\mu = \lambda$ [28].

Then the general solution process of EA considering replacement can be illustrated as follows.

[45] μ best individuals are selected deterministically from $(\mu+\lambda)$ individuals or λ individuals in $(\mu+\lambda)$-ES and (μ,λ)-ES, respectively. So we often call it a *deterministic replacement*.

[46] Here we assume that crossover and mutation will change the old population. There is a small probability that p_c and p_m will also provide some overlap.

General Solution Process of EA

Phase 1: Initialization.

Step 1.1: Assign the parameters for EA, such as $\mu, \lambda, p_c, p_m, popsize$, stop criteria (such as *maxgen*), etc.

Step 1.2: Generate μ uniformly distributed individuals randomly to form the initial population and evaluate their fitness values. $gen = 0$.

Phase 2: Main loop. Repeat the following steps until stop criteria are satisfied (such as $gen > maxgen$).

Step 2.1: Generate the mating pool using some selection process introduced in Sect. 3.3.

Step 2.2: Generate λ new individuals using variation operators introduced in Sect. 3.2.

Step 2.3: Evaluate the fitness values of new individuals.

Step 2.4: Replace the $G\mu$ old individuals with new individuals according to replacement rules introduced in Sect. 3.4.1. $gen = gen + 1$.

Phase 3: Submitting the final μ individuals as the results of the EA.

If $G = 1$, the algorithm is called *generational* EA. By contrast, if $G = 1/\mu$ or $G = 2/\mu$, and $\lambda = 1$ or $\lambda = 2$, which means that only one or two new individuals will be generated and one or two old individuals will be replaced by the new ones in one generation, then the algorithm is called *incremental*, or *steady state*.[47] After one or two new individuals replace the old one(s), we say this algorithm goes to the next generation.

In steady state EA, the replacement rules can be generally divided into three categories as follows:

- **Fitness-based replacement.** Worse individuals have a greater chance of being replaced.
- **Random-based replacement.** The individuals being replaced are selected randomly.
- **Age-based replacement.** Older individuals have a greater chance of being replaced.

There are many rules for fitness-based replacement. The most straightforward one is to replace the worst individual. We can also use a tournament selection from the old population[48] to determine the one being replaced.

In random-based replacement, we just pick up a random individual among the old individuals. In age-based replacement, we need to define the age of an individual as the generation of its existence. The most often used age-based replacement

[47] Most of the population will remain unchanged in incremental EA. It seems that the algorithm is in a steady state, hence the name.

[48] Here the unfitter one wins the tournament. So it is sometimes called a *kill tournament*.

rule is to replace the oldest individual, also known as the first-in-first-out (FIFO) rule. The initial population has the same age, so we need to perform random-based replacement until all the individuals have different ages.

Replacement has a strong influence on the selective pressure, which will subsequently influence the search result. So we need some more guidelines for the selective pressure introduced by replacement.

We analyze the problem as in genetic drift. Suppose there is only one global optimal solution in μ individuals. The parental selection is based on binary tournament selection and there are no variation operators. We define the generation the population consisting of only the global optimal solution as the *takeover* time. This is also a Markov chain and the state is the number of global optimal solutions in the population. Probabilities of state transfer can be analyzed and the takeover time can be calculated mathematically [29, 30]. Also numerical experiments are available. We just summarize the sequence of the selective pressure as follows:

$$\textit{Replace worst} > \textit{Kill tournament} > \textit{FIFO} \approx \textit{Replace random}$$

The *replace worst* rule ensures the improvement of the population if the new offspring is not worse than the worst one in the old population. So takeover will absolutely happen. The *kill tournament* rule has the tendency to keep the better one, the loser of the kill tournament, so that it contains some selective pressure. The *FIFO* rule and the *replace random* rule are similar on a statistical basis. They have the weakest selective pressure.

There are many other techniques to adjust the replacement rule. Some of them are listed as follows:

- **Elitism.** If the best one is selected to be replaced according to the replacement rule, the replacement will not happen.[49] So elitism increases the selective pressure. Elitism can be combined with the *kill tournament* rule, the *FIFO* rule, and the *random* rule.
- **Conservative way.** If one individual is selected to be replaced according to the replacement rule, it will participate in the parental tournament selection. In this way, if the best one is selected to be replaced, it will be in the offspring. So the conservative technique keeps the best one in another way and thus increases the selective pressure. A conservative FIFO rule and a conservative random rule are often used.
- **Competition between parents and offspring.** The competition could be between the direct parents and offspring in a steady state EA, or between μ parents and λ offspring. In this way, the best individuals will never be lost and thus increase the selective pressure.

[49] Elitism could also be used in generational EA, where the best solution up till now, *best*, is maintained in another memory place. After the new population replaces the old one in a generational way, check whether *best* is in the population. If not, replace any one in the population with *best*. The memory to store the elitist is called the *archive*.

- **Replacement happens with probability.** We can define a probability for replacement so that the one being selected according to the replacement rule will have a chance to survive. This technique decreases the selective pressure.

Smith's research in 2007 reveals that the sequence of the selective pressure of these four techniques is in the follow way and these four techniques are weaker than the *replace worst* rule [30].

elitism > *conservative* > *p vs. o competition* >> *probabilistic replacement*

Since we have introduced so many types of parental selection, survivor selection, crossover operator, and mutation operator, a very straightforward question must be asked by alert readers: How do I determine the most suitable technique for my problem? Unfortunately, the answer is we do not know! Parental selection and survivor selection determine the selective pressure among *popsize* individuals. The crossover operator and mutation operator determine the explorative ability. The convergence force and the exploration force need to be adjusted elaborately to enable adequate search over the solution landscape. We can only give the suggestion that you may select the components of the algorithms more precisely when you know more about the problem. For example, if we know that the solution landscape of the problem is large, isolated, and multimodal according to previous experiences or trial sampling, then highly explorative variation operators and techniques to maintain the population diversity for a relatively long time is necessary.

Now we introduce two EA examples with replacement process.

Boltzmann Selection

Boltzmann selection can be implemented in both generational and an incremental way. Here we discuss steady state Boltzmann selection for a maximum problem.

After two individuals, p_1 and p_2, have been selected randomly with replacement from a population, crossover and mutation are used to generate two offspring, c_1 and c_2. c_1 will compete with p_1 and c_2 will compete with p_2 for survival in the population.[50] Offspring win the competition with the following probability:

$$p_{c_i} = \frac{1}{1 + \exp\left(\frac{f_{p_i} - f_{c_i}}{t}\right)} \tag{3.48}$$

where f_{p_i} and f_{c_i} are the fitness values of p_i and c_i, respectively (the larger, the better), and t is the temperature, which will be decreased with the evolving process.

If p_i is better than c_i, the denominator will be greater than 2, which makes c_i have a lower probability of winning the competition, and *vice versa*. Higher temperature tends to diminish the difference and lower temperature tends to magnify the difference.

[50] Here we just form the comparison pair randomly. Special concerns will be discussed in Chap. 5.

Species Adaptation Genetic Algorithm

Species adaptation genetic algorithm (SAGA), proposed by Harvey in 2001, is a very simple yet effective search algorithm for large-scale operation domain searching [31].[51]

In a large-scale operation domain, the code might have large redundancy, and the landscape might contain *ridges* between different local optimal solutions. The fitness values on the ridge is the same or similar,[52] but the ridges provide the path for searching for the global optimal solution.

For every generation, two individuals are randomly selected without replacement. The fitter one is called W, i.e., winner, and the other one is called L, i.e., loser. The user needs to assign two probabilities, i.e., probability of copying the gene from W to L p_r and the probability of bit-flip mutation p_m.[53] L changes every gene with probability $p = p_r + p_m$. If a change happens, then gene L_j copies the relevant gene W_j with probability p_r or flips it with probability p_m. Then W and L are reinserted into the population. That is one generation in SAGA.

There is no crossover operator in SAGA. In the large-scale searching domain SAGA faces,[54] for a very long time, i.e., several thousand generations, its best fitness remains unchanged until a sudden improvement happens. Harvey uses the concept of ridge and neutral network to explain this phenomenon.

3.4.2 Stop Criteria

The selection, crossover, mutation, and replacement discussed above constitute one generation in EA. Then the evolving process continues to the next generation. As mentioned above, selection grants fitter individuals a higher opportunity of propagating its high-quality genes, crossover and mutation explore the solution landscape, and replacement might increase or decrease the population diversity with different rules. We could have the rational expectation that the population will become better and better.

When do we stop?

Evolution in nature will never stop. But we need to define some *stop criteria*, or *termination criteria*, for the optimization or learning problems we face. Generally speaking, defining stop criteria is a hard task, like defining *popsize*, because we have no idea of the true performance of the EA on a specific problem. We only have the individuals' chromosomes and fitness values in all generations, and even though they are getting better, we do not know whether they are good enough.

There are three categories of stop criteria often used by EA participants.

---

[51] The concept of species will be discussed in Chap. 5. Not knowing what a specie is will not influence one's understanding of SAGA.

[52] So Harvey also called it a *neutral network* or *neutral plateaus*.

[53] Harvey only uses SAGA in binary code. It could be expanded to other codes.

[54] Whose size is about 2^{1900} in the example.

1. **Stop with fitness value.** If we know the fitness value of the optimal solution, then we could define a threshold so that EA stops when the current best, or the recorded best, fitness value is within the threshold of the optimal one. Another way to perform a stop with fitness value is to verify the variance of the fitness values in the current population. If most individuals in the current population are similar,[55] we call this situation *convergence* If the population is convergent, the variance of the fitness values is small. Then a threshold could be defined so that EA stops when the variance in the current population is below the threshold. The third way to stop the algorithm according to fitness value is to use the ratio of maximum fitness value to minimum fitness value or the relative error between them. The closer the ratio is to 1 (or the smaller the relative error is), the more convergent the current population is. A threshold could then be used to stop the algorithm.
2. **Stop with fitness change.** We could record the changes of the best fitness values for every generation. If the change is below the threshold of predefined values, EA stops. The changes in the average fitness value could also be used. The change rate is another option.
3. **Stop with time.** We could record the generation number. If it reaches the predefined maximum generation, expressed by *maxgen*, EA stops. We could also record the number of objective function evaluation. If this number reaches the predefined value, EA stops. We prefer the latter way to *maxgen* for the following two reasons. The most time-consuming step in real-world problem is often objective evaluation (which might include a decoding procedure). If the calculating ability is limited or if there are requirements for online performance, a stop with objective evaluation could reveal more information in evaluating the speed of searching high-quality solutions. The second reason is that the generation might be totally different with different algorithms.[56] Then it is hard to compare them using generation.

In a real implementation, several criteria could be combined to achieve a flexible yet effective rule to terminate an EA.

[55] The comparison might be carried out on an operation/definition domain or an objective/fitness value domain. If the problem is unimodal and the three maps illustrated in Fig. 3.1 all maintain the neighborhood relationship, then the domain on which the comparison is carried out is irrelevant. If not, then special considerations are necessary. Even though the variances of chromosomes are possible in practical use, we just discuss the variances of fitness values here.

[56] Consider the generation of steady state GA and SGA.

3.5 Parameter Control

3.5.1 Strategy Parameter Setting

Thus far we have introduced many variation operators and selection techniques. These options are double-edged swords. They give you the possibility of acquiring better solutions with the assumption that you have adequate knowledge about the solution landscape and the ability of these techniques. It basically violates the original intention of designing a robust, universal, and easy-to-use optimization and learning algorithm enlightened by Darwin's natural selection principle. But that's the truth of the real world: no free lunch.[57]

Right now, we are actually facing two problems: finding the optimal solution of the original problem, referred to as the *problem parameter*, and finding the optimal operators and their optimal parameters, referred to as the *strategy parameter*. Due to the nonlinear intrinsic characteristics of EAs, sometimes the latter problem is even harder than the former one! That's why there are so many papers discussing how to optimize the performance of optimization algorithms.

We can roughly divide the elements affecting the optimization results into two groups: local factors and global factors. Those options and parameters that will only have individual-level effects are *local factors*, e.g., crossover and mutation operators and their parameters, and those options and parameters that will have population-level effects are *global factors*, e.g., selection type and parameters, population size, stop criteria. Global factors have a stronger influence on the selective pressure and population diversity than local factors. So they are harder to set. In this section, we will introduce various techniques, from local factors to global factors, to set EAs with the intention that these techniques will release us, at least to some extent, from the later optimization problem mentioned above.

Eiben *et al.* presented a wonderful survey on strategy *parameter setting* [32]. We will follow the definition of the terms in that paper.

There exist some theoretical analyses on the optimal strategy parameter. But since EAs themselves are complicated systems and given the various intrinsic characteristics of the problems, these analyses all make strong assumptions that are hard to satisfy in real-world problems.

Much effort has been devoted to fine-tuning the strategy parameters for most problems in a numerical way. The most famous one is De Jong's experiments in his Ph.D. thesis [33]. He suggested that $popsize = 50$, $p_c = 0.6$, $p_m = 0.001$, $G = 1$,[58] and elitism are proper strategy parameters for his test functions.

Grefenstette used a meta-level GA to handle the strategy parameter optimization problem [34]. Every individual of the meta-level GA is a set of strategy parameters of lower-level GA by which a full optimization process on the real problem could be carried out. The best objective value of the lower-level GA was then regarded as the fitness value of the individual in the meta-level GA. The optimal solution of the

57 We will discuss this paradox in Sect. 3.6 again.

58 Generation gap.

meta-level GA, i.e., the best strategy parameters for online (offline) performance,[59] were $popsize = 30(80)$, $p_c = 0.95(0.45)$, $p_m = 0.01(0.01)$, $G = 1(0.9)$, and elitism.

As can be seen from the above two efforts,[60] suggestions from different researchers are achieved by different test functions and the performance criteria are different, which put GA users in an embarrassing situation, i.e., which suggestion is best suit for my problem?

So sometimes GA users need to do the *parameter tuning* discussed above themselves. Whether they use the trial-and-error method or utilize the meta-GA on their problems, both of them are very time consuming because the possible value number of the strategy parameters might be large and the coupling between strategy parameters might be tight.

To make things worse, you may have already realized that different problems might require different optimal strategy parameters. Even though you have already found the best parameters for your problem, you may feel uncomfortable if you want to adjust the population diversity in the evolving process of the EA, which means that the strategy parameters need to be changed on the fly. Eiben *et al.* gave the taxonomy of strategy parameter setting as follows and thus divided on the fly *parameter control* into the following three groups.[61]

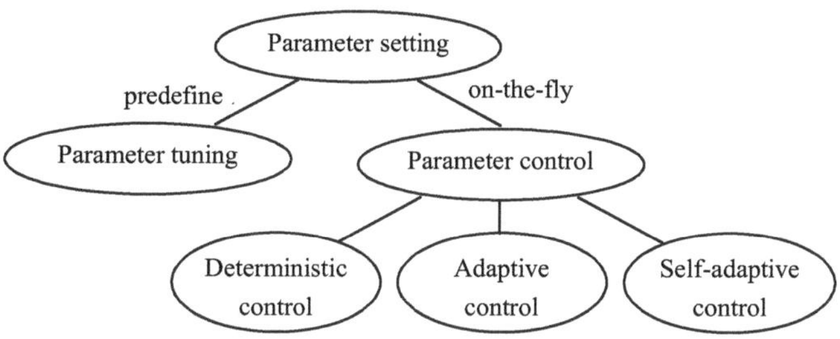

Fig. 3.18 Taxonomy of strategy parameter setting

1. **Deterministic control.** Strategy parameters are adjusted with heuristic rules generally only depend on *gen*.
2. **Adaptive control.** Strategy parameters are adjusted with heuristic rules that depend on feedback from the current or previous population.
3. **Self-adaptive control.** Strategy parameters are encoded into chromosomes and optimized at the same time as problem parameters using the same EA.

[59] We will mention these two performance criteria in Sect. 3.6. Here you can just consider them as two sets of optimal strategy parameters for different requirements.

[60] There are many other suggestions. One thing that needs to be mentioned here is we sometimes set $p_m = 1/n$ in real code, where n is the number of variables, with the intention of changing a variable by mutation in a blind environment.

[61] Kita and Deb considered the self-adaptive ability of GAs to be the ability to generate offspring according to the distribution of parents, which is a different concept from Eiben's definition [3, 12].

In the following part of this subsection, we take a mutation parameter as an example to illustrate these three groups.

Deterministic Control over Normal Mutation

The very straightforward consideration is that we want to explore the solution landscape in the early stage of EAs with high standard deviation normal mutation and exploit the optimal solution basin in the late stage using low standard deviation normal mutation, i.e., σ needs to be decreased with gen. One simple way to implement this idea is

$$\sigma\left(gen\right) = 1 - 0.9\frac{gen}{maxgen}$$

where $\sigma(gen)$ is the strength parameter, i.e., standard deviation, at generation gen in Eq. 3.15. Equations 3.13–3.48 discussed in the above sections are other deterministic controls for parameters.

If our heuristic rule is correct for the problem, deterministic control is easily defined as the change tendency of the strategy parameter, but special considerations on the implementation details, e.g., the speed and shape of the annealing curve, are needed. Furthermore, the robustness of deterministic control on different problems is generally weak because it has no idea of the online performance of the heuristic rule.

Adaptive Control over Normal Mutation

In (1+1)-ES, a mutation is considered to be successful if its mutant is better than the current individual. Rechenberg suggested the "1/5 success rule" to adaptively control σ [35]. The heuristic rule is that the ratio of the success mutation[62] to all mutations, p_s, should be 1/5.[63] If we want to adjust σ every τ generations, we just need to count the success mutation rate in the recent τ generations and define the following procedure.

if ($gen \bmod \tau = 0$) then

$$\sigma\left(gen\right) = \begin{cases} \sigma\left(gen-\tau\right)/c, & p_s > 1/5 \\ \sigma\left(gen-\tau\right)c, & p_s < 1/5 \\ \sigma\left(gen-\tau\right), & p_s = 1/5 \end{cases} \tag{3.49}$$

[62] The mutant is fitter.

[63] We guess that Rechenberg's rule came from the Pareto principle, also called the 80-20 rule, which states that, for many events, roughly 80% of the effects come from 20% of the endeavor. We will encounter Mr. Pareto again in Chap. 6.

else

$$\sigma(gen) = \sigma(gen-1)$$

end

where $0.871 \leq c \leq 1$ is a control parameter. If the success mutation rate is lower than 1/5, σ decreases so as to limit the search scale near the current solution, and *vice versa*.

Three questions need to be answered before designing an adaptive parameter control scheme: (1) What is the feedback information and how does one collect them? (2) What is the controlled parameter? (3) What is the heuristic rule whose input is the feedback information and output is the controlled parameter? In the above 1/5 rule, the success mutation rate in the recent τ generations is the feedback information, σ is the controlled parameter, and Eq. 3.49 is the heuristic control rule. To implement one type of heuristic control rule, there are many specific details to consider, e.g., c in Eq. 3.49. So a comprehensive study on the robustness of the rule and its parameters over different problems is necessary.

Self-adaptive Control over Normal Mutation

For the self-adaptive control of σ in Eq. 3.15, we just need to code it into the chromosome as $(x_1, x_2, \cdots, x_n, \sigma)$. Bäck further suggested that the mutation for σ should use lognormal distribution instead of normal distribution [36]. If we get a $\eta \sim N(\mu, \sigma^2)$, then $\xi = \mathrm{e}^{\eta}$ is a lognormal distribution random number. The density function of lognormal distribution for $\xi > 0$ is as follows:

$$p(\xi) = \frac{1}{\xi\sigma\sqrt{2\pi}} \exp\left(-\frac{(\ln(\xi)-\mu)^2}{2\sigma^2}\right) \tag{3.50}$$

where μ and σ are, respectively, the mean and the standard deviation of the variable's natural logarithm $\ln\xi = \eta$. We can draw the density function of standard normal distribution and lognormal distribution ($\mu = 0$) in Fig. 3.19.

Compared to a normal distribution, with the same σ, a lognormal distribution has a smaller variance. Apart from that, with smaller σ, a lognormal distribution will more likely be 1. Bäck wanted small perturbation for the standard deviation of normal distributions and used multiplication to mutate it as follows:

$$\sigma' = \sigma \mathrm{e}^{N(0,\tau)} \tag{3.51}$$

where τ is a user-defined parameter. Bäck suggested that $\tau \propto 1/\sqrt{n}$, where n is the number of variables, i.e., dimensions, of the problem. After σ gets its mutant, other genes, i.e., variables, mutate according to the new σ'. In this way, we first get the mutant of the stepsize and then use it to mutate the chromosome. This is called a *mutative step-size control* (MSC).

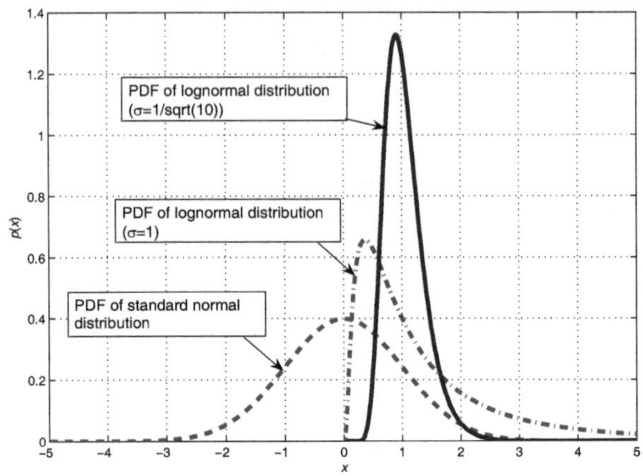

Fig. 3.19 Lognormal distribution and normal distribution

Besides uncorrelated mutation with one step size, uncorrelated mutation with n step sizes and correlated mutation could also be implemented in a similar self-adaptive way.[64]

A self-adaptive parameter control uses EAs to optimize problem parameters and strategy parameters simultaneously, which is a fascinating idea, but this increases the requirement for the exploration and exploitation ability of EAs.

Another thing that needs to be mentioned is that not all parameters can be self-adaptively controlled. Generally speaking, only those that have no direct influence on the fitness value can be self-adaptively controlled. For example, if we want to self-adaptively control parameter a in Eq. 3.38, the only possible result is that all individuals will have the same a gene, its upper bound.

In the following subsection, we will introduce some recently published interesting parameter control examples.

3.5.2 Examples of Variation Operator Control

3.5.2.1 Conditional Variation Operators

Wang and Okazaki suggested a conditional, not probabilistic, way to control variation operators in a deterministic way [39].

[64] Interested readers are referred to [37]. In addition, Schwefel discusses the initialization and the control parameter update of ES in detail [38].

We introduced the concept of mating restriction on page 63. Positive assortative mating promotes similar parents to cross over and negative assortative mating has the reverse effect. Both cases need a distance threshold to determine whether parents are permitted to mate or not.

After generating the mating pool and finishing parent pair matching, the *difference degree* (d_i) of the ith pair of parents could be calculated as follows:

$$d_i = \frac{N_d}{N_g} \tag{3.52}$$

where N_d is the number of different genes between the two parent chromosomes and N_g is the length of the chromosome. Wang and Okazaki gave the example with binary code or integer code, but this idea could be expands to other codes with modification.

If d_i is higher than a predefined threshold, i.e., *setting difference degree* D_s, which means that two parents are far away, they crossover. After cross over, if the number of offspring is smaller than *popsize*, those parents whose $d_i < D_s$ will undergo mutation until the new population has *popsize* individuals. This is a negative assortative mating rule introduced in Sect. 3.2

Wang and Okazaki also control the setting difference degree D_s by an annealing like process with the following rule:

$$D_s\left(gen+1\right) = \mu D_s\left(gen\right) \tag{3.53}$$

where $0 < \mu < 1$ is the predefined cooling ratio. With the evolving process, D_s is getting smaller, which makes crossover easier. On the other hand, if the population is getting convergent, then individuals are becoming similar and crossover is becoming harder. These two facets constitute the balance between exploration and exploitation of their EAs.

3.5.2.2 Adaptive Control on the Intensity of Assortative Mating

In 2008 Fernandes and Rosa proposed an adaptive way to control the threshold of negative assortative mating [40]. Their GA was also implemented with binary code, so the Hamming distance is used for distance metric.

In every generation, after generating the mating pool using a selection process and forming the parent pairs randomly, if the Hamming distance in one pair is larger than a *threshold*, this pair is allowed to cross over and it is a *successful mating*. If not, it does not cross over, but each of the parents mutates with probability p_{m}, and it is a *failed mating*.

At the end of every generation, new individuals and old ones are combined together, and their fitter *popsize* will form the new population.[65]

---

[65] It could be regarded as a $(\mu + \lambda)$-ES.

Also at the end of every generation, the number of successful matings and the number of failed matings are compared. If more matings failed, then $threshold = threshold - 1$ in order to promote crossover and *vice versa*.

The initial $threshold_0 = L - 1$, where L is the length of the chromosome. It is a rather high threshold so that there will be little crossover and the algorithm has a strong explorative effect by mutation at that time. If the population is approaching convergence, the Hamming distances between parents are getting smaller, and the *threshold* will decrease adaptively.[66] If the *threshold* is too low so that the number of successful matings is very large, then it will increase. The adaptive control of the *threshold* gives the algorithm good balance.

3.5.2.3 Fuzzy Logic Controller for P_c and P_m

There are many *fuzzy logic controllers* (FLC) for P_c and P_m. All of them use a *fuzzification interface* to change real feedback information into fuzzy numbers and a *defuzzification interface* to change fuzzy numbers into real strategy parameters. Here we just give examples on the heuristic control methods, i.e., *fuzzy inference*, and refer the interested reader on fuzzification and defuzzification to [41].

Yun and Gen suggested an FLC for P_c and P_m [42]. They took the improvement of the average fitness value in the successive two generations as the feedback information, i.e., $\Delta f_{avg}(gen) = f_{avg}(gen) - f_{avg}(gen-1)$ and $\Delta f_{avg}(gen-1) = f_{avg}(gen-1) - f_{avg}(gen-2)$, where $f_{avg}(gen)$ is the average fitness value in generation gen.[67] After fuzzification interface, $\Delta f_{avg}(gen)$ and $\Delta f_{avg}(gen-1)$ were mapped into two fuzzy numbers with value NR (negative larger), NL (negative large), NM (negative medium), NS (negative small), ZE (zero), PS (positive small), PM (positive medium), PL (positive large), PR (positive larger). Yun and Gen considered that successive improvement of a population meant that the current population is evolving toward a promising area so that we need to enlarge p_c and P_m, and *vice versa*. The control strategy is illustrated by Table 3.1.

After fuzzy number $\Delta p_c(gen+1)$ has been determined by the fuzzy rules, defuzzification can be carried out to transform it into a real value. Then the following equation could be used:[68]

$$p_c(gen+1) = p_c(gen) + \Delta p_c(gen+1) \tag{3.54}$$

Lin and Gen further improved the above FLC controller for strategy parameters [43]. They defined three parameters: (1) p_c is the ordinary crossover probability, (2) p_m is the probability of using heuristic local search methods to improve the current individual,[69] and (3) p_i is the probability of introducing randomly generated

[66] Why?

[67] Here the larger fitness means better.

[68] p_m can also be controlled in a similar way.

[69] Although it is written as p_m, this operator is a heuristic local search method instead of a probabilistic mutation.

Table 3.1 Fuzzy decision rule for Δp_c

$\Delta p_c\,(gen+1)$		$\Delta f_{avg}\,(gen-1)$								
		NR	NL	NM	NS	ZE	PS	PM	PL	PR
$\Delta f_{avg}\,(gen)$	NR	NR	NL	NL	NM	NM	NS	NS	ZE	ZE
	NL	NL	NL	NM	NM	NS	NS	ZE	ZE	PS
	NM	NL	NM	NM	NS	NS	ZE	ZE	PS	PS
	NS	NM	NM	NS	NS	ZE	ZE	PS	PS	PM
	ZE	NM	NS	NS	ZE	ZE	PS	PS	PM	PM
	PS	NS	NS	ZE	ZE	PS	PS	PM	PM	PL
	PM	NS	ZE	ZE	PS	PS	PM	PM	PL	PL
	PL	ZE	ZE	PS	PS	PM	PM	PL	PL	PR
	PR	ZE	PS	PS	PM	PM	PL	PL	PR	PR

new individuals. These three operators will be carried out on the current population and their results form the offspring population. Then the best *popsize* individuals are selected from the union of these two populations.[70] These three probabilities should satisfy the constraint $p_c + p_m + p_i = 1$ so that there is a tradeoff between exploration, caused by crossover and random initialization, and exploitation, caused by local search-based mutation. Then the FLC-based adaptive control rules for p_c and p_m are as follows.

First the weighted sum of the improvement in the recent τ generations is calculated:

$$\Delta f = \sum_{i=1}^{\tau} 2^{\tau-i} \lambda\,(\Delta f_{avg}\,(gen-i)), \quad gen \geq \tau \tag{3.55}$$

where τ is a user-predefined parameter, $\lambda\,(x) = 1$ if $x \geq 0$ or $\lambda\,(x) = 0$ if $x < 0$, and $\Delta f_{avg}(gen-i) = f_{avg}(gen-i) - f_{avg}(gen-i-1)$. According to Eq. 3.55, improvements in the recent generations are paid more attention. Lin and Gen think that a large weighted improvement means that the current population is in the global optimal attraction basin so that we need to employ more heuristic local search methods to improve the quality of the current population and limit crossover in order to avoid wasting computation resources. Thus, an FLC is used to increase Δp_m and decrease Δp_c with linear proportion to Δf.

The above two methods have different considerations on control rules, which might be utilized at different stages of EAs. In this direction, Zhang *et al.* suggested a clustering-based adaptive control method for p_c and p_m [44].[71]

They first cluster individuals into several groups using their chromosomes. After clustering, every individual has its own cluster and we know the size of each cluster,

[70] It can be regarded as a kind of $(\mu+\lambda)$-EA.

[71] Clustering is a method to classify data into groups. It is often used in optimization, pattern recognition, and machine learning. The simplest clustering method might be K-means clustering, which is utilized by Zhang *et al.* Interested readers are referred to [45–49].

i.e., the individual number belonging to that cluster, and denote G_{B} as the size of the cluster containing the best individual in the current population and G_{W} as the size of the cluster containing the worst individual in the current population. Then they are normalized in the scale $[0,1]$ as follows:

$$\hat{G}_{\mathrm{B}} = \frac{G_{\mathrm{B}} - G_{\min}}{G_{\max} - G_{\min}}$$

$$\hat{G}_{\mathrm{W}} = \frac{G_{\mathrm{W}} - G_{\min}}{G_{\max} - G_{\min}}$$

where $G_{\min}$ and $G_{\max}$ are the size of the smallest and the largest cluster, respectively.

Then the fuzzification procedure is carried out to map $\hat{G}_{\mathrm{B}}$ and $\hat{G}_{\mathrm{W}}$ into fuzzy numbers PB (positive big) or PS (positive small) depending on how close they are to 1 and 0, respectively.

Zhang *et al.* considered that there are four different stages in EAs' evolving procedure that need to be considered separately.

- If $\hat{G}_{\mathrm{B}}$ is PB and $\hat{G}_{\mathrm{W}}$ is PS, the population is in the mature stage, so we need to decrease both p_{c} and p_{m}.
- If $\hat{G}_{\mathrm{B}} = \hat{G}_{\mathrm{W}}$ and are all PB, the population is in the maturing stage. The global optimal solution has not been found and we want the EA to explore the solution space using large p_{c} and maintain the good solution with small p_{m}. So we need to increase p_{c} and decrease p_{m}.
- If $\hat{G}_{\mathrm{B}} = \hat{G}_{\mathrm{W}}$ and are all PS, the population is in the submaturing stage. The best solution has not been surrounded by many individuals, so we need to increase both p_{c} and p_{m} to accelerate the search.
- If $\hat{G}_{\mathrm{B}}$ is PS and $\hat{G}_{\mathrm{W}}$ is PB, the population is in a bad situation. The population might be in the initial stage. There are many individuals surrounding the worst points, so we need to decrease p_{c} in order to decrease the chance of generating similar worse individuals. At the same time, p_{m} needs to be increased to do the global exploration.

The above heuristic rules can be implemented with fuzzy numbers and then the defuzzification procedure is used to get real Δp_{c} and Δp_{m}.

3.5.2.4 Adaptive Control on Mutation Types

We introduced on page 59 the idea that mutation with Cauchy distribution occurs with a relatively larger variance to promote global search and normal distribution occurs with a relatively smaller variance to promote local search. So Yao *et al.* suggested a simple way to adaptively control the mutation type [16]. Every individual undergoes two mutations, normal mutation and Cauchy mutation, and the better mutant is the offspring.

Qin and Suganthan solved the problem of adaptive mutation type selection in DE with a probability [50, 51]. When an individual $\mathbf{x}_i$ requires mutation, it might be mutated using "rand/1," Eq. 3.28, with probability p_1 and "current to best/1," Eq. 3.30, with probability $p_2 = 1 - p_1$.

The initial value p_1 for every individual is 0.5, i.e., both strategies have an equal chance to produce mutants.

During the selection, Eq. 2.20, we count the successes of the trial vector and denote the numbers as ns_1 and ns_2 according to its mutation type, i.e., ns_1 is the number of trial vectors that are mutated by "rand/1," successfully entering the next generation, and *vice versa*. Also the number of failed trial vectors is counted and denoted as nf_1 and nf_2. $ns_1 + nf_1$ is the number of trial vectors generated by "rand/1" and $ns_2 + nf_2$ is the number of trial vectors generated by "current to best/1."

These numbers are accumulated in several generations (Qin and Suganthan used 50), and p_1 is adjusted according to

$$p_1 = \frac{\frac{ns_1}{(ns_1+nf_1)}}{\frac{ns_1}{(ns_1+nf_1)} + \frac{ns_2}{(ns_2+nf_2)}} = \frac{ns_1\,(ns_2+nf_2)}{ns_1\,(ns_2+nf_2) + ns_2\,(ns_1+nf_1)} \tag{3.56}$$

where $\frac{ns_1}{(ns_1+nf_1)}$ and $\frac{ns_2}{(ns_2+nf_2)}$ are the success rates of "rand/1" and "current to best/1," respectively. It is quite certain that higher success rates cause a larger possibility of being used later.

Then $p_2 = 1 - p_1$, rest ns_1, ns_2, nf_1, nf_2 to 0, and continue the iteration.

In this way, competition is carried out between two mutation types statistically so that the proper type has more chances of producing mutants.

3.5.2.5 Evolutionary Gradient Search

Classical gradient methods provide fast and reliable search on a differentiable solution landscape. But sometimes the differentiation assumption on the solution landscape cannot be satisfied or the gradient-based search will be trapped in a local optimal solution if we cannot provide a proper initial solution. Salomon suggested a method, *evolutionary gradient search* (EGS), using EAs to construct gradient information on a nondifferential landscape and later developed it for noisy environment optimization [52, 53].

EGS uses self-adaptive control for mutation strength, i.e., the chromosome is coded as $\mathbf{x} = (x_1, x_2, \cdots, x_n, \sigma)$.

EGS has the sense of $(1,\lambda)$-ES. So it only works on one individual. From current point $\mathbf{x}$, EGS generates λ new individuals $\mathbf{t}_1, \cdots, \mathbf{t}_\lambda$ using normal mutation, Eq. 3.15, and calculates their fitness values as $f(\mathbf{t}_1), \cdots, f(\mathbf{t}_\lambda)$. The estimated gradient is as follows:

$$\mathbf{g} = \sum_{i=1}^{\lambda} \left(f(\mathbf{t}_i) - f(\mathbf{x})\right)(\mathbf{t}_i - \mathbf{x}) \tag{3.57}$$

$(\mathbf{t}_i - \mathbf{x})$ is the direction from $\mathbf{x}$ toward $\mathbf{t}_i$. For maximum optimization problems, if this is an improvement direction, $f(\mathbf{t}_i) - f(\mathbf{x}) > 0$, it contributes to the estimation of the gradient. Larger improvements account for larger weights. By contrast, if this is a worsen direction, $f(\mathbf{t}_i) - f(\mathbf{x}) < 0$, it also contributes to the estimation of the gradient because $f(\mathbf{x}) - f(\mathbf{t}_i)$ is the improvement direction. For minimum optimization, Eq. 3.57 is also the estimation of the gradient.[72]

Then the estimated gradient is normalized as follows:

$$\mathbf{e} = \frac{\mathbf{g}}{\|\mathbf{g}\|} \tag{3.58}$$

After acquiring the estimation for the gradient, EGS utilizes an adaptive technique to change the mutation strength as on page 90. For maximum optimization problems,[73] two trial points could be generated according to σ and estimated gradient $\mathbf{e}$, i.e., $\mathbf{x} + (\sigma\zeta)\mathbf{e}$ and $\mathbf{x} + (\sigma/\zeta)\mathbf{e}$,[74] and the one with the larger fitness value determines the mutant of σ, i.e.,

$$\sigma' = \begin{cases} \sigma\zeta, & f(\mathbf{x} + (\sigma\zeta)\mathbf{e}) > f(\mathbf{x} + (\sigma/\zeta)\mathbf{e}) \\ \sigma/\zeta, & f(\mathbf{x} + (\sigma\zeta)\mathbf{e}) \le f(\mathbf{x} + (\sigma/\zeta)\mathbf{e}) \end{cases} \tag{3.59}$$

Then we get the new individual

$$\mathbf{x}' = \mathbf{x} + \sigma'\mathbf{e} \tag{3.60}$$

With the random properties of normal mutation, EGS could in the meantime estimate the gradient with the ability to escape from the local optimal attraction basin.

3.5.2.6 Covariance Matrix Adaptation

Salomon's EGS uses EAs to construct gradient information that is used to direct the search efficiently. Hansen and Ostermeier suggested *covariance matrix adaptation* (CMA) ES in 2001 [54] to further accelerate the search efficiency. CMA-ES supposes that the local solution space of the current point has a quadratic shape, i.e., the Taylor series of $f(\mathbf{x})$ around an $\mathbf{x}_k$ is as follows:

$$f(\mathbf{x}_k + \Delta\mathbf{x}) \approx f(\mathbf{x}_k) + \nabla f(\mathbf{x}_k)^{\mathrm{T}}\Delta\mathbf{x} + \frac{1}{2}\Delta\mathbf{x}^{\mathrm{T}}\mathbf{H}(\mathbf{x}_k)\Delta\mathbf{x} \tag{3.61}$$

where $\nabla f(\mathbf{x}_k)$ and $\mathbf{H}(\mathbf{x}_k)$ are the gradient and the Hessian matrix at $\mathbf{x}_k$, respectively. There are many techniques for convex quadratic numerical optimization, such as the Newton's method, the conjugate gradient method, and the quasi-Newton method [55]. Convex quadratic programming uses a positive definite Hessian ma-

---

[72] Why?

[73] We leave the minimum optimization problem as an exercise.

[74] Salomon suggested that $\zeta \approx 1.8$ is a good option.

trix to adjust the search direction so that these algorithms converge very fast, i.e., n-dimensional quadratic programming could be finished with at most n steps.

Real-world problems are often multimodal and isolated, so using quadratic programming directly could only guarantee the local optimal solution.

Recall Fig. 3.4. We use it to demonstrate that problems with nonseparable variables are hard. PCX, UNDX, UNDX-m, and correlated normal mutation could provide a multi dimensional search, which increases the search efficiency. As mentioned above, in self-adaptive ES, the standard deviation and the covariance (or the corresponding rotation angle) of multi dimensional normal distribution could be encoded into chromosome to be optimized by the algorithm.

In CMA-ES, the λ new individuals generated with normal distribution are regarded as samplings on the solution space. The density function, Eq. 3.17, is a quadratic function of $\mathbf{x}$. If we could simplify the local area of the solution space as a convex quadratic surface, the μ best individuals among λ might form a better density function by which we could generate better individuals. We can use a numerical experiment to illustrate the above idea.

Consider the problem with nonseparable variables illustrated by Fig. 3.4. Let us suppose that $(3,3)$ is the start point. We use the technique introduced in Sect. 3.2 to generate $\lambda = 100$ 2-D normal distribution random points with $\mathbf{a}^{(g)} = (3,3)^{\mathrm{T}}, \mathbf{B}^{(g)} = \begin{bmatrix} 1 & 0 \\ 0 & 1 \end{bmatrix}$, where $g = 0$. The individuals are illustrated in Fig. 3.20a.

We rank these 100 points according to their objective values in ascending order and pick the first $\mu = 20$ points $\mathbf{x}^{(g+1)}$ to represent the hopeful search direction and use them to construct the new $\mathbf{B}^{(g+1)}$ as follows:

$$\mathbf{B}^{(g+1)} = \frac{1}{\mu} \sum_{i=1}^{\mu} \left(\mathbf{x}_i^{(g+1)} - \mathbf{a}^{(g)}\right) \left(\mathbf{x}_i^{(g+1)} - \mathbf{a}^{(g)}\right)^{\mathrm{T}} \tag{3.62}$$

where $\mathbf{x}_i^{(g+1)}$ is an individual i of generation $(g+1)$, and $\mathbf{a}^{(g)}$ is the mean of generation (g). The new covariance matrix is $\mathbf{B}^{(1)} = \begin{bmatrix} 0.9870 & 0.6597 \\ 0.6597 & 1.4560 \end{bmatrix}$, which suggests that they are strongly correlation between two variables.

Then the expectance of the new generation could be calculated as follows:

$$\mathbf{a}^{(g+1)} = \frac{1}{\mu} \sum_{i=1}^{\mu} \mathbf{x}_i^{(g+1)} \tag{3.63}$$

The new expectance is $\mathbf{a}^{(1)} = (2.19, 2.02)$. After that we could use the new expectance $\mathbf{a}^{(g+1)}$ and the new covariance matrix $\mathbf{B}^{(g+1)}$ to generate $\lambda = 100$ new individuals, illustrated by Fig. 3.20b. Then $g = g+1$.

Using the above procedure again and again, we can see that the estimated density function could estimate the solution surface in the local area of the current mean, cf. Fig. 3.20c–e, the covariance matrices are $\mathbf{B}^{(2)} = \begin{bmatrix} 1.6171 & 1.5699 \\ 1.5699 & 2.0693 \end{bmatrix}$, $\mathbf{B}^{(3)} = \begin{bmatrix} 1.1142 & 0.7717 \\ 0.7717 & 0.6916 \end{bmatrix}$, $\mathbf{B}^{(4)} = \begin{bmatrix} 0.0659 & 0.0092 \\ 0.0092 & 0.0388 \end{bmatrix}$, respectively. In this way, CMA

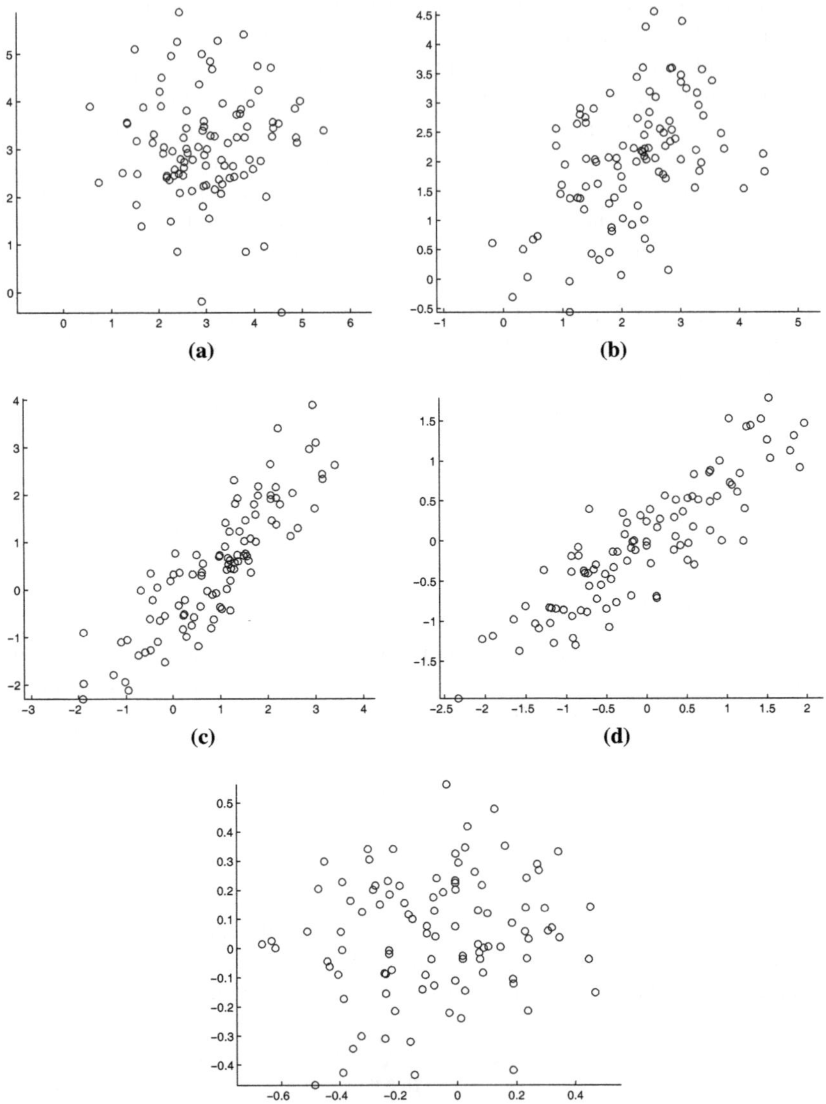

Fig. 3.20 A simple example of CMA: (a) initial distribution [$\mathbf{a} = (3,3)$], (b) first distribution [$\mathbf{a} = (2.19, 2.02)$], (c) second distribution [$\mathbf{a} = (1.05, 0.72)$], (d) third distribution [$\mathbf{a} = (0.08, 0.00)$], and (e) fourth distribution [$\mathbf{a} = (-0.04, 0.04)$]

can use the current best μ individuals (feedback information) and Eqs. 3.62 and 3.63 (control rule) to adaptively change the density function for generating λ new individuals (controlled parameter). From these iterative covariance matrices we can see that CMA cannot only point the main search in the most hopeful direction but also adjust the step size, i.e., variance, adaptively.[75] The power of CMA can be illustrated clearly by comparing Figs. 3.21 and 3.4b.[76]

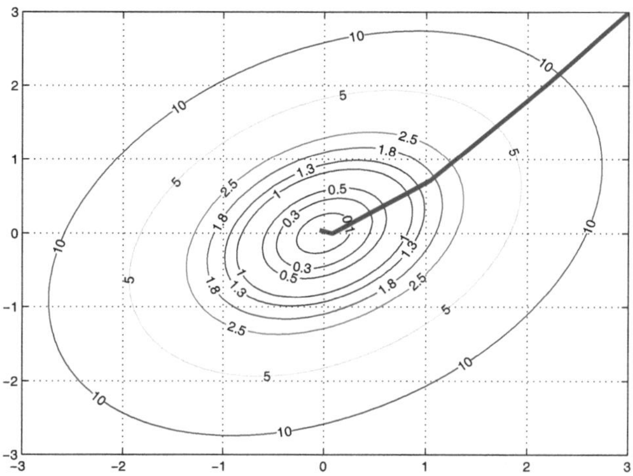

Fig. 3.21 Searching the nonseparable variables with CMA

Here we just illustrate the basic idea of CMA. The mutation step size is changed adaptively in CMA, which could be regarded as a *derandomized mutative step-size control* compared to the randomized σ control in Eq. 3.51. Hansen *et al.* also suggested many other ways to iteratively control **a** and **B**. Interested readers are referred to [54, 56].[77]

Another thing that needs to be mentioned is that CMA could be regarded as an *estimation of distribution algorithm* (EDA), which considers the solution landscape as a probability density function space and uses the population to estimate that probability distribution. EDAs are beyond the scope of this textbook, and interested readers are referred to [57–60].

Those algorithms that use the previously found solutions in such a way that the search will concentrate on the regions containing high-quality solutions and information gathered during the search are denoted *model-based algorithms*.[78] Zlochin

[75] Readers are encouraged to compare the axis scales of Fig. 3.20b and e.

[76] Readers are now encouraged to reread about normal mutation on page 54 to gain an in-depth understanding of the geometrical interpretation of different covariance matrices **B** using Fig. 3.20.

[77] The MATLAB® source code of CMA-ES can be downloaded at `http://www.bionik.tu-berlin.de/user/niko/cmatutorial.pdf`.

[78] The counterpart of model-based algorithm is *instance-based algorithm*, which generate new solutions using only current solutions. Examples of instance-based algorithms are SGA, simulated

et al. give an excellent survey on model-based search for combinatorial optimization [61].

3.5.2.7 Self-adaptive Crossover Probability Control in Differential Evolution

Many parameter control methods have been suggested for DE recently. Here we mention just two different self-adaptive considerations based on their crossover probability Cr.[79]

In DE, Cr determines the probability that the offspring will inherit genes from the mutant, cf. Eq. 2.21. The self-adaptive control for Cr needs to code it into chromosomes, i.e., $(x_1, x_2, \cdots, x_n, Cr)$.

Qin and Suganthan considered that the crossover probability Cr^i of individual i is initially generated with $N(0.5, 0.1)$ [50, 51]. The value remains unchanged in five generations. After that, a new Cr is generated with the same distribution, i.e., $N(0.5, 0.1)$. If a trial vector successfully enters the new generation, we record its Cr.

Then for every 25 generations, every Cr has been changed $25/5 = 5$ times with the same distribution, and the centroid of all recorded successful Cr is calculated as Cr_m. Then Cr is generated with $N(Cr_m, 0.1)$ and the success record is reset to zero.

In this way, the better value of Cr could have more and more influence on the whole population.

Brest *et al.* suggested another simple way to mutate Cr [62]. The mutation is carried out as follows:

$$Cr' = \begin{cases} rand_1, & rand_2 < 0.1 \\ Cr, & \text{otherwise} \end{cases} \tag{3.64}$$

where $rand_1$ and $rand_2$ are all $U(0,1)$ numbers. Equation 3.64 means that Cr changes to a new uniform distribution random number within $(0,1)$ with probability 0.1 and remains unchanged otherwise.

3.5.3 Examples of *popsize* Control

As mentioned above, *popsize* has strong influences on the population level, so adequate control on *popsize* might noticeably increase EAs' search. Many studies have been done on this topic, and we will select some of them according to our interests.

---

annealing, etc. As discussed previously, sometimes it is hard to discriminate between model-based algorithms and instance-based algorithms when the algorithm absorbs several powerful techniques from both of them.

[79] The papers cited below also have adaptive or self-adaptive control over other parameters. We just chose Cr as an example.

3.5.3.1 Saw-tooth-type Deterministic Control on *popsize*

Koumousis and Katsaras suggested a deterministic control on *popsize* [63]. A larger *popsize* is good for exploration in the early stage, while a smaller *popsize* is good for exploitation in the late stage. So Koumousis and Katsaras decreased the *popsize* with *gen*.

Another aspect of *popsize* is that population reinitialization, with the help of elitism, has been reported to have ascendancy in optimizing multimodal problems. Koumousis and Katsaras combined the idea of variable population size and population reinitialization. The main control rule for *popsize* is illustrated by Fig. 3.22.

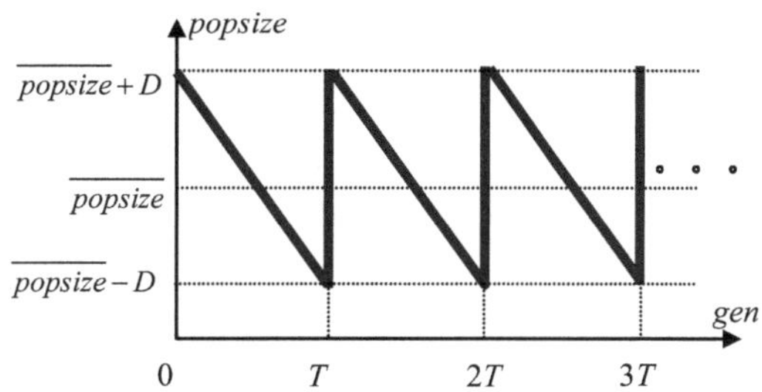

Fig. 3.22 Saw-tooth deterministic control on *popsize*

As can be seen from Fig. 3.22, the initial population is $\overline{popsize}+D$. It decreases linearly with *gen* to $\overline{popsize}-D$ after T generations. Then reinitialization is carried out with elitism to ensure that the best solution will never be lost. Detailed discussions have been done on the sensitivity of $\frac{T}{\overline{popsize}}$ and $\frac{D}{\overline{popsize}}$ to different types of benchmark problems. One result is that T is not sensitive to the performance of EAs if $T > 20$. Another conclusion is that a larger D seems to have more advantages, e.g., $\frac{D}{\overline{popsize}} \in [0.90, 0.98]$.

3.5.3.2 Adaptive Control on *popsize* Using an Age Concept

As far as we know, Arabas *et al.* were the earliest researchers focusing on the adaptive control of *popsize* using an age concept. They proposed a GA with varying population size (GAVaPS) [64].

In GAVaPS, every individual has two extra properties: *lifetime*, which is proportional to its relative fitness value, and *age*, which will be increased every generation. Whenever an individual's *age* is greater than its *lifetime*, it dies. Otherwise it will remain in the population. In this way *popsize* becomes an observable parameter, rather than a strategy parameter. The general solution process of GAVaPS could be illustrated as follows:

General Solution Process of GAVaPS

Phase 1: Initialization.

Step 1.1: Assign the parameters for GAVaPS, such as reproduction ratio ρ, maximum and minimum lifetime $MaxLT$ and $MinLT$, initial population size $popsize(0)$, and stop criteria.

Step 1.2: Generate $popsize(0)$ uniformly distributed individuals randomly to form the initial population. Evaluate the fitness value f_i, calculate $lifetime_i$, and set $age_i = 0$ for every individual i separately. $gen = 0$. $popsize(1) = popsize(0)$.

Phase 2: Main loop. Repeat the following steps until stop criteria are satisfied.

Step 2.1: $age_i = age_i + 1$ for every individual. $gen = gen + 1$.

Step 2.2: Select $\rho \times popsize(gen)$ individuals randomly, independently of f_i or age_i, and generate new individuals using crossover and mutation based on these individuals.

Step 2.3: Evaluate the fitness value for every new individual and calculate its lifetime. The age of the new individual is 1.

Step 2.4: Combine the old individuals with the new individuals and remove those whose age_i is greater than their $lifetime_i$.

Phase 3: Submitting the final individuals as the results of GAVaPS.

The population size of GAVaPS is

$$popsize\,(gen+1) = popsize\,(gen) + newpopsize\,(gen) - D\,(gen) \tag{3.65}$$

where $newpopsize(g)$ is the number of the new individuals and $D(gen)$ is the number of the removed individuals in generation gen.

To assign $lifetime_i$, Arabas *et al.* suggested three rules: proportional allocation, linear allocation, and bilinear allocation for maximum problems as follows:

$$lifetime_i = \min\left(MinLT + \eta\frac{f_i}{f_{\mathrm{avg}}}, MaxLT\right) \tag{3.66}$$

$$lifetime_i = MinLT + 2\eta\frac{f_i - f_{\min(Abs)}}{f_{\max(Abs)} - f_{\min(Abs)}} \tag{3.67}$$

$$lifetime_i = \begin{cases} MinLT + \eta\frac{f_i - f_{\min}}{f_{\mathrm{avg}} - f_{\min}}, & f_i \le f_{\mathrm{avg}} \\ \frac{1}{2}(MinLT + MaxLT) + \eta\frac{f_i - f_{avg}}{f_{\max} - f_{avg}}, & f_i > f_{\mathrm{avg}} \end{cases} \tag{3.68}$$

where $MaxLT$ and $MinLT$ are user-defined parameters corresponding to the maximum and minimum lifetime of an individual, f_{avg}, $f_{\max}$, and $f_{\min}$ are, respectively, the average, maximum, and minimum fitness values of the current population, $f_{\max(Abs)}$ and $f_{\min(Abs)}$ are, respectively, the maximum and minimum fitness values found so far, and $\eta = \frac{1}{2}(MaxLT - MinLT)$. As can be seen from Eqs. 3.66–3.68, all of these lifetime appointing schemes are with the intuitive idea that individuals with better relative fitness values have longer lifetimes.

There is no obvious selection process in GAVaPS. The selective pressure is adjusted by age_i and $lifetime_i$. Better individuals have longer lifetimes, so that they could have more chances to produce offspring.

Four parameters, $popsize(0)$, $MaxLT$, $MinLT$, and ρ, are necessary to eliminate the static $popsize$ in GAVaPS. Arabas *et al.* studied the sensitivity of these parameters and suggested that $\rho = 0.4$ might be the optimal value and found that $popsize(0)$ has little influence on the solution quality. They used $MinLT = 1$ and $MaxLT = 7$ in their numerical experiments. On the three lifetime allocation methods, their conclusion was that Eq. 3.67 had the best performance but with high computation cost, Eq. 3.68 had the cheapest computation cost but lowest performance, and Eq. 3.66 was in the middle position.

The main drawback of GAVaPS is that, according to Eq. 3.65, in the worst case, the population size will be doubled in the successive generations so that there might be a population explosion. So Bäck *et al.* suggested an adaptive population size GA (APGA) to handle this issue [65].

There are only three differences between APGA and GAVaPS:

1. APGA uses $lefttime_i$ instead of age_i. The initial $lefttime_i = lifetime_i$. In each generation, $lefttime_i = lefttime_i - 1$.
2. APGA is a steady state GA. Only two new individuals will be generated and inserted into a population. Those whose $lefttime_i = 0$ will be removed.
3. APGA utilizes an elitism mechanism, i.e., the $lefttime$ of the best individual in the current population will not be decreased.

The population size of APGA is

$$popsize(gen+1) = popsize(gen) + 2 - D(gen) \tag{3.69}$$

where $D(gen)$ is the number of the removed individuals in generation gen. According to Eq. 3.69, APGA has a rather “stable” population size control mechanism. Its maximum population size is $2MaxLT + 1$ and its approximate population size is $MinLT + MaxLT + 1$ [66].

In order to make $popsize$ change but not change too much, Fernandes and Rosa suggested the self-regulated population size EA (SRP-EA) [67]. Similar to the mating restriction introduced on page 87, they utilized negative assortative mating and counted the number of successful matings and the number of failed matings. If more matings fail, $threshold = threshold - Dec$. By contrast, $threshold = threshold + Inc$.

Newly generated individuals are added to the population. SRP-EA uses the same $lifetime_i$ allocation mechanism and age_i adding mechanism as GAVaPS, but has its probabilistic killing rule. All of the individuals, except the best one, die with the probability

$$p_{i,die} = 1 - \frac{lifetime_i - age_i}{maxLT} \tag{3.70}$$

Equation 3.70 means that better individuals, with larger $lifetime_i$, and younger individuals, with smaller age_i, have a smaller probability of dying.

The main mechanism to inhibit excessive population size in SRP-EA is that if the number of new individuals is larger than that of dying individuals in this generation, then $threshold = threshold + Inc$, which decreases the possibility of generating new individuals in the next generation. So in SRP-EA, there is a self-regulated population control mechanism.

Five parameters, $popsize_0$, $MaxLT$, $MinLT$, Dec, and Inc, are necessary to eliminate the static $popsize$ in SRP-EA. Fernandes and Rosa did the parameter sensitivity analysis on benchmark problems. They claimed that $Inc = Dec \in (1\%L, 10\%L)$, where L is the length of chromosome, are good options for SRP-EA.

3.5.3.3 Population Competition-based Adaptive Control on *popsize*

In 1999 Harik and Lobo considered the problem of *popsize* control as the competition between different population sizes and suggested a very interesting method for controlling *popsize* [68]. They call the algorithms *parameter-less GA*.

Their rationale for population competition is that smaller population should have more chance to exploit the solution to save computational cost. But if the average fitness value of a smaller population is worse than that of a larger population, the smaller population will be overtaken by the larger population and will be deleted. Apart from that, if all the chromosomes are the same in one population, meaning it is stuck in the local optimal solution, it will be deleted.

The mechanism to control the competition is very straightforward: a simple counter of base 4.

In a 4-base counter, we get (0), (1), (2), (3), (10), (11), (12) , (13), (20), (21), (22), (23), (30), (31), (32), (33), (100), (101), etc.

Every counter in the above sequence means a generation for one population.

We start with the first population with $popsize = 4$, and the first four counters in the above sequence mean that we run this population four generations. The carry, from (3) to (10), means the generation of the second population, which is twice the size of the first population, i.e., $popsize = 8$ in this example. The new population will run for a generation and return the control to the first population, i.e., the population with four individuals will run for the following three generations. Then the change in the high position, from (13) to (21), means a generation for the second population, i.e., the population with eight individuals will run for one generation and then return to the first population.

To sum up, every carry into the higher position means an initialization of a new population whose *popsize* is twice that of its previous population, every change in the high position means a generation for the larger population, and after these two operations the first population will continue to run until the next change in the high position or carry. If one population has been deleted, the counter will be reset.

In this way, parameter-less GA has a mechanism of increase and decrease *popsize* with very few predefined parameters.[80] If we are dealing with some problem and have no idea of the complexity of its solution landscape, i.e., have no idea of the magnitude of *popsize*, a parameter-less GA might be a powerful adaptive way to search.

Some interesting numerical comparisons have been done on the above *popsize* control mechanisms [66, 69]. Lobo and Lima gave a very comprehensive survey on this subject [70]. We adopt some guidelines discussed in the survey as follows:

- **No upper bound for adaptive *popsize* control.** We have no idea on the difficulty of the real-world problem. Specific *popsize*, selection methods, and variation operators constitute the specific search ability of the algorithm over the solution landscape. So setting the upper bound for *popsize* might not be a good option.
- **Do not forget the excuse of adaptive control.** We want to alleviate the pressure of strategy parameter tuning for EA users. So do not transfer the pressure of strategy parameter tuning to the pressure of control parameter tuning.
- **Consider the scalability of the algorithm.** Try to test how the algorithm scales up with problems of different sizes.
- **Compare in both directions.** Both the solution quality after the same objective function evaluation time and the objective function evaluation time to reach the same solution quality should be considered.

We have no content to introduce the population size control in DE. Interested readers are referred to three papers published in 2006, 2008, and 2009, respectively [71–73].

3.6 Performance Evaluation of Evolutionary Algorithms

3.6.1 General Discussion on Performance Evaluation

If you have read some research papers on EAs according to our suggestion or your interests, you must have noticed that at least half the content of the papers is about the performance evaluation of an algorithm or an operator. That means we need to focus ourselves on evaluating the improvements we made fairly and statistically to confirm that our initial idea, which sounds reasonable, is correct. But unfortunately, few papers and textbooks discuss such an important issue.[81]

---

[80] The initial $popsize = 4$ might be regarded as one parameter.

[81] Eiben and Smith's textbook has a chapter on it. We strongly encourage our readers to read [37].

3.6.1.1 No Free Lunch Theorem for Optimization

The first thing you need to keep in mind about the performance evaluation of algorithms is that the performance depends on problems.

Wolpert and Macready published a paper with a very strong title: "No Free Lunch Theorems for Optimization" [74]. The key contents of the paper can be quoted as follows:

> For both static and time dependent optimization problems, the average performance of any pair of algorithms across all possible problems is identical.

That's really a very interesting yet annoying statement. The interesting part is the fact that the performance evaluation is on specific (benchmark) problems, while the annoying part is that we do not know what the real-world problem we will face is while researching EAs. So we generally have no idea on the real performance of our deliberately designed algorithm.[82] We can roughly illustrate the No Free Lunch theorem with Fig. 3.23.

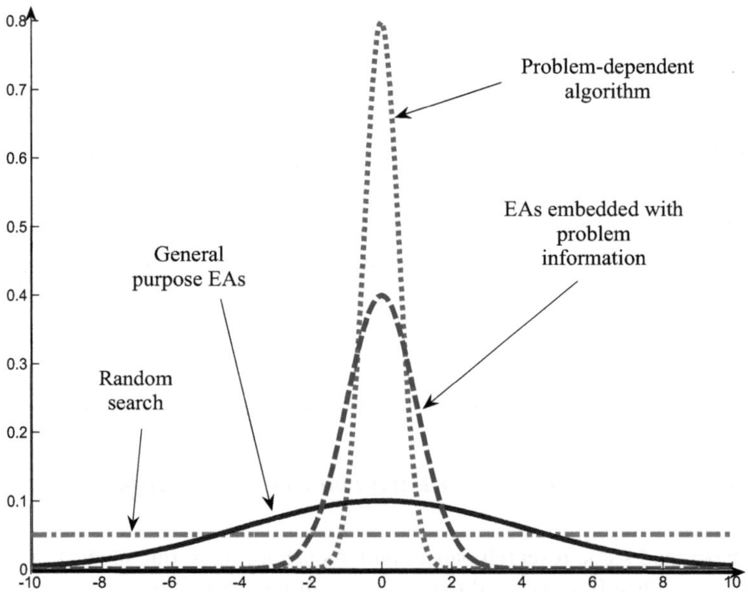

Fig. 3.23 One kind of interpretation of No Free Lunch theorem

The horizontal axis is various types of problems and we suppose that they have been elaborately arranged so that problems with similar properties have been put

[82] There are many other agreements and disagreements on this paper and they may strengthen your understanding of performance evaluation [75].

closer, and the vertical axis is one performance criterion.[83] We use one uniform distribution density function and three normal distribution density functions with different standard deviations to illustrate four algorithms. Several points need to be mentioned about Fig. 3.23.

- The more we understand the problem, the more specific technique we could design for solving it, and the better performance it will have, but the less robust it will be for other problems.
- We need to demonstrate that our algorithms are better than random search on the problem we face.[84]
- General purpose EAs are reliable methods when you are doing a blind or near-blind search in most cases.
- If problem information could be embedded into the encoding and decoding process and into operators, together with a problem-dependent local search method, the performance of the algorithm would be improved at the expense of lower adaptability for other problems.

To sum up, the No Free Lunch theorem is the sword of Damocles when you want to prove that your algorithm is an ideal one.

3.6.1.2 Considerations on Benchmark Problems

A set of *benchmark problems* is often used by EA researchers to do the performance evaluation and comparison. But the alert reader must wonder why we need to do the performance comparison if there exists the No Free Lunch theorem. We have four answers to this question.

- The No Free Lunch theorem addresses all kinds of problems including many random or artificially generated freak examples. But most of the real-world problems we face are some kind of "normal" problems, which prompts us to use other "normal" benchmark problems to test the performance of our algorithm on the problem we are concerned with.
- We can use different types of benchmark problems and different performance criteria to find out which algorithm (with which operator) is good at which problem type. In fact, we do not do a totally blind search on a given problem. We know some properties about the problem through previous experience or the trial calculation. Thus we can choose our weapons, if we know what they are good at in advance, to handle it.
- We can use different types of benchmark problems to test the robustness, or the sensitivity, of the algorithm, the operator, or the parameter we are researching so that we can further improve it.

---

[83] We will discuss several criteria later. Here we suppose that a larger performance criterion value means better performance.

[84] This is not a trivial requirement and we will show you how to find competitors for your algorithm later.

- Generally, we know the optimal value of the benchmark problem or at least know the most recently reported best value. There is great value in doing performance evaluation and comparison.

We can roughly classify benchmark problems into three groups as follows:

1. Problems separately suggested by different researchers. These researchers have different backgrounds and interests, so their suggestions cover a large scale of horizontal axis in Fig. 3.23, but it is hard to decide which one to use for test. Famous benchmark problems are introduced in [16, 17, 33, 36]. Starting in 2005, the IEEE Congress on Evolutionary Computation has given special sessions on numerical optimization competitions, including real parameter optimization (2005), constrained real parameter optimization (2006), multiobjective optimization (2007), large-scale global optimization (2008), dynamic optimization (2009), and multiobjective optimization with constraints (2009). Interested readers may consult its Web site.[85]
2. Benchmark problem generators suggested by different researchers. We can generate problems rather easily using these generators. But generally these generators generate problems with similar properties so it is not good to put all your eggs in one basket. Famous generators are introduced in [76–78]. Interested readers might refer to the Repository of Test Problem Generators Web site.[86]
3. Real-world problems. To solve real-world problem is the real value of EAs. But generally every detail of real-world problems canot be published because of commercial, confidential, too complicated excuses, so it is hard for other researchers to replicate the numerical experiments and results, which is a basic requirement for comparison.

Selecting the test suite from the above benchmark problems for your performance evaluation depends on what you want to test. Generally, efficacy, i.e., quality, efficiency, i.e., speed, and reliability, i.e., success rate, are the three main considerations that we will discuss later. The following guidelines were suggested by Bäck [36] and enhanced by Eiben in [37]; we add some other considerations to help you form the test suite.

1. The test suite should include a few unimodal functions to test efficiency.
2. The test suite should include several multimodal functions with a large number of local optima to test efficacy.[87]
3. The test suite should include functions with uncertain elements, such as noisy and dynamic problems to test the robustness of the algorithms.
4. The test suite should contain scalable problems, whose dimension can be changed easily, to test the scalability of the algorithm for large problems.
5. The test suite should contain problems with nonseparable variables to avoid a tricky decomposition.[88]

[85] `http://www3.ntu.edu.sg/home/EPNSugan/`

[86] `http://www.cs.uwyo.edu/~wspears/generators.html`

[87] One special type of multimodal problem is the so-called deceptive problem [79].

[88] Rotation provides a way to introduce nonseparability [4].

3.6.2 Performance Evaluation and Comparison

In 1995 Barr *et al.* published a very helpful survey on designing and reporting numerical experiments with heuristic methods [80], which could also be used in EAs. According to Huband *et al.*'s suggestion, the typical scenario of EA comparison is as follows [81]:

1. Select the EAs to compare.
2. Form the proper test suite.
3. Select adequate performance indices to evaluate the algorithms.
4. Obtain results for all competitors, either from the published results or generate the results by users and then generate the performance indices for each algorithm.
5. Compare the metrics statistically.
6. Draw conclusions.

The procedure of selecting competitors depends on the current state of your research. If little algorithm research has been done on your model, then at least a comparison between your contribution and the random search is necessary. If other (meta-)heuristic algorithms have been used to solve the model, such as SGA, you need to make them the baseline. If the problem you are researching is rather "hot" so that there are several "classical" algorithms, such as SR in constrained optimization and NSGA-II in multiobjective optimization, you need to compare with them. If your research is the direct improvement of one algorithm, the comparison should be done between the new one and the old one.

In the above scenario, different test suites, performance indices, and stop criteria might suggest different conclusions of the comparison.[89] So doing fair performance evaluation and comparison is rather complicated. We will discuss these three elements and their interactions in the following part of this section.

3.6.2.1 Performance Indices of Numerical Optimization

The *performance indices* (PIs)[90] strongly depend on the application, i.e., what you want from the algorithm. De Jong suggested that the online and offline performance imitated the performance requirements in on-line and off-line situations, respectively, separately for EAs [33]. Then in 2003, Eiben and Smith suggested three indices that have in recent years been used most often [37]. In 2005, at a special session on real parameter optimization of the IEEE Congress on Evolutionary Computation, they were developed further [82].

EAs are random-based algorithms, so they might provide different results at different runs. In order to evaluate and compare them fairly, we need to run them many times and take the results of all runs to generate a PI. There is no rule regarding how

---

[89] Sometimes even controversial conclusions.

[90] Other names include *performance assessment*, *performance metric*, *performance indicator*, and *performance measure*.

many runs are sufficient or necessary for an algorithm on a problem, which might be limited by the computing environment, i.e., the objective function evaluation cost. But we believe a larger number of runs could provide more objective performances and the following statistical procedure requires large samples to draw conclusions. So at least 20 to 30 runs on one problem for each algorithm are required. Different runs should start with different initial populations, but we suggest that the different EAs have the same initial populations in one run in order to make the comparison fairer.[91]

We have introduced many techniques to dynamically control *popsize* and adaptively stop the computation. So for fair comparison, it is better to stop the algorithms with either of the following two conditions.

1. The difference between the fitness value of the current best individual f_{best} and a predefined value f^* is under a predefined threshold ε. f^* is the optimal value of the benchmark problem or the most recently reported best value and ε depends on user requirements. We want to know whether the algorithm can reach the threshold or not. If so, we further want to know how fast the algorithm can reach it.
2. The *number of objective function evaluation* (NOFE) has reached the predefined value Max_{NOFE}. We want to know the solution quality after Max_{NOFE} iterations of the objective function.

Smaller ε and larger Max_{NOFE} mean more search effort. It is necessary to mention that determining ε and Max_{NOFE} is not a trivial task because the convergence speeds of different algorithms vary distinctly, which will be demonstrated later.

Generally speaking, there are two categories of PI: the *overall PI*, which describes the overall performance with a number, and the *evolving PI*, which describes the evolving process with a series of numbers.

We can group the overall PI into efficacy, efficiency, and reliability.

1. **Efficacy.** We want to evaluate the quality of the results the algorithm provides and do not care about the speed. The *mean best fitness* (MBF) is defined as the average of the best fitness in the last population over all runs. MBF could be used in totally offline situations. Apart from the best fitness in the last population, the best fitness values thus far could be used as a more absolute evaluation for efficacy.
2. **Reliability.** We want to know the extent to which the algorithm can provide acceptable results. *Success rate* (SR) is defined as the percentage of runs terminated with success. We define a successful run as the difference between the best fitness value in the last generation f_{best} and a predefined value f^* under a predefined threshold ε, i.e., the algorithm ends with stop criterion 1. SR, combined with MBF, could be used in situations with time requirements, so we want to get good results with limited runs. Low SR and high MBF might suggest that the

[91] Fogel and Beyer suggested that the initialization should be biased against the global solution of the benchmark problems, i.e., do not include the global solution in the initial generation domain of the variables, so that we can test the true explorative ability of the algorithms [83].

algorithm converges slowly, and high SR and low MBF might suggest that the algorithm is basically reliable but might provide very bad results accidentally.

3. **Efficiency.** In general, we want to find the global optimal solution as soon as possible. So the *average number of evaluations to a solution* (AES) is defined as the number of evaluations it takes on average for the successful runs to find an optimum, i.e., stop with criterion 1. If an algorithm has no successful runs, its AES is undefined. AES can be used in online situations.
4. **Reliability and efficiency.** We want smaller AES and larger SR, so why not combine them as smaller AES/SR. This criterion considers reliability and efficiency simultaneously.

Each of the above PIs provides a number to represent the overall performance regarding one aspect for an algorithm on a benchmark problem with many runs. Sometimes we are also interested in considering the evolving process of the algorithms. Thus many other generation-based evolving PI are necessary, which are illustrated as follows:

1. **Best-so-far (BSF).** We record the best solution found by the algorithm thus far for each generation in every run.[92] Obviously the BSF index is monotonic. The *final BSF* is the BSF when the algorithm satisfies the stop criteria.
2. **Best-of-current-population (BCP).** We record the best solution in each generation in every run. MBF is the average of final BCP or final BSF over multiple runs.
3. **Average-of-current-population (ACP).** We record the average solution in each generation in every run.
4. **Worst-of-current-population (WCP).** We record the worst solution in each generation in every run.

The answer to the problem of which evolving PI to take depends on the application environment. If we want only one best solution by the end of the run, BSF might be the proper selection. But if all the *popsize* individuals will be used by the end of the run, BCP, ACP, and WCP describe the performance of the algorithm from various viewpoints.

3.6.2.2 Performance Description and Comparison of Evolutionary Algorithms

The overall PI might suggest a concise description for the performance, but the evolving PI might provide more detailed information. The problem of the evolving PI lies in the fact that after many runs with random initial populations, we might get so many evolving PI numbers that sometimes we are at a loss to describe and compare the performance of EAs using these data. In this part, we will show you three

---

[92] The best solution so far might participate in the following evolving process of the algorithm or not, depending on the requirement of various algorithms. If it involves the evolving process, the algorithm has an elitism mechanism.

types of statistical description, i.e., statistical visualization, descriptive statistics, and statistical inference, as the tools to draw conclusions about your algorithm.

Statistical Visualization

Statistical visualization uses graphs to describe and compare EAs, which is very illustrative.

The box plot is the most useful way to graphically illustrate the performance of EAs. Suppose we run an algorithm on a problem 100 times and get 100 numbers for one PI, e.g., final BSF.[93] We can rank these 100 numbers in ascending order. The *lower quartile* is the $0.25 \times 100 + 1 = 26$th smallest BSF $x_{.25}$, the *median* is the $0.50 \times 100 + 1 = 51$st smallest BSF $x_{.50}$, and the *higher quartile* is the $0.75 \times 100 + 1 = 76$th smallest BSF $x_{.75}$. Then we could draw a box to include numbers between $x_{.75}$ and $x_{.25}$ to illustrate the main part of the data[94] and draw a line at $x_{.50}$ to illustrate the median,[95] just like in Fig. 3.24. The *interquartile range* (IQR) is $x_{.75} - x_{.25}$. Any data that lie more than $1.5 \times$ IQR lower than the lower quartile or $1.5 \times$ IQR higher than the higher quartile is considered an *outlier*. We draw two lines to indicate the smallest number that is not a lower outlier and the largest number that is not a higher outlier,[96] denoted as *whisker*, and connect the whiskers with the box. Other outliers are illustrated by "+". We can do the above procedure for every algorithm and draw their box plots together. Many softwares provide a function to draw box plots easily.[97] An example of the box plots of four algorithms is illustrated in Fig. 3.24.

From the *performance graph* in Fig. 3.24 we can easily say that algorithm 4 is the best one because its median BSF is large and its IQR is small, which can be interpreted as meaning that the average performance of algorithm 4 is good, along with its small variance. But there are difficulties in differentiating the performance between algorithms 1 and 3. Statistical visualization is not strong enough to do this. We will deal with it later.

We often want to represent the evolving process of many runs. So the *convergence graph* illustrating the performance over NOFE is quite useful.

Suppose we need to compare two groups of results, e.g., 100 runs by algorithm A and another 100 runs by algorithm B, with BSF,[98] then a box plot is a very clear and straightforward way to illustrate the convergence procedure. We can record BSF

---

[93] Suppose we are dealing with the maximum problem, i.e., larger final BSF means better performance.

[94] The vertical location of the lower and upper horizontal lines of the box are the true numbers of $x_{.25}$ and $x_{.75}$, respectively.

[95] The vertical location of the line is the true number of $x_{.50}$.

[96] Here the numbers are the true numbers in the data set, not the number calculated by $1.5 \times$ IQR.

[97] MATLAB® has a "boxplot" function. There are other variations of box plot. Interested readers are referred to [84].

[98] In this example, the smaller the better.

every 1000 NOFEs.[99] We run every algorithm 100 times and record the 100 performance values on each record time. Then we draw the convergence box plot graph illustrated in Fig. 3.25.

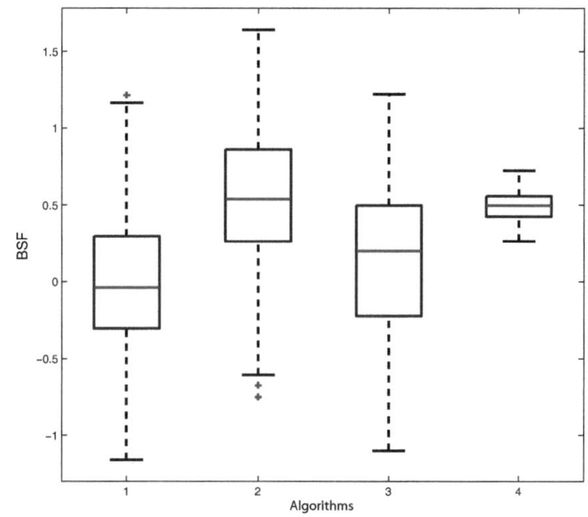

Fig. 3.24 Performance graph using box plot for four algorithms

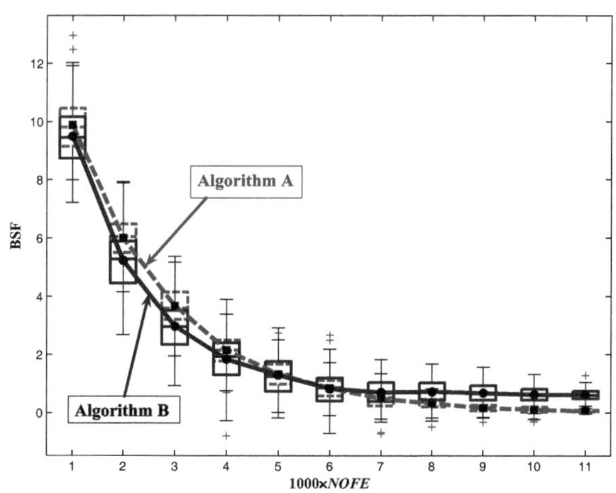

Fig. 3.25 Convergence graph using box plot for two algorithms

[99] This means that we take 1000 NOFEs as a virtual common generation if we want to compare two algorithms with *popsize* variation techniques. Here every 1000 NOFEs is called a record time.

We also calculate the mean of 100 runs at every record time for every algorithm and connect them in Fig. 3.25, i.e., the squares are for algorithm A and the circles are for algorithm B. Several useful conclusions can be drawn from Fig. 3.25, such as algorithm B converges fast but the final quality is not as good as algorithm A, and the variance in the final best solutions of algorithm A is small, which means that A is more reliable, etc.

With the help of Fig. 3.25, we can further discuss drawing conclusions. The various stop criteria introduced above might influence the comparison results. For example, if ε is very small or Max_{NOFE} is very large, we can say that algorithm A is better than algorithm B. But with larger ε or smaller Max_{NOFE}, opposite conclusions might be drawn. But that does not mean that you can cheat others by selecting stop criteria suitable for your proposed algorithms. Our suggestion is to run the algorithms with the smallest affordable ε or the largest affordable Max_{NOFE}, and then compare them with various PIs in various record times and find out the advantages and disadvantages, using the various statistical descriptions described above and below, of your algorithm over others.

Descriptive Statistics

Graphs are easy to understand, but sometimes the difference between different algorithms is small. Then we need specific numbers to describe and compare the performance.

The most often used *descriptive statistics* are mean and variance (or standard deviation). The mean of n data is

$$\bar{X} = \frac{1}{n}\sum_{i=1}^{n} x_i$$

where x_i is datum i. The variance of n data is

$$S^2 = \frac{1}{n}\sum_{i=1}^{n}\left(x_i - \bar{X}\right)^2 = \frac{1}{n}\sum_{i=1}^{n} x_i^2 - \bar{X}^2$$

where x_i is datum i and $\bar{X}$ is the mean of these n data. The standard deviation is the square root of variance. Then we calculate and tabulate the mean and the variance of all PIs and compare them.

Statistical Inference

Sometimes descriptive statistics is also not strong enough to differentiate between two results, in which case we need *statistical inference*. Statistical inference includes parameter estimation, hypothesis testing, and many other techniques. Here we focus on hypothesis testing to verify whether the difference between two results is statistically significant.

Let us take the results of algorithms 1–3 in Fig. 3.24 for example.

The mean BSF of algorithm 1 is −0.0764, that of algorithm 2 is 0.5750, and that of algorithm 3 is 0.1429. We can see from Fig. 3.24 or the mean of these algorithms that algorithm 2 is the best one. But how do we draw conclusions about algorithms 1 and 3? Does the difference come from the nature of these algorithms, i.e., the two groups of results by algorithms 1 and 3 are two samples on two different density functions, so the difference comes from different expectancies of density functions, or from the sampling noises, i.e., the two groups of results by algorithms 1 and 3 are two samples on the same distribution functions so the difference comes from sampling noises? We need to use hypothesis testing to verify whether there exists *statistical significance* or not.

The most often used hypothesis testing method in EA community to verify the above hypothesis is the *t test*, which assumes that the two groups of results all obey normal distribution and have same variance. Then there is a theorem.

Theorem 3.2. *For two groups of normal distribution random numbers with the same standard deviation, $N(\mu_1, \sigma^2)$ and $N(\mu_2, \sigma^2)$, we independently sample n_1 points from group 1 and n_2 points from group 2. Suppose that their means are $\bar{X}_1$ and $\bar{X}_2$ and variances are S_1^{*2} and S_2^{*2}, respectively. Then*

$$T = \frac{(\bar{X}_1 - \bar{X}_2) - (\mu_1 - \mu_2)}{\sqrt{\frac{1}{n_1} - \frac{1}{n_2}} S^*} \tag{3.71}$$

obeys a t distribution with $n_1 + n_2 - 2$ degrees of freedom, where

$$S^* = \sqrt{\frac{(n_1 - 1)\, S_1^{*2} + (n_2 - 1)\, S_2^{*2}}{n_1 + n_2 - 2}}$$

The proof of the theorem can be found in any statistics textbook. The density function of a t distribution with $n = 1$ degrees of freedom is illustrated in Fig. 3.26. Higher n makes the t distribution approach the standard normal distribution.

After the above "tedious"preparation, we can start the *hypothesis testing*. One way to do it is to hypothesize that $\mu_1 = \mu_2$, i.e., the two groups of samples are from the same normal distribution. This suggests a *null hypothesis* of $H_0 : \mu_1 = \mu_2$.[100] The counterpart of this null hypothesis is an *alternative hypothesis*, i.e., $H_1 : \mu_1 \neq \mu_2$.

If our hypothesis is right, $\mu_1 = \mu_2$, according to Eq. 3.71, the following random number T satisfies t distribution with $n_1 + n_2 - 2$ degrees of freedom, i.e., T is a sample on the t distribution function.

$$T = \frac{\bar{X}_1 - \bar{X}_2}{\sqrt{\frac{1}{n_1} - \frac{1}{n_2}} \sqrt{\frac{(n_1-1)S_1^{*2} + (n_2-1)S_2^{*2}}{n_1+n_2-2}}} \tag{3.72}$$

[100] There are other options, such as $H_0 : \mu_1 < \mu_2$. We just introduce the simplest one here.

We can calculate the number T by Eq. 3.72 using the $n_1 + n_2$ numbers and we want to verify whether T is a t distribution random number or not. If so, our hypothesis $H_0 : \mu_1 = \mu_2$ is approved, otherwise it is rejected.

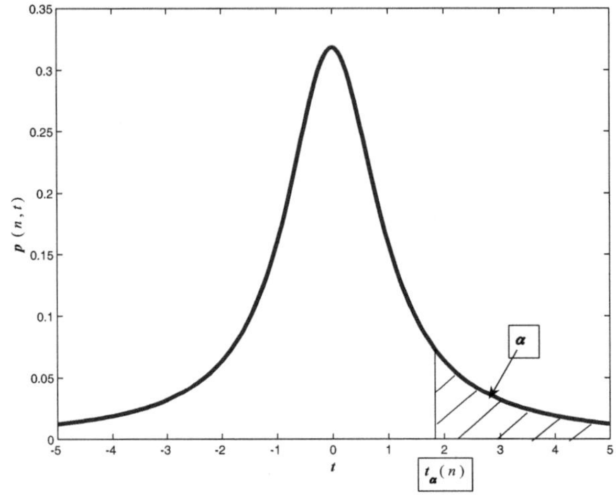

Fig. 3.26 t distribution density function ($n = 1$) and its upper α percent number

But a t distribution number can have any value according to Fig. 3.26. So we need to find out the scale in which the t distribution mainly resides. We can define the *upper α percent number* of t distribution with n degrees of freedom $t_\alpha(n)$ as

$$P\{t(n) > t_\alpha(n)\} = \int_{t_\alpha(n)}^{\infty} p(n,t)\,\mathrm{d}t = \alpha \tag{3.73}$$

where $t(n)$ is a random number obeying a t distribution with n degrees of freedom and $p(n,t)$ is the density function of the t distribution with n degrees of freedom. The upper percent number is also illustrated in Fig. 3.26.

From the definition of the upper α percent number and the symmetry property of the t distribution, we know that if we generate a t distribution random number, it has probability $1 - \alpha$ of being within the interval $(-t_{\alpha/2}(n), t_{\alpha/2}(n))$.

Then we can predefine a threshold α to verify whether the calculated number T, according to Eq. 3.72, is a t distributed number or not, i.e., T is within the interval $(-t_{\alpha/2}(n), t_{\alpha/2}(n))$ or not, where $n = n_1 + n_2 - 2$. If so, we claim that the probability of T being a t distribution random number is larger than $1 - \alpha$ and then accept H_0, else reject H_0.

Actually α determines the probability of regret when T is a t distribution number but we claim it is not, i.e., the hypothesis is true but we claim that it is wrong. A larger α makes the same T harder to claim as a t distribution number, i.e., the accepted T is more like to be a t distribution number.[101] So we call α the *significance level*.

In practice, we need to first make the null hypothesis, e.g., $\mu_1 = \mu_2$, and assign the significance level α.[102] Then we can calculate T with two groups of numbers, n_1 and n_2, and find the $\alpha/2$ upper percent number in the table of t distribution with $n_1 + n_2 - 2$ degrees of freedom.[103] If $|T| \geq t_{\alpha/2}(n_1 + n_2 - 2)$, we reject H_0 and claim that these two groups of numbers are samples from two normal distribution functions with different expectancies, i.e., the means of these two groups of numbers have a statistical difference with significance level α. Otherwise, we accept H_0 and claim that these two groups of numbers are samples from the same normal distribution function, i.e., the means of these two groups of numbers do not have a statistical difference with significance level α.

Many softwares provide a function to do a t test.[104] The results of a t test between the data of algorithms 1 and 2 is that the means of algorithms 1 and 2 have a statistical difference with significance level $\alpha = 0.05$. According to the t test result, we can say that the means of algorithms 1 and 3 do not have a statistical difference with significance level $\alpha = 0.05$. In this way, we can statistically claim whether the difference in mean between two groups of numbers is from their internal differences or from the sampling noise.

The t test has the assumption that the two groups of results all obey the normal distribution and have the same variance. There is a Mann–Whitney test to do a similar task without this assumption. Also the difference between two variations could be verified using an F test.[105]

ANOVA and Orthogonal Experiment Design

The last, but not least, thing about performance evaluation is how to design numerical experiments to verify the influence of one parameter and the interaction between parameters. This type of statistical inference is called *ANalysis Of VAriance* (ANOVA).

Recall the taxonomy of parameter control discussed in Sect. 3.5; one might argue that we do not need to do the parameter tuning because there are so many deterministic, adaptive, and self-adaptive control methods for strategy parameter control. But if you look at these techniques closely, most of them have other control parameters to "intelligently" control the strategy parameters. So at least we need to prove that these control parameters are robust, i.e., different values of these parameters have less influence on the performance of EAs. This could also be transferred to a hy-

---

[101] It also means the higher confidence for the accepted H_0 but higher possibility to reject H_0 with the same T.

[102] The most commonly used significance level is $\alpha = 0.05$.

[103] Every textbook on statistics has such a table.

[104] MATLAB® has a "ttest2" function to do a t test.

[105] MATLAB® has a "ranksum" function to do a Mann–Whitney test and a "vartest2" function to do an F test. Interested readers are referred to [85].

pothesis testing problem with H_0: the mean value of results with different control parameters do not have a statistical difference with significance level α.[106]

When there are two control parameters, we want to know not only their robustness but also their interactions, which means we cannot tune these control parameters one by one. ANOVA can also solve this problem.[107]

Under the condition that there are many control parameters, each having many possible values, it will be very hard to test their robustness and interactions fully. So this is where the experiment design comes in. We just introduce the *orthogonal experiment design* based on Latin square; other methods can be found in [86, 87].

In experiment design, the control parameters are called *factors*. Each factor can have different *levels*. Suppose there are r factors, each with s levels; the full experiment number is s^r. This number is sometimes too large to be implemented. So people have already designed orthogonal experiment tables, such as $L_4(2^3)$, $L_8(2^7)$, $L_{16}(2^{15})$, $L_{12}(2^{11})$, $L_9(3^4)$, and $L_{27}(3^{13})$, to decrease the experiment number while still testing these factors and levels fully and fairly. $L_n(s^r)$ means for r factors, each with s levels, we can do n experiments to test their influence on the performance. The $L_9(3^4)$ table, where the full experiment number is $3^4 = 81$, is as follows:

Table 3.2 Orthogonal experiment table $L_9(3^4)$

Experiment	Factor			
	1	2	3	4
1	1	1	1	1
2	1	2	2	2
3	1	3	3	3
4	2	1	2	3
5	2	2	3	1
6	2	3	1	2
7	3	1	3	2
8	3	2	1	3
9	3	3	2	1

The first column in Table 3.2 represents the experiment number and every other column represents one factor. The contents in the table represent the level of the factor in one experiment. In nine experiments, every level of every factor was used three times. Every level of every factor encounters every level of every other factor, so that we can use ANOVA to analyze the interaction between them.[108]

[106] MATLAB® has a "anova1" function to do this.

[107] MATLAB® has a "anova2" function to do this.

[108] MATLAB® has a "anovan" function to do this.

Aggregate Score on All Test Problems

Previously we gave suggestions for fair comparisons between algorithms with one benchmark problem. With different benchmark problems, conclusions might be different. How do we draw the aggregate conclusion on all test problems? Kennedy has suggested a way [88].

Let us discuss Kennedy's method with a simple example. Suppose we have tested algorithms A, B, and C on two minimum benchmark problems, Each algorithm optimizes each problem ten times, and we use the final BSF as the PI. Then we have ten numbers of each algorithm on each problem (Table 3.3).

Table 3.3 The final BSF of three algorithms on two minimum problems

	Problem 1			Problem 2		
Run	A	B	C	A	B	C
1	103.1149	98.4529	108.8182	−28.9812	−28.0573	−22.4254
2	101.3634	94.1934	125.6188	−28.6885	−27.2502	−13.3915
3	108.1008	104.5227	90.0454	−26.6605	−28.4597	−28.3950
4	98.2710	108.7934	77.6257	−28.9092	−40.0238	−31.8664
5	88.2701	101.6964	116.1530	−28.4129	−35.8542	−25.0883
6	93.5067	100.0010	100.8240	−28.5062	−35.7957	−33.4131
7	70.5098	104.6402	84.8758	−26.3803	−33.7700	−16.9626
8	99.7329	103.2524	98.2174	−27.9191	−37.2893	−22.2520
9	96.1695	95.4273	59.8230	−29.0811	−38.9106	−38.5930
10	89.0890	117.9255	121.6784	−29.1245	−36.7872	−34.6892

The bottleneck of the aggregate score on different problems lies in the different magnitude of the results (Table 3.3).

To scale them into comparable magnitudes, we can first calculate the mean and the standard deviation of the total results for each problem (30 numbers for each problem in this example), i.e., $\mu_1 = 98.6904$, $\sigma_1 = 13.9389$, $\mu_2 = -29.7313$, and $\sigma_2 = 6.1936$.

Then, by $\mathrm{BSF}_1' = (\mathrm{BSF}_1 - \mu_1)/\sigma_1$ and $\mathrm{BSF}_2' = (\mathrm{BSF}_2 - \mu_2)/\sigma_2$,we can transform all the data into zero mean and unit standard deviation,[109] thereby solving the problem of different magnitudes.

Now we can calculate the mean of each algorithm on each problem. For example, for algorithm A, we get $\mu_{A1} = -0.2782$ on problem 1 and $\mu_{A2} = 0.2365$ on problem 2. Since for the total 30 results of problem 1 (and problem 2) the transformed BSF_1' are with zero mean and unit standard deviation, we can say that algorithm A is good at problem 1 (because $-0.2782 < 0$) but bad at problem 2 (because

[109] Here BSF_1' and BSF_2' represent the transformed final BSF for each datum.

$0.2365 > 0$). Then we can sum these two data up to represent the overall performance of algorithm A as $\mu_A = \mu_{A1} + \mu_{A2} = -0.0417$. In the same way, we can get that $\mu_B = -0.4234$ and $\mu_C = 0.4650$.[110] Because all the benchmark problems have a minimum objective, we can then come to the conclusion that algorithm B is the best one among them on the two benchmark problems with respect to final BSF.

3.7 Brief Introduction to Other Topics

3.7.1 Coevolution

In biology, coevolution is the mutual evolutionary influence between two species. There are many types of coevolutions, and the two most famous are cooperative and competitive coevolution.

An example of *cooperative coevolution* is the relationship between bees and flowers. Bees get food from flowers while flowers get propagation of their seeds from bees. In optimization, when the dimension of the problem is very large, the chromosomes are very long. The search ability of our algorithm might not be strong enough to handle it. If we can decompose the full solution into different parts adequately, effective searches over smaller landscapes are possible [89].

Let us just discuss a simple example: designing a device whose best structure is not clear, not to mention the best parameters under that structure. A simple structure might require a small number of parameters and *vice versa*. If we use one population whose individual is represented as the most complicated structure possible with its parameters, we will waste a lot of computing resources in searching the hopeless region. So we would like to decouple the population into two subpopulations, one to represent and code the structure and the other to represent and code the parameter. For every individual i in each subpopulation, we combine it with k randomly selected individuals from another subpopulation to form k full solutions to the problem. The fitness value of individual i could be the average (or the best) fitness values among those k full solutions' fitness values. Other operators of cooperative coevolution are the same as standard EAs. In this way, simple structures only search small regions and complicated structures are also able to find the optimal solution [90].

An example of *competitive coevolution* is the relationship between wolves and rabbits. Wolves take rabbits as food. Their relationship is competitive, which prompts both of them to develop speed and flexibility. While cooperative coevolution could be regarded as a decomposer for a complicated optimization problem, competitive coevolution could be regarded as a best objective value estimator for complicated learning problems, especially in a competitive environment [91, 92].

Game playing is an example. Generally we do not know the best strategy, which is what we need to learn, unless the game is too simple. So we cannot use the best strategy, which could be seen as the objective function in the optimization problem,

[110] $\mu_A + \mu_B + \mu_C = 0$.

to evaluate strategy i's fitness value. So we can only randomly select k individuals, representing other strategies, to compete with i. The fitness value of i is proportional to its winning rate in k games. Competitive coevolution has a self-regulating selective pressure. In the early stage, most strategies are weak, so the bad ones have a chance to survive. With the evolving process, strategies get smarter, so the selective pressure increases.

3.7.2 Memetic Algorithms

The general consideration for EAs' performance on a solution landscape search is that they have strong global search ability but are relatively weak at local fine tuning. So why not combine the already existing effective *local search* (LS) methods, such as conjugate gradient method in differentiable problems, with EAs so that they can complement each other? Here's where the term *hybrid EA* or *genetic local search* comes in. In the EA literature, we call this combination *memetic algorithm* (MA).

The word "meme," from biology, describes a type of behavior that is passed from one member of a group to another, not in the genes but by other means such as people copying it.[111] Memes are like LS methods in MAs.

When designing or analyzing an MA, we suggest that the following six issues need to be considered.

1. **When?** On page 77, we illustrate the general solution process for EAs. An LS can be used in the initialization phase to improve the initial individuals' quality. But the most frequent situation is between steps 2.2 and 2.3, i.e., rewrite step 2.3 as follows:

> Step 2.3: Select new individuals to undergo LS and evaluate the fitness values for all individuals.

2. **Who?** Which new individuals will under go LS? All? The bad ones? The good ones? The balance of computational costs between LS and the main EAs need to be maintained elaborately.
3. **What?** What kind(s) of LS methods will be used in step 2.3? We want to improve the quality of individuals using LS. So these methods are called *hill climbers*.[112] For a specific problem, generally there exist many of these hill climbers with different speeds and efficiencies but suffers from local optimum.[113]
4. **How long?** Do we need to wait until LS methods can no longer improve or until we stop them with some heuristic rules? This issue is also related to the balance of computational costs.

---

[111] Oxford Advanced Learner's Dictionary.

[112] The maximum problem is considered here.

[113] We need to mention that EA operators, such as crossover, can be used in LS [93].

5. **How should individuals be treated after a local search?** One way is to replace one individual, both chromosome and fitness value, before LS with the one after it. This is called *Lamarckian learning*.[114] The other way is to only use the fitness value after LS to represent the individual before it without the replacement of chromosomes. This way uses the ability to learn to represent an individual. It is called *Baldwin learning*.[115] Many discussions on the comparison of these two learning strategies have been made [94].
6. **How does one replace?** How does one appropriately design step 2.4 of MAs? This will strongly affect the tradeoff between selective pressure and population diversity.

Due to the introduction of MAs, a reconsideration on the tradeoff between exploration and exploitation is necessary because individuals after LS methods will be improved and thus have selective advantages later. The evolving process might be accelerated with higher chances of premature. So there are many discussions on the balance between LS and other elements of MAs. Some of them are listed as follows:

- Using steady state EAs whose replacement method has a smaller selective pressure. Sect. 3.4.1 discussed this aspect.
- Using variation operators who can preserve population diversity. Sect. 3.2.2 discussed this.
- Using large *popsize*, variable *popsize*, and a reinitialization mechanism [15]. Sect. 3.5.3 discussed this.
- Using adaptive control on the probability of undergoing LS. You may consider that individuals with higher fitness values represent the hopeful search direction so that they should have a greater chance of undergoing LS. Likewise the less crowded individuals represent the undeveloped area so that they have a greater chance of undergoing LS. With different ideas, adaptive control rules could be designed with techniques introduced in Sect. 3.5.2.
- Using clustering methods and only picking one individual in each cluster to undergo LS so that the balance of computational costs between LS and the main EAs can be maintained.
- Using negative assortative mating to maintain population diversity [93, 95]. Sect. 3.5.3 discussed this.
- Using self-adaptive control for LS methods, i.e., every individual has its own LS methods [96].
- Using a niche technique, which will be introduced in Chap. 5, to preserve population diversity.
- Using parallel/distribution environments to maintain diversity. In some parallel/distribution implementations, different subpopulations will evolve in a parallel way so that they can exploit different areas of the solution landscape.

---

[114] Lamarck was a French biologist whose idea that an organism can pass on characteristics that it acquired during its lifetime to its offspring was widely accepted before Darwin's natural selection.

[115] Baldwin was an American psychologist with the idea that offspring would tend to have an increased capacity for learning new skills rather than being confined to genetically coded knowledge.

The research on MAs is an interdisciplinary field that has strong ties with OR and machine learning. Springer Press launched a new journal, *Memetic Computing*, in 2009. An example of MA, suggested by Chelouah and Siarry [97], is to combine Nelder–Mead's simplex search, which is introduced in Sect. 2.4.1, with EA. Krasnogor and Smith suggested an intensive survey on the taxonomy of MAs and discussed important issues for designing a competent MA [98].

Looking from EAs' perspective, LS techniques could also be considered one type of operator apart from crossover and mutation. So sometimes EA participants consider these operators a component of EAs and do not mention MAs.

3.7.3 Hyper-heuristics

In the above subsection, MA uses a hill climber with the intention of improving mutants. So it is an improving mechanism for EAs.

Hyper-heuristics is a relative new concept, it is a mechanism to use different heuristics, so the whole algorithm might have remarkable search ability in many different instances of a problem or even many types of problems.

In order to expand the robustness of the algorithm, hyper-heuristics divides itself into two levels: upper and lower. The lower level contains heuristics, including *mutational heuristics*, whose intention is to introduce perturbation, and a *hill climber*, whose intention is to improve the quality. Generally we suppose that these heuristics are available and readily prepared.

The upper level contains the rule(s) to schedule the lower level heuristics using feedback information from the lower level such as the improvement of the objective value using some heuristics and the CPU time on it, etc. In this way, the upper level is isolated from the specific problem, which makes hyper-heuristics a robust problem-solver.

There are two problems to be answered when designing a hyper-heuristics: the rules of scheduling lower level heuristics, i.e., *heuristic selection*, and the rules to accept the result of the heuristics, i.e., *move acceptance*. Various implementations have different technical details. We will introduce the idea suggested by Cowling *et al.* [99], perhaps the first one, and refer interested readers to [100–102].

After generating a solution using any method, we spend some time calling ζ heuristics n times, i.e., $(N_1, \cdots, N_n)$, in a random way in order to collect the initial performances of these heuristics. This time is called the *warm-up period.*[116]

At the end of the warm-up period, we calculate three properties for every heuristics as follows:

$$f_1(N_j) = \sum_{i=1}^{m} \alpha^{i-1} \frac{I_i(N_j)}{T_i(N_j)} \tag{3.74}$$

[116] Generally we want $n >> \zeta$ so that we can evaluate the performance of each heuristic and their combinations as thoroughly as possible. But too long a warm-up time is impossible. In Cowling *et al.*'s implementation, the warm-up period is $1/3$ of the total run time.

where $I_i(N_j)$ and $T_i(N_j)$ are, respectively, the performance improvement and the CPU time of heuristic j in the recent i's call, m is the number of recent calls of heuristic i, and $0 < \alpha < 1$. Equation 3.74 means that we want greater performance with smaller CPU time from heuristic j. We count the recent m calls for heuristics and put more focus on the recent calls with weight α^{i-1}. A larger $f_1(N_j)$ means a better heuristic j.

$$f_2(N_j,N_k) = \sum_{i=1}^{l} \beta^{i-1} \frac{I_i(N_j,N_k)}{T_i(N_j,N_k)} \tag{3.75}$$

where $I_i(N_j,N_k)$ and $T_i(N_j,N_k)$ are, respectively, the performance improvement and the CPU time of the calls in which heuristic k is called immediately after heuristic j in the recent i's call, l is the number of the recent calls in which heuristic k is immediately after heuristic j, and $0 < \beta < 1$. We need to calculate f_2 for every pair and also put more focus on the recent calls. A larger $f_2(N_j)$ means better combinations of heuristics j and k.

$$f_3(N_j) = \tau(N_j) \tag{3.76}$$

where $\tau(N_j)$ is the CPU time that has elapsed since heuristic j was last called.

Then the hyper-heuristics starts. We calculate the overall performance of heuristic k as follows:

$$F(N_k) = \alpha f_1(N_k) + \beta f_2(N_j,N_k) + \delta f_3(N_k) \tag{3.77}$$

where δ is a control parameter and N_j is the last heuristic in the warm-up period. Then RWS can be used to select one heuristic for the next call.[117]

After the selection, the winner, heuristic k, is called. Its fitness values is modified according to the following equations.

$$f_1'(N_k) = \frac{I(N_k)}{T(N_k)} + \alpha f_1(N_k) \tag{3.78}$$

where $I(N_k)$ and $T(N_k)$ represent the performance improvement and CUP time of this call, respectively.

$$f_2'(N_k) = \frac{I(N_j,N_k)}{T(N_j,N_k)} + \beta f_2(N_k) \tag{3.79}$$

where heuristic j is heuristic k's immediately preceding call.

We also count the time for every heuristic since its last call, i.e., $f_3(N_k)$.

Then we can use Eq. 3.77 to get the fitness for all heuristics and use RWS to pick another heuristic and continue the hyper-heuristics until the predefined stop criteria are satisfied.[118]

---

[117] Cowling *et al.* suggested the technique to implement RWS when the overall performance of some heuristics is smaller than zero.

[118] According to Eq. 3.77, we know that larger $f_3(N_k)$ means higher probability of being called in the next run, which embodies the idea that even though a heuristic might be bad currently, it still

The results of sensitivity analysis based on numerical experiments done by Cowling *et al.* showed that hyper-heuristics is insensitive either to the value of the strategy parameters, e.g., α, β, δ etc., or to problem instance and problem types (so long as the appropriate low-level heuristics are available).

EAs can certainly be used in the upper level of hyper-heuristics [103] and hyper-heuristics can be embedded in the LS of MAs to make them more effective, i.e., *meta-Lamarckian learning* [104].

3.7.4 Handling Uncertain Environments

Up till now, we have been discussing optimization and learning problems in an ideal environment, i.e., there is no sampling noise, no fabrication errors, and the problem has been formulated in a not-so-difficult explicit expression and will never change. These assumptions would never be valid in a real-world situation. So the last, but not least, topic in this chapter is the *uncertain environment* and how EAs can handle it. Jin and Branke published a comprehensive survey in 2005, and we adopt their taxonomy [105].

Noise

We consider the situation where the evaluation process of EAs suffers from additive noise, i.e., we can only get the contaminated fitness values for individuals. Whenever we need to sample from the real world, there will be various additive noises. Generally we assume that the additive *noise* is normally distributed. For individual $\mathbf{x}$, its real fitness value is $f(\mathbf{x})$ and the contaminated fitness value is $F(\mathbf{x}) = f(\mathbf{x}) + \xi$, where $\xi \sim N(0, \sigma^2)$. We need to work on $F(\mathbf{x})$ and try to find the real optimal solution for $f(\mathbf{x})$. The essence behind all possible solutions is that the mean of the contaminated fitness value is the real fitness value, i.e.,

$$\int_{-\infty}^{+\infty} [f(\mathbf{x}) + \xi] p(\xi) \mathrm{d}\xi = f(\mathbf{x}) \tag{3.80}$$

where $p(\xi)$ is the density function of ξ. So the simplest way is to make multiple samplings for one solution and use the average contaminated fitness value as the real fitness value, i.e.,

$$f(\mathbf{x}) \approx \frac{1}{n} \sum_{i=1}^{n} F_i(\mathbf{x}) \tag{3.81}$$

has the chance to be called later because of its increasing f_3. This property might be helpful in the condition that different heuristics are adequate in different search stages.

The method of average sum could be regarded as a kind of filtering technique. There exist many other advanced filtering techniques that can be utilized in treating noise in EAs.

Robustness

Robustness refers to the fact that a variable suffers from additive noise. It is very normal in manufacturing. Every element of a device contains tolerances, i.e., the same exact global optimal solution $\mathbf{x}$ we got from EAs could never be implemented with the same exact value in real conditions. We can only implement $\mathbf{x}+\xi$, where $\xi \sim N(0,\sigma^2)$.

Even though the global optimal solution is very good, if the fitness values close to it decrease rapidly,[119] we do not like it because there might be a low-quality production due to the fabrication tolerance and "sharp" nature around the global optimal solution.

In this situation, we generally want EAs to find a good and robust solution, which means that the fitness value of it is high and the fitness values close to it are also high. This brings up another closely related term, *sensitivity analysis*. We want our best solution to be insensitive to fabrication tolerance.

The basic solution for this problem is to add $\xi \sim N(0,\sigma^2)$ *artificially* to $\mathbf{x}$ when we are optimizing the problem. For every individual $\mathbf{x}$, we use the average sum of the fitness values of n samplings of $\mathbf{x}+\xi$ as its fitness value:

$$f(\mathbf{x}) \approx \frac{1}{n}\sum_{i=1}^{n} f(\mathbf{x}+\xi_i) \tag{3.82}$$

Equations 3.81 and 3.82 look similar but are different in nature. The random number in Eq. 3.81 is inevitable and we want to cancel its effects as far as possible and find the global optimal solution. By contrast, the random number in Eq. 3.82 is added artificially, and we want to find a robust good solution instead of the global optimal solution.

Fitness Approximation

If the model of the optimization problem comes from mechanics, electromagnetics, and thermodynamics, generally its fitness function cannot be expressed in an explicit form due to the partial differential equations behind the model. It is just an example of a real condition we often face, i.e., the fitness value is very "expensive" to evaluate so that we need some approximation techniques to generate a simplified model and use the value from the simplified model as the fitness value of the real model.

---

[119] We consider the maximum problem here.

So there inevitably exist estimation errors. For solution $\mathbf{x}$, suppose its real fitness value is $f(\mathbf{x})$; we will use $F(\mathbf{x}) = f(\mathbf{x}) + \varepsilon(\mathbf{x})$ to do the optimization, where $\varepsilon(\mathbf{x})$ is the estimation error.[120]

The general consideration for the *fitness approximation* is to use better models and adaptively modify the model parameters during the optimization so that we can depend on the approximation model more and more and reduce the cost of evaluation with acceptable errors. Many techniques have been suggested to handle the fitness approximation, including the topics we just introduced in this section, i.e., MA-based approximation [106] and coevolution-based approximation [107].

Time-varying Fitness Functions

Suppose you want to solve an optimization problem. The variable is the rotation angles of a photoelectric conversion plate, which can transfer solar power into electric power, and the objective is to maximize the generated electric power. For a specific time, e.g., 9:00 AM, there is an optimal rotation angle to make the plating face east. But the bad thing is that the Sun moves, which means the optimal solution changes and the optimal value also changes with time. Then the function is called a *time-varying fitness function* and the problem is called a *dynamic* or *nonstationary* problem.

We can certainly do the optimization every hour and get the optimal angle again. The feasibility behind this technique is that the Sun moves slowly and we compute fast so that we can afford the total reoptimization. In many other dynamic optimization problems with more rigorous online requirements, we have no time to totally reoptimize.

In these situations, we hope that the change in the fitness function is not so violent that previous good solutions could be used to accelerate the optimization of the new function.

So the perseverance of population diversity is the critical consideration in optimizing dynamic functions. Many techniques discussed in Sect. 3.7.2 can be used. Apart from that, we can use *redundant encodings* [108], *diploid* [109], or *evolvability* [108].

3.8 Summary

This is a rather long, perhaps tough, chapter.[121] Many of the key elements for solving optimization and learning problems have been discussed. Suppose the problem you are facing is the raw material and the programming environment is the kitchen. We hope these topics provide enough flavoring to cook your own gourmet meal.

---

[120] We write the expression in this way only with the intention of emphasizing the existence of the estimation errors. The real $F(\mathbf{x})$ is calculated using the simplified estimated model.

[121] But we do not want to make it tedious.

Different types of problems may require different adequate code scheme and related variation operators. We only briefly discuss the binary code, put more focus on the real code, and will discuss permutation code and tree code in the following chapters. In real-world problems, an efficient code scheme, i.e., how to encode and decode quickly in a not-so-large operation domain, always needs to be considered as the first priority.

Variation operators explore and exploit the solution landscape, together with the selection process, constitute the search ability of an algorithm, and maintain the tradeoff between selective pressure and population diversity, which might be the key element after the code scheme has been determined.

For the above two reasons, Sects. 3.2–3.4 are basic requirements for EA designers. As to the specific crossover, mutation, and selection method, we just provided the flavoring and you can select them according to your problem requirement and preference.

One type of problem, perhaps one instance of a certain problem, has its own solution landscape property, and thus requires a different search ability in different evolving stages. So a strategy parameter control, whether deterministic, adaptive, or self-adaptive, should be considered as a practical requirement for a robust problem solver. We hope that the techniques discussed in Sect. 3.5 will inspire you to develop innovative ideas for your algorithm.

Then you need to demonstrate that your innovative ideas really work, so numerical experiments are absolutely necessary.[122] Due to their stochastic nature, EAs may provide different results for different runs. So the experiment design and results analysis will directly influence the conclusions. We hope that readers can draw rigorous conclusions using methods introduced in Sect. 3.6.

After reading this chapter, you should skillfully grasp several real code variation operators, understand the selective pressure of different parental selections and survival selections, comprehend the advantages and disadvantages of various strategy parameter control methods, and know how to compare different operators and algorithms using numerical experiments.

In all, designing an EA is the art of tradeoff between selective pressure and population diversity.

Suggestions for Further Reading

Eiben and Smiths' textbook [37] and Haupt and Haupt's textbooks [110] give clear discussions on the basic contents of this chapter. More detailed information on related topics can be found in Bäck *et al.*'s textbooks [92, 111].

This book does not discuss the theoretical part of EAs, such as building block hypothesis, schema theorem, Markov chain analysis, linkage, epistasis, deceptive problems, etc. Interested readers are referred to [37, 112]. There is a biannual inter-

[122] Unless in the rare case, you can prove the absolute advantage of one way over others.

national workshop on the foundations of genetic algorithms, and recent proceedings are of great value for this topic [113, 114].

The real code crossover operators discussed in this chapter, such as SPX,SBX, PCX, UNDX, and UNDX-m, do not consider the quality of parents. Nakanishi *et al.* use biased probability distribution functions to direct the search with preference on the best individual in the current population [115].

We have discussed many crossover operators. Each of them has a different search ability on different problems. If we are dealing with a blind problem, i.e., we know almost nothing about the solution space, it is an interesting idea to verify the synergy effects of various crossover operators. Yoon and Moon discuss this issue in a paper published in 2002 [116]. Even though there was still no selection rule for specific crossover operators or a combination of operators, they observed strong synergy effects.

We introduced rank as a way to scale fitness values and adjust selective pressure in Sect. 3.3.4, but Cervantes and Stephens use ranks to adjust p_{m} by granting fitter individuals a smaller p_{m} and to promote mating restriction by allowing individuals with similar ranks to cross over [117].

If readers have a strong interest in adaptive and self-adaptive parameter control,[123] a book edited by Lobo *et al.* in 2007 [118] and one written by Kramer in 2008 [119] are suggested.

The selection process has a very strong influence on the evolving process, so designing an adequate self-adaptive selection scheme is a rather elaborate task. Interested readers are referred to a paper published in 2006 [120].

The orthogonal experiment design introduced in Sect. 3.6.2 is quite useful in analyzing the sensitivity of control parameters. Leung *et al.* developed the idea of orthogonal design into a crossover operation and initialization to search the solution landscape in a statistically sound manner [121, 122].

Yuen and Chow used a binary space partitioning tree to generate and query all the searched points in a solution space so that their algorithm will never revisit the searched points, which improves the search efficiency significantly [123].

Lobo and Goldberg suggested an empirical way to measure the difficulty of a given benchmark problem [124]. Their paper is recommended to those with an interest in evaluating problems, not algorithms.

We did not discuss two very interesting and useful topics – *evolvable hardware* and *parallel evolutionary algorithm.*

Generally we design an equipment, implement it, and use it. We do not know in advance the possible change of equipment functions due to environmental changes or the failure of its inner parts. A robust design and dynamic problem handling techniques may relieve the pain. But we have more powerful tools – evolvable hardware. Evolvable hardware uses field-programmable gate array (FPGA) or other field-programmable devices to realize an equipment. There are environmental change monitors and inner part failure monitors in the equipment so that whenever the change or failure happens, the EAs in the equipment can automatically solve the

[123] We guess more than 50% readers have.

new optimal design problem and use the remnant gates in FPGA to realize the new functions. Then the new part replaces the failed one. Interested readers are referred to [125–127]; Springer Press publishes a related journal: *Genetic Programming and Evolvable Machines*; and IEEE hosts an annual conference called the *IEEE Workshop on Evolvable and Adaptive Hardware*.

EAs use populations to implement optimization and learning, which means there must be a lot of samplings on the solution landscape. So the ideas of using parallel/distributed computing infrastructure to implement EAs and designing the EAs suitable for parallel/distributed computing environment are quite attractive. Interested readers are referred to [128].

The must-read papers of this chapter are [7] for SBX, [12] for UNDX, [32] for strategy parameter control,[124] [64] for dynamic *popsize* control (but mostly for writing a short yet informative paper), [63, 93] for strategy parameter sensitivity analysis, [80] for performance evaluation, and [62] for an example of using hypothesis testing to draw statistical conclusions.

Exercises and Potential Research Projects

3.1. Why are single-point crossover, multiple-point crossover, and uniform crossover not suitable for real code? (Hint: consider the search ability of these crossovers on real code space.)

3.2. Prove or verify that Eqs. 3.5, 3.23, 3.25, and 3.46 are probability density functions.

3.3. Implement BLX-0.5 and simplex crossover in your programming environment.

3.4. Implement SBX and UNDX crossover in your programming environment, use the benchmark problems in Appendix and the techniques introduced in Sect. 3.6 to do a fair comparison and draw statistical conclusions according to your numerical experiments.

3.5. How does one generate a positive definite covariance matrix using rotation transform? Present your method and verify it by MATLAB® using the method discussed on page 57.

3.6. Implement nonuniform mutation and one kind of self-adaptive control normal mutation in your programming environment, and use the benchmark problems in Appendix and the techniques introduced in Sect. 3.6 to do a fair comparison and draw statistical conclusions according to your numerical experiments.

3.7. Find the maximum selective error of RWS and SUS.

[124] This paper won the IEEE Transactions on Evolutionary Computation Outstanding Paper Award.

3.8. How do we do fitness transferral if we want to maximize a function with negative objective value?

3.9. Why do we say that the probability of being selected decreases from q to $\frac{2}{popsize} - q$ linearly in ranking by Eq. 3.45?

3.10. Implement SUS and binary tournament selection in your programming environment.

3.11. Implement at least one method of fitness scaling and one method of ranking in your programming environment, and use the benchmark problems in Appendix and the techniques introduced in Sect. 3.6 to do a fair comparison and draw statistical conclusions according to your numerical experiments.

3.12. Analyze parameter c's influence on the selective pressure in sigma truncation.

3.13. Select and implement at least one method of deterministic, adaptive, and self-adaptive control over p_{m} or p_{c} using the benchmark problems given in Appendix. Comparisons should be made using statistical methods suggested in Sect. 3.6.

3.14. Why for minimum optimization is Eq. 3.57 also used to estimate the gradient? Explain it with sentences similar to those in the textbook. Do we need to change the trial point generation, Eq. 3.59, and Eq. 3.60 for a minimum optimization problem? If so, how?

3.15. Start from any adaptive strategy parameter control method and try to use techniques introduced in this chapter to improve it. Report your basic idea, your implementation details, your numerical experimental results on the benchmark problems in Appendix, and the statistical conclusions using methods introduced in Sect. 3.6.

3.16. Different variables might have different influences on the objective, i.e., the sensitivity of different variables might vary. Can you suggest a way to adaptively find the most sensitive variables and focus the search on these variables? Report your basic idea, your implementation details, your numerical experimental results on the benchmark problems in Appendix, and the statistical conclusions using methods introduced in Sect. 3.6.

References

1. Blum C, Roli A (2003) Metaheuristics in combinatorial optimization: overview and conceptual comparison. ACM Comput Surv 35(3):268–308
2. Kita H, Ono I, Kobayashi S (1999) Multi-parental extension of the unimodal normal distribution crossover for real-coded genetic algorithms. In: Proceedings of the IEEE congress on evolutionary computation, p 1588
3. Deb K, Beyer H (2001) Self-adaptive genetic algorithms with simulated binary crossover. Evol Comput 9(2):197–221

4. Salomon R (1996) Reevaluating genetic algorithm performance under coordinate rotation of benchmark functions. BIOSYSTEMS 39:263–278
5. Eshelman LJ, Schaffer JD (1993) Real-coded genetic algorithms and interval schemata. In: Proceedings of foundation of genetic algorithms, pp 187–202
6. Tsutsui S, Yamamura M, Higuchi T (1999) Multi-parent recombination with simplex crossover in real coded genetic algorithms. In: Proceedings of the genetic and evolutionary computation conference, pp 657–664
7. Deb K, Agrawal R (1995) Simulated binary crossover for continuous search space. Complex Syst 9:115–148
8. Papoulis A, Pillai S (2002) Probability, random variables and stochastic processes, 4th edn. McGraw-Hill Higher Education, New York
9. Ross S (2009) A first course in probability, 8th edn. Prentice Hall, Englewood Cliffs, NJ
10. Deb K, Joshi D, Anand A (2002) Real-coded evolutionary algorithms with parent-centric recombination. In: Proceedings of the IEEE congress on evolutionary computation, 1:61–66
11. Ono I, Tatsuzawa Y, Kobayashi S (1997) A real-coded genetic algorithm for function optimization using unimodal normal distribution crossover. In: Proceedings of the 7th international conference on genetic algorithms, pp 81–84
12. Kita H (2001) A comparison study of self-adaptation in evolution strategies and real-coded genetic algorithms. Evol Comput 9(2):223–241
13. Ono I, Kita H, Kobayashi S (2003) A real-coded genetic algorithm using the unimodal normal distribution crossover. In: Ghosh A and Tsutsui S (eds) Advances in evolutionary computing: theory and applications. Springer, Berlin Heidelberg New York, pp 213–237
14. Kirkpatrick S, Gelatt CD, Vecchi MP (1983) Optimization by simulated annealing. Science 220(4598):671–680
15. Fukunaga A (1998) Restart scheduling for genetic algorithms. In: Proceedings of the international conference on parallel problem solving from nature, pp 357–366
16. Yao X, Liu Y, Lin G (1999) Evolutionary programming made faster. IEEE Trans Evol Comput 3(2):82–102
17. Lee C, Yao X (2004) Evolutionary programming using mutations based on the levy probability distribution. IEEE Trans Evol Comput 8(1):1–13
18. Deb K, Goyal M (1996) A combined genetic adaptive search (GeneAS) for engineering design. Comput Sci Informat 26:30–45
19. Goldberg D, Deb K, Korb B (1989) Messy genetic algorithms: motivation, analysis, and first results. Complex Syst 3:493–530
20. Harvey I, Harvey I (1992) The SAGA cross: the mechanics of recombination for species with variable-length genotypes. In: Proceedings of the international conference on parallel problem solving from nature, pp 269–278
21. Hutt B, Warwick K (2007) Synapsing variable-length crossover: meaningful crossover for variable-length genomes. IEEE Trans Evol Comput 11(1):118–131
22. Ross SM (2006) Introduction to probability models, 9th edn. Elsevier, Singapore
23. Baker JE (1987) Reducing bias and inefficiency in the selection algorithm. In: Proceedings of the 2nd international conference on genetic algorithms on genetic algorithms and their application, pp 14–21
24. Sokolov A, Whitley D (2005) Unbiased tournament selection. In: Proceedings of the conference on genetic and evolutionary computation, pp 1131–1138
25. Sokolov A, Whitley D, Barreto AMS (2007) A note on the variance of rank-based selection strategies for genetic algorithms and genetic programming. Genet Programm Evolv Mach 8(3):221–237
26. Goldberg D (1990) A note on Boltzmann tournament selection for genetic algorithms and population-oriented simulated annealing. Complex Syst 4(4):445–460
27. Vavak F, Fogarty T (1996) Comparison of steady state and generational genetic algorithms for use in nonstationary environments. In: Proceedings of the IEEE international conference on evolutionary computation, pp 192–195
28. Jong KAD, De KA, Sarma J (1992) Generation gaps revisited. In: Proceedings of international workshop on foundations of genetic algorithms, pp 19–28

29. Smith J, Vavak F (1998) Replacement strategies in steady state genetic algorithms: Static environments. In: Proceedings of international workshop on foundations of genetic algorithms, pp 219–234
30. Smith J (2007) On replacement strategies in steady state evolutionary algorithms. Evol Comput 15(1):29–59
31. Harvey I (2001) Artificial evolution: a continuing SAGA. In Gomi T (ed) Evolutionary robotics: from intelligent robotics to artificial life. Springer, Berlin Heidelberg New York, pp 94–109
32. Eiben A, Hinterding R, Michalewicz Z (1999) Parameter control in evolutionary algorithms. IEEE Trans Evol Comput 3(2):124–141
33. de Jong KAD (1975) An analysis of the behavior of a class of genetic adaptive systems. Ph.D. thesis, University of Michigan
34. Grefenstette J (1986) Optimization of control parameters for genetic algorithms. IEEE Trans Sys, Man Cybern 16(1):122–128
35. Rechenberg I (1973) Evolutionsstrategie: Optimierung technischer Systeme nach Prinzipien der biologischen Evolution. Fromman-Hozlboog, Stuttgart
36. Bäck T (1996) Evolutionary algorithms in theory and practice: evolution strategies, evolutionary programming, genetic algorithms. Oxford University Press, Oxford, UK
37. Eiben AE, Smith JE (2003) Introduction to evolutionary computing. Springer, Berlin Heidelberg New York
38. Schwefel H (1995) Evolution and optimum seeking. Wiley-Interscience, New York
39. Wang R, Okazaki K (2007) An improved genetic algorithm with conditional genetic operators and its application to set-covering problem. Soft Comput 11(7):687–694
40. Fernandes C, Rosa AC (2008) Self-adjusting the intensity of assortative mating in genetic algorithms. Soft Comput 12(10):955–979
41. Zimmermann H (2001) Fuzzy set theory and its applications, 4th edn. Springer, Berlin Heidelberg New York
42. Yun Y, Gen M (2003) Adaptive hybrid genetic algorithm with fuzzy logic controller. In: Verdegay J(ed) Fuzzy sets based heuristics for optimization. Springer, Berlin Heidelberg New York, pp 251–263
43. Lin L, Gen M (2009) Auto-tuning strategy for evolutionary algorithms: balancing between exploration and exploitation. Soft Comput 13(2):157–168
44. Zhang J, Chung HS, Lo W (2007) Clustering-based adaptive crossover and mutation probabilities for genetic algorithms. IEEE Trans Evol Comput 11(3):326–335
45. Xu R, Wunsch D (2008) Clustering. Wiley-IEEE, New York
46. Xu R, Wunsch D (2005) Survey of clustering algorithms. IEEE Trans Neural Netw 16(3):645–678
47. Duda RO, Hart PE, Stork DG (2000) Pattern classification, 2nd edn. Wiley-Interscience, New York
48. Bishop CM (2007) Pattern recognition and machine learning. Springer, Berlin Heidelberg New York
49. Alpaydin E (2004) Introduction to machine learning. MIT Press, Cambridge, MA
50. Qin A, Suganthan P (2005) Self-adaptive differential evolution algorithm for numerical optimization. In: Proceedings of the congress on evolutionary computation, 2:1785–1791
51. Qin A, Huang V, Suganthan P (2009) Differential evolution algorithm with strategy adaptation for global numerical optimization. IEEE Trans Evol Comput 13(2):398–417
52. Salomon R (1998) Evolutionary algorithms and gradient search: similarities and differences. IEEE Trans Evol Comput 2(2):45–55
53. Arnold D, Salomon R (2007) Evolutionary gradient search revisited. IEEE Trans Evol Comput 11(4):480–495
54. Hansen N, Ostermeier A (2001) Completely derandomized self-adaptation in evolution strategies. Evol Comput 9(2):159–195
55. Nocedal J, Wright S (2006) Numerical optimization, 2nd edn. Springer, Berlin Heidelberg New York

56. Hansen N, Mller SD, Koumoutsakos P (2003) Reducing the time complexity of the derandomized evolution strategy with covariance matrix adaptation (CMA-ES). Evol Comput 11(1):1–18
57. Pelikan M (2005) Hierarchical Bayesian optimization algorithm: toward a new generation of evolutionary algorithms. Springer, Berlin Heidelberg New York
58. Larranaga P, Lozano JA (2002) Estimation of distribution algorithms: a new tool for evolutionary computation. Springer, Berlin Heidelberg New York
59. Lozano JA, Larraaga P, Inza I *et al* (2006) Towards a new evolutionary computation: advances on estimation of distribution algorithms. Springer, Berlin Heidelberg New York
60. Kern S, Müller SD, Hansen N *et al* (2004) Learning probability distributions in continuous evolutionary algorithms - a comparative review. Nat Comput 3(3):355–356
61. Zlochin M, Birattari M, Meuleau N *et al* (2004) Model-based search for combinatorial optimization: a critical survey. Ann Operat Res 131(1):373–395
62. Brest J, Greiner S, Boskovic B *et al* (2006) Self-Adapting control parameters in differential evolution: a comparative study on numerical benchmark problems. IEEE Trans Evol Comput 10(6):646–657
63. Koumousis V, Katsaras C (2006) A saw-tooth genetic algorithm combining the effects of variable population size and reinitialization to enhance performance. IEEE Trans Evol Comput 10(1):19–28
64. Arabas J, Michalewicz Z, Mulawka J (1994) GAVaPS-a genetic algorithm with varying population size. In: Proceedings of the first IEEE conference on evolutionary computation, 1:73–78
65. Bäck T, Eiben AE, van der Vaart NAL (2000) An empirical study on GAs "without" parameters. In: Proceedings of the international conference on parallel problem solving from nature, pp 315–324
66. Lobo FG, Lima CF (2006) Revisiting evolutionary algorithms with on-the-fly population size adjustment. In: Proceedings of the annual conference on genetic and evolutionary computation, pp 1241–1248
67. Fernandes C, Rosa A (2006) Self-regulated population size in evolutionary algorithms. In: Proceedings of the international conference on parallel problem solving from nature, pp 920–929
68. Harik GR, Lobo FG (1999) A parameter-less genetic algorithm. In: Proceedings of the genetic and evolutionary computation conference, pp 258–265
69. Eiben AE, Marchiori E, Valk VA (2004) Evolutionary algorithms with on-the-fly population size adjustment. In: Proceedings of the international conference on parallel problem solving from nature, pp 41–50
70. Lobo FG, Lima CF (2005) A review of adaptive population sizing schemes in genetic algorithms. In: Proceedings of the 2005 workshops on genetic and evolutionary computation, pp 228–234
71. Teo J (2006) Exploring dynamic self-adaptive populations in differential evolution. Soft Comput 10(8):673–686
72. Brest J, Maucec MS (2008) Population size reduction for the differential evolution algorithm. Appl Intell 29(3):228–247
73. Teng N, Teo J, Hijazi M (2009) Self-adaptive population sizing for a tune-free differential evolution. Soft Comput 13(7):709–724
74. Wolpert D, Macready W (1997) No free lunch theorems for optimization. IEEE Trans Evol Comput 1(1):67–82
75. Whitley D, Watson J (2005) Complexity theory and the no free lunch theorem. In: Burke EK and Kendall G (eds) Search methodologies. Springer, Berlin Heidelberg New York, pp 317–339
76. Spears WM (2004) Evolutionary algorithms: the role of mutation and recombination. Springer, Berlin Heidelberg New York
77. Liang J, Suganthan P, Deb K (2005) Novel composition test functions for numerical global optimization. In: Proceedings of the IEEE swarm intelligence symposium, pp 68–75

78. Gallagher M, Yuan B (2006) A general-purpose tunable landscape generator. IEEE Trans on Evol Comput 10(5):590–603
79. Goldberg DE, Deb K, Horn J (1992) Massive multimodality, deception, and genetic algorithms. Tech. rep., Illinois Genetic Algorithms Laboratory, UIUC
80. Barr R, Golden B, Kelly J *et al* (1995) Designing and reporting on computational experiments with heuristic methods. J Heurist 1(1):9–32
81. Huband S, Hingston P, Barone L *et al* (2006) A review of multiobjective test problems and a scalable test problem toolkit. IEEE Trans Evol Comput 10(5):477–506
82. Suganthan P, Hansen N, Liang J *et al* (2005) Problem definitions and evaluation criteria for the CEC 2005 special session on real parameter optimization. Tech. rep., Nanyang Technological University and IIT Kanpur, Singapore
83. Fogel DB, Beyer H (1995) A note on the empirical evaluation of intermediate recombination. Evol Comput 3(4):491–495
84. Mcgill R, Tukey J, Larsen W (1978) Variations of box plots. Am Statistic 32(1):12–16
85. Marques JP (2007) Applied statistics using SPSS, STATISTICA, MATLAB and R, 2nd edn. Springer, Berlin Heidelberg New York
86. Montgomery DC (2004) Design and analysis of experiments, 6th edn. Wiley, New York
87. Box GEP, Hunter JS, Hunter WG (2005) Statistics for experimenters: design, innovation, and discovery, 2nd edn. Wiley-Interscience, New York
88. Kennedy J (2003) Bare bones particle swarms. In: Proceedings of the 2003 IEEE swarm intelligence symposium, pp 80–87
89. Potter M, Jong KD (1994) A cooperative coevolutionary approach to function optimization. In: Proceedings of the international conference on parallel problem solving from nature, pp 249–257
90. Yu Y, Xinjie Y (2007) Cooperative coevolutionary genetic algorithm for digital IIR filter design. IEEE Trans Ind Electron 54(3):1311–1318
91. Angeline PJ, Pollack JB (1993) Competitive environments evolve better solutions for complex tasks. In: Proceedings of the 5th international conference on genetic algorithms, pp 264–270
92. Bäck T, Fogel D, Michalewicz Z (2000) Evolutionary computation 2: advanced algorithms and operations. Taylor & Francis, London, UK
93. Lozano M, Herrera F, Krasnogor N *et al* (2004) Real-coded memetic algorithms with crossover hill-climbing. Evol Comput 12(3):273–302
94. Whitley LD, Gordon VS, Mathias KE (1994) Lamarckian evolution, the Baldwin effect and function optimization. In: Proceedings of the international conference on parallel problem solving from nature, pp 6–15
95. Garcła-Martłnez C, Lozano M (2008) Local search based on genetic algorithms. In: Rozenberg G (ed) Advances in metaheuristics for hard optimization. Springer, Berlin Heidelberg New York, pp 199–221
96. Krasnogor N, Gustafson S (2004) A study on the use of self-generation in memetic algorithms. Nat Comput 3(1):53–76
97. Chelouah R, Siarry P (2003) Genetic and Nelder–Mead algorithms hybridized for a more accurate global optimization of continuous multiminima functions. Eur J Operat Res 148:335–348
98. Krasnogor N, Smith J (2005) A tutorial for competent memetic algorithms: model, taxonomy, and design issues. IEEE Trans Evol Comput 9(5):474–488
99. Cowling PI, Kendall G, Soubeiga E (2001) A hyperheuristic approach to scheduling a sales summit. In: Selected papers from the third international conference on practice and theory of automated timetabling, pp 176–190
100. Burke E, Kendall G, Newall J *et al* (2003) Hyper-heuristics: an emerging direction in modern search technology. In: Glover FW, Kochenberger GA (eds) Handbook of metaheuristics. Springer, Berlin Heidelberg New York, pp 457–474
101. Ross P (2005) Hyper-heuristics. In: Burke EK and Kendall G (eds) Search methodologies. Springer, Berlin Heidelberg New York, pp 529–556

102. Ozcan E, Bilgin B, Korkmaz EE (2008) A comprehensive analysis of hyper-heuristics. Intell Data Anal 12(1):3–23
103. Cowling P, Kendall G, Han L (2002) An investigation of a hyperheuristic genetic algorithm applied to a trainer scheduling problem. In: Proceedings of the IEEE congress on evolutionary computation, 2:1185–1190
104. Ong YS, Keane A (2004) Meta-Lamarckian learning in memetic algorithms. IEEE Trans Evol Comput 8(2):99–110
105. Jin Y, Branke J (2005) Evolutionary optimization in uncertain environments-a survey. IEEE Trans Evol Comput 9(3):303–317
106. Zhou Z, Ong YS, Lim MH *et al* (2007) Memetic algorithm using multi-surrogates for computationally expensive optimization problems. Soft Comput 11(10):957–971
107. Schmidt MD, Lipson H (2008) Coevolution of fitness predictors. IEEE Trans Evol Comput 12(6):736–749
108. Choi S, Moon B (2008) Normalization for genetic algorithms with nonsynonymously redundant encodings. IEEE Trans Evol Comput 12(5):604–616
109. Yang S (2006) On the design of diploid genetic algorithms for problem optimization in dynamic environments. In: IEEE congress on evolutionary computation, pp 1362–1369
110. Haupt RL, Haupt SE (2004) Practical genetic algorithms, 2nd edn. Wiley, New York
111. Bäck T, Fogel D, Michalewicz Z (2000) Evolutionary computation 1: basic algorithms and operators. Taylor & Francis, London, UK
112. Goldberg DE (1989) Genetic algorithms in search, optimization, and machine learning. Addison-Wesley, Reading, MA
113. Wright AH, Vose MD, Jong KAD *et al* (2005) Proceedings of 8th international workshop on foundations of genetic algorithms. Springer, Berlin Heidelberg New York
114. Stephens CR, Toussaint M, Whitley D *et al* (2007) Proceedings of 9th international workshop on foundations of genetic algorithms. Springer, Berlin Heidelberg New York
115. Nakanishi H, Kinjo H, Oshiro N *et al* (2007) Searching performance of a real-coded genetic algorithm using biased probability distribution functions and mutation. Artif Life Robot V11(1):37–41
116. Yoon H, Moon B (2002) An empirical study on the synergy of multiple crossover operators. IEEE Trans Evol Comput 6(2):212–223
117. Cervantes J, Stephens C (2009) Limitations of existing mutation rate heuristics and how a rank GA overcomes them. IEEE Trans Evol Comput 13(2):369–397
118. Lobo FG, Lima CF, Michalewicz Z (2007) Parameter setting in evolutionary algorithms. Springer, Berlin Heidelberg New York
119. Kramer O (2008) Self-adaptive heuristics for evolutionary computation. Springer, Berlin Heidelberg New York
120. Eiben AE, Schut MC, de Wilde AR (2006) Boosting genetic algorithms with self-adaptive selection. In: Proceedings of IEEE congress on evolutionary computation, pp 477–482
121. Zhang Q, Leung Y (1999) An orthogonal genetic algorithm for multimedia multicast routing. IEEE Trans Evol Comput 3(1):53–62
122. Leung Y, Wang Y (2001) An orthogonal genetic algorithm with quantization for global numerical optimization. IEEE Trans Evol Comput 5(1):41–53
123. Yuen SY, Chow CK (2009) A genetic algorithm that adaptively mutates and never revisits. IEEE Trans Evol Comput 13(2):454–472
124. Lobo F, Goldberg DE (2004) The parameter-less genetic algorithm in practice. Inf Sci 167(1-4):217–232
125. Zebulum RS, Pacheco MA, Vellasco MMB (2001) Evolutionary electronics: automatic design of electronic circuits and systems by genetic algorithms. CRC, Boca Raton, FL
126. Gokhale M, Graham PS (2005) Reconfigurable computing: accelerating computation with field-programmable gate arrays. Springer, Berlin Heidelberg New York
127. Greenwood GW, Tyrrell AM (2006) Introduction to evolvable hardware: a practical guide for designing self-adaptive systems. Wiley-IEEE Press, New York
128. Nedjah N, Alba E, Mourelle LM (2006) Parallel evolutionary computations. Springer, Berlin Heidelberg New York

Part II
Dealing with Complicated Problems

Chapter 4
Constrained Optimization

Abstract In previous chapters, we only searched the space defined by variables' upper and lower bounds. But real-world problems are always with constraints. One important question that needs to be answered when applying EAs in constrained optimization is how to evaluate a solution that violates some constraints. Generally, we want the final results of our EAs to satisfy all the constraints. But discarding the those that violate some constraints and generating new ones again is very inefficient. Several wonderful ideas for constraint handling will be discussed in this chapter.

4.1 Introduction

4.1.1 Constrained Optimization

The real world is constrained by various rules, so that the mathematical models representing real-world optimization and learning problems will have various constraints. The general form of *constrained optimization problems* (COPs) can be illustrated as follows:[1]

$$\begin{aligned} &\min f(\mathbf{x}),\ \mathbf{x} \in \mathbb{R}^n \\ &\text{s.t.}\quad g_i(\mathbf{x}) \le 0, \quad i = 1, \cdots, q \\ &\qquad\ \ h_j(\mathbf{x}) = 0, \quad j = q+1, \cdots, k \\ &\qquad\ \ L_l \le x_l \le U_l, \quad l = 1, \cdots, n \end{aligned} \tag{4.1}$$

where L_l and U_l are the lower and upper bounds of variable x_l, respectively, which forms the *search space* S.[2] q inequality constraints (linear or nonlinear) and $k-q$ equality constraints (linear or nonlinear) need to be satisfied, which forms the

[1] If objective or constraint function is nonlinear with $\mathbf{x}$, it is also called *nonlinear programming* (NLP).

[2] Generally, these n inequalities are not regarded as constraints but form the definition domain.

feasible region F. $F \subseteq S$. If point $\mathbf{x} \in F$, we say $\mathbf{x}$ is *feasible*, else if point $\mathbf{x} \in S \backslash F$, we say $\mathbf{x}$ is *infeasible*.[3] If the purpose of OR is to find feasible solutions, i.e., the objective is not considered, then such problems are called *constrained satisfaction problems* (CSPs).

If a point $\mathbf{x}$ satisfies $g_i(\mathbf{x}) = 0$ for inequality constraint i, we say that constraint i is *active* at point $\mathbf{x}$. All equality constraints are considered to be active at feasible region F.

Due to the possible complex form of constraints, the relationship between F and S might be complicated. Figure 4.1 is one example.

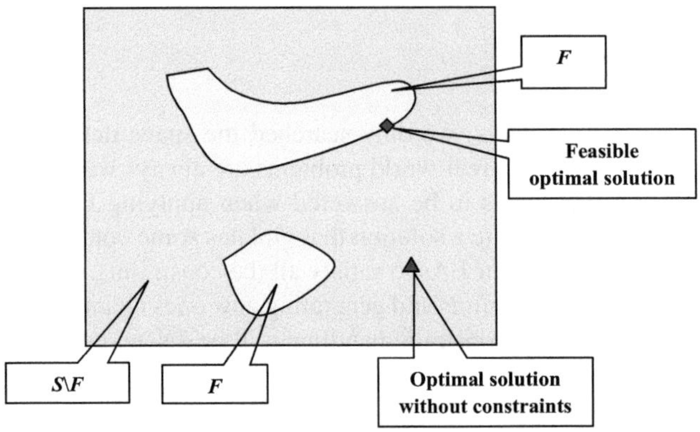

Fig. 4.1 Feasible region and search space

In Fig.4.1, F is a nonconvex disconnected set and the feasible optimal solution is at the edge of the feasible region, which is rather far away from the optimal solution without constraints. All of these characteristics of F add difficulties to COPs.

Due to the above facets and considering the No Free Lunch theorem, generally it is impossible to develop a deterministic method for COPs that might be better than an exhaustive search, i.e., a COP is an intractable model for OR methods.

Equality constraints $h_j(\mathbf{x}) = 0$ in Eq. 4.1 might be the most difficult part of NLP because they make F extremely small compared to S. So generally, for almost all NLP solvers, we need to relax the equality constraints to inequality constraints as follows:

$$\left|h_j(\mathbf{x})\right| \leq \delta, \quad j = q+1, \cdots, k \tag{4.2}$$

where δ is the tolerance value predefined by users.[4] In this way, we can transfer Eq. 4.1 into an NLP with k inequality constraints.

---

[3] If $\mathbf{x} \notin S$, we say $\mathbf{x}$ is *illegal*.

[4] $\delta = 0.001$ or $\delta = 0.0001$ are commonly suggested fixed values. Techniques to control δ in the evolving process will be introduced later.

4.1.2 Constrained Optimization Evolutionary Algorithms

EAs can maintain a group of individuals, either feasible or infeasible, that could promote evolution in different feasible regions, the finding of new feasible regions, and approaching a feasible optimal solution from different directions. So EAs have some advantages in COPs especially for those whose feasible regions are disconnected or whose feasible optimal solution is at the edge of the feasible region, like the example in Fig. 4.1.

EAs with special considerations for COPs are called *constrained optimization EAs* (COEAs). Because we often want COEAs to have the above-mentioned exploration abilities, population diversity needs to be preserved carefully.

The following points need to be emphasized when designing and analyzing a COEA; the first one is of most importance.

1. How does one compare a feasible solution with a larger objective value and an infeasible solution with a smaller objective value?[5] This consideration often affects the selection or replacement part of EAs.
2. Standard EAs are generally ill-suited for solving CSPs or COPs because standard variation operators might generate infeasible offspring from feasible parents, i.e., these operators are "blind" to constraints. So are there any special considerations on the variation operators that can promote the search on the constrained solution landscape effectively?
3. How does one control the tolerance δ in Eq. 4.2 for equality constraints?
4. How does one maintain the feasibility of individuals if variation operators generate illegal or infeasible offspring?

The taxonomy of COEAs is given in Fig 4.2.

If a COEA can always limit its search in feasible region F with special encoding and decoding techniques, it belongs to the "feasibility maintenance" group. We will introduce an example of real-code COEA in Sect. 4.2 and discuss other topics in Chap. 7.

If users can indicate a preference between the constraint violation and objective value, i.e., using predefined or adaptive controlled weights to change COPs into optimization problems without constraints, the COEA belongs to the "penalty function" group. We will introduce several methods to set or control the weight parameters in Sect. 4.3.[6]

COEAs generally treat objective and constraint violation separately, i.e., they maintain elaborate balance between infeasible solutions with relatively small objective values and feasible solutions with relative large objective values, which belongs to the "separation of constraint violation and objective value" group. We will introduce three examples in Sect. 4.4.

---

[5] In this chapter, we will stick to the minimum constrained problem unless otherwise specified.

[6] In Chap. 3 we sometimes ignored the difference between objective value and fitness value because many selection processes can handle minimum requirements directly. In this chapter, we will use objective value to represent $f(\mathbf{x})$ in Eq. 4.1 and fitness value to represent the weighted penalty function value.

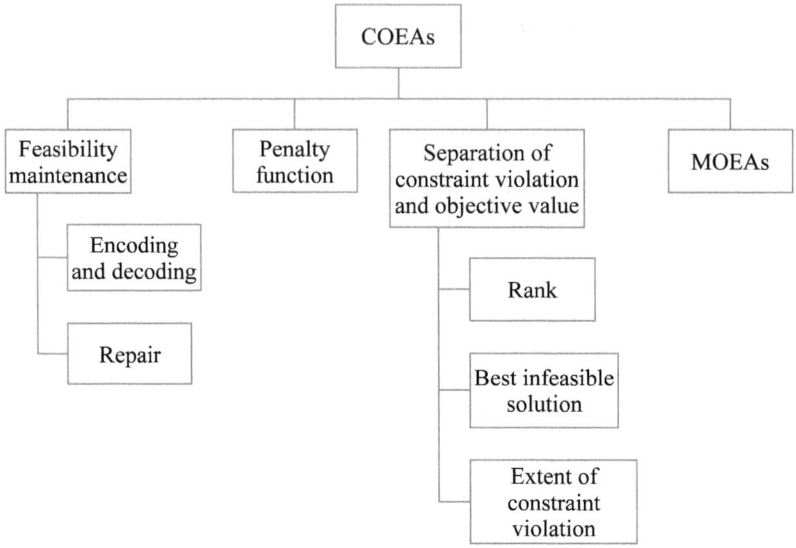

Fig. 4.2 The taxonomy of COEAs

Also, we can treat k constraints as k objectives or the total constraint violation as an objective, thus change the COPs into multiobjective problems and solve them using multiobjective EAs, which belongs to the "MOEAs" group. Sect. 6.8.2 will introduce constraint handling techniques developed from MOEAs.

Figure 4.3 illustrates the number of papers indexed by the SCI on COEAs.[7]. As can be seen from Fig. 4.3, COEAs have attracted interest in recent years.

4.2 Feasibility Maintenance

4.2.1 Genetic Algorithm for Numerical Optimization of Constrained Problems

In 1998 Michalewicz suggested GENOCOP (genetic algorithm for numerical optimization of constrained problems) as an example COEA of "feasibility maintenance" [1]. GENOCOP can handle COPs with convex feasible region F. We can illustrate GENOCOP by a simple linear constrained problem as follows:

[7] TS = (("constrained optimization") AND ("genetic algorithm" OR "genetic algorithms" OR "evolutionary computation" OR "evolutionary computing" OR "evolutionary algorithms" OR "evolutionary intelligence")). The SCI index "TS" is for the search topic in the title, the keywords, and the abstract.

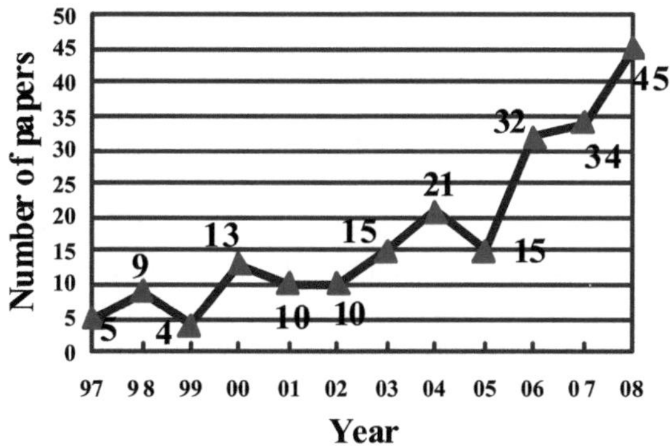

Fig. 4.3 Number of papers indexed by SCI on COEAs

$$
\begin{aligned}
&\min f(x_1,x_2,x_3,x_4,x_5,x_6) \\
&\text{s.t.}\quad 2x_1+x_2+x_3=6 \\
&\qquad x_3+x_5-3x_6=10 \\
&\qquad x_1+4x_4=3 \\
&\qquad x_2+x_5\le 120 \\
&\qquad -40\le x_1\le 10,\quad 50\le x_2\le 75 \\
&\qquad 0\le x_3\le 10,\quad 5\le x_4\le 15 \\
&\qquad 0\le x_5\le 20,\quad -5\le x_6\le 5
\end{aligned}
\tag{4.3}
$$

In Eq. 4.3, we can first eliminate all the equality constraints as follows:

$$
\begin{aligned}
x_1 &= 3-4x_4 \\
x_2 &= -10+8x_4+x_5-3x_6 \\
x_3 &= 10-x_5+3x_6
\end{aligned}
\tag{4.4}
$$

By substituting Eq. 4.4 into Eq. 4.3, we get the following NLP with only linear inequality constraints, which constitute the convex F.

$$
\begin{aligned}
&\min\ g(x_4,x_5,x_6)=f(3-4x_4,-10+8x_4+x_5-3x_6,10-x_5+3x_6,x_4,x_5,x_6) \\
&\text{s.t.}\quad -10+8x_4+2x_5-3x_6\le 120 \\
&\qquad -40\le 3-4x_4\le 10 \\
&\qquad 50\le -10+8x_4+x_5-3x_6\le 75 \\
&\qquad 0\le 10-x_5+3x_6\le 10 \\
&\qquad 5\le x_4\le 15, 0\le x_5\le 20, -5\le x_6\le 5
\end{aligned}
\tag{4.5}
$$

Provided that we have *popsize* feasible individuals,[8] whole arithmetic crossover, introduced in Sect. 3.2.2, can maintain the feasibility of the offspring.

As for mutation, feasible individual (x_4, x_5, x_6) might generate infeasible mutants by the methods discussed in Sect. 3.2.2. Since Eq. 4.5 only has linear inequality constraints, we can easily generate the possible mutation region for each variable.

Suppose we want to mutate $(x_4 = 10, x_5 = 8, x_6 = 2)$. All we need to do is fix two variables and get the possible mutation region for the third variable. For example, if we want to mutate gene x_4, then $x_5 = 8$ and $x_6 = 2$ will not change. Substituting $x_5 = 8$ and $x_6 = 2$ into the constraints of Eq. 4.5, we can get several linear inequality constraints for x_4. The possible mutation region for x_4 is the intersection of the solutions of these linear inequality equations. In this example, if $x_5 = 8$ and $x_6 = 2$, then the mutation region for x_4 is $[7.25, 10.375]$; if $x_4 = 10$ and $x_6 = 2$, the mutation region for x_5 is $[6, 11]$; and if $x_4 = 10$ and $x_5 = 8$, the mutation region for x_8 is $[1, 2.666]$.[9] Then the techniques discussed in Sect. 3.2.2 can be used.

GENOCOP can only maintain the feasibility of individuals within convex F, which limits its application. But the idea of eliminating linear equality constraints is used by almost all COEA designers.

4.2.2 Homomorphous Mappings

In 1999 Koziel and Michalewicz further developed GENOCOP for nonconvex and disconnected F by *homomorphous mappings* (HM) [2].

In abstract algebra, a homomorphism is a structure-preserving map between two algebraic structures. If we can generate a bijective map between a point in the cube of n-dimensional space $[-1, 1]^n$ and a point in the feasible region F, i.e., every point in cube $[-1, 1]^n$ has an exact image in F and *vice versa*, and the images in F of the close points in cube $[-1, 1]^n$ are close, this map an is *isomorphism* and cube $[-1, 1]^n$ and F are *isomorphic*.

Then we can just evolve in the convex cube $[-1, 1]^n$ with real-code crossover and mutation operators and maintain the feasibility of the population. To determine the original image in the cube and its image in F, their relative distance to the origin are the same, which is the structure being preserved in the map.

We will discuss HM in the 2-D situation. Higher dimensions will increase the computational cost and complexity, but with the same idea. The first example is the HM between a rectangle in 2-D and the convex F in S as in Fig. 4.4.

For point A in Fig 4.4a, we can determine a radial line starting from the origin and going toward A and denote this radial line as follows:

---

8 If the cardinality of F is small compared to S, generating *popsize* feasible individuals initially is not a trivial task. Here we overlook this difficulty and just simplify the problem by discarding all the randomly generated infeasible solutions until we get *popsize* feasible ones.

9 For the reason of simplicity, we use linear constraints. If the constraints are nonlinear but we know that F is convex, mutation region for each variable could be found by numerical methods, such as binary search.

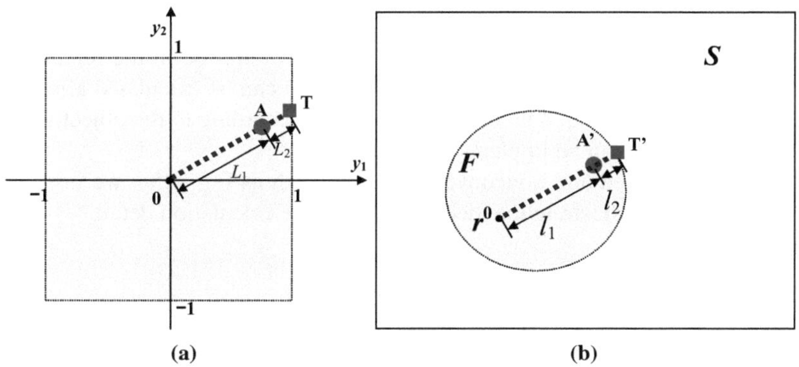

Fig. 4.4 HM between the rectangle in 2-D and the convex F in S: (a) rectangle in 2-D, and (b) convex F in S

$$\mathbf{y} = \mathbf{y}^0 t \tag{4.6}$$

where $\mathbf{y}^0$ is the coordinates of point A and t is a parameter. We can calculate the distance between point A and the origin as L_1. The radial line intersects with the boundary of the rectangle one time at one point because of the convexity of the rectangle. We can get the intersection point T, i.e., get the maximum value of t as $t_{\max} = \frac{1}{\max(|y_1^0|,|y_2^0|)}$.[10] So the distance between T and point A can be calculated as L_2. The structure being preserved by HM is $\frac{L_1}{L_1+L_2}$.

For the points in F in Fig. 4.4b, we need to appoint a *reference point* $\mathbf{r}^0$ artificially corresponding to the origin in Fig 4.4a. After that we can determine another radial line starting from $\mathbf{r}^0$ and with the same direction of $\mathbf{y}^0$ as follows:

$$\mathbf{x} = \mathbf{r}^0 + \mathbf{y}^0 \tau \tag{4.7}$$

where τ is a parameter. For convex F, we can find only one intersection, i.e., point T', of the radial line and the boundary of F with numerical methods, i.e., find $\tau_{\max}$. So the distance between the intersection and the reference point can be calculated as $l_1 + l_2$.

To keep the relative distance to the origin (the reference point), the image of point A in Fig 4.4a, i.e., point A' in Fig 4.4b, should satisfy

$$\frac{L_1}{L_1+L_2} = \frac{l_1}{l_1+l_2} \tag{4.8}$$

So A' can be calculated using

$$\mathbf{x}^0 = \mathbf{r}^0 + \mathbf{y}^0 \frac{\tau_{\max}}{t_{\max}} \tag{4.9}$$

[10] Why the maximum coordinate value determines the maximum t?

where $\mathbf{x}^0$ is the coordinates of point A'.

After we generate the initial individuals in the rectangular area of Fig 4.4a, variation operators that guarantees that the offspring will still be in the rectangle can be used freely. The objective value for an individual can be calculated after we get its real coordinates by Eq. 4.9. Then we can select according to the objective value. Thus the whole EA can be implemented.

For a disconnected and nonconvex situation, such as Fig. 4.5b, we just use the distance concept to illustrate the idea and neglect the calculation detail.

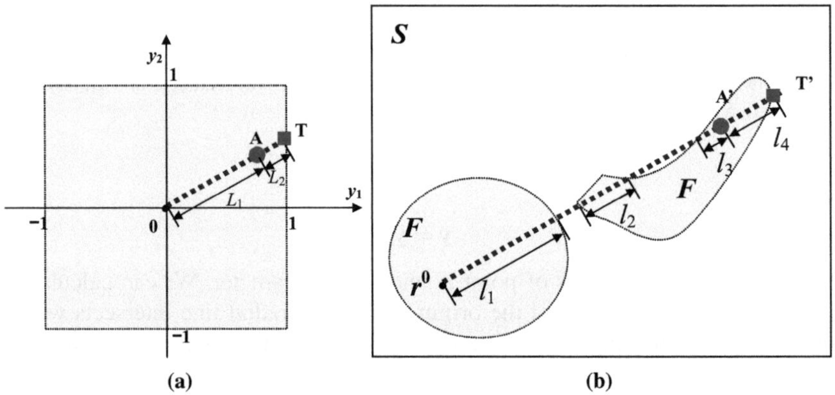

Fig. 4.5 HM between the rectangle in two-dimension and the nonconvex F in S: (a) rectangle in 2-D, and (b) nonconvex F in S

In nonconvex and disconnected F, we can calculate multiple intersections between the radial line and the feasible boundary by numerical methods.[11] After calculating all the intersections, we can determine the farthest intersection point T'. To keep the relative distance to the origin (the reference point) in Fig. 4.5b, point A' can be calculated by

$$\frac{L_1}{L_1+L_2} = \frac{l_1+l_2+l_3}{l_1+l_2+l_3+l_4} \tag{4.10}$$

where the definition of l_1, l_2, l_3, and l_4 can easily be understood with Fig. 4.5b. Other considerations are the same as with convex F.

The advantages and characteristics of HM can be summarized as follows:

- HM will always preserve the feasibility of a population, so there is no requirements for evaluating infeasible solutions.
- Although users need to provide a reference point $\mathbf{r}^0$ in F before the map, theoretically any point in F could be selected and would not affect the map.[12]

[11] This might be hard to do if F has a complex shape.

[12] But different $\mathbf{r}^0$ might affect the optimization results.

- No additional parameters need to be provided by users.

The disadvantages of HM can be listed as follows:

- The numerical calculation for finding all the intersections between the radial line and the boundary of F is hard.
- Basically, HM cannot handle equality constraints.[13]
- Sometimes finding the initial feasible solutions is not a trivial task.
- A bad selection of $\mathbf{r}^0$ might yield bad results.

Another topic that needs to be mentioned is that sometimes the variation operators might generate *illegal* solutions, i.e., $x_l \geq U_l$ or $L_l \geq x_l$ in Eq. 4.1. Several methods can be used that are basically based on a repair idea. We need to "repair" the current illegal solution so that it becomes feasible. The simplest way to repair $x_l \geq U_l$ is to assign $x_l = U_l$ directly. If you know that a feasible solution is on the boundary of F, this simple technique might work. Otherwise, we can reflect the illegal variable by $x_l^{'} = 2U_l - x_l$, where $x_l^{'}$ is the repaired legal variable. This symmetrical reflection method can also be controlled using an annealing idea. Similar methods can be used with $L_l \geq x_l$.[14] Although we discuss illegal repairing approaches in this section, they can also be used in other COEAs.

4.3 Penalty Function

The penalty function is the ordinary OR method for COPs and is the most often used technique in COEAs. For the COP model illustrated by Eq. 4.1, we can define the *violation of constraint* j, i.e., $v_j(\mathbf{x})$, as follows:[15]

$$v_j(\mathbf{x}) = \begin{cases} \max\{0, g_j(\mathbf{x})\} & 1 \leq j \leq q \\ |h_j(\mathbf{x})| & q+1 \leq j \leq k \end{cases} \tag{4.11}$$

If we know that the constraint violations are of the same order of magnitude or we have already normalized them, the overall constraint violation can be generated by either Eq. 4.12 or Eq. 4.13:[16]

$$v(\mathbf{x}) = \sum_{j=1}^{k} v_j(\mathbf{x}) \tag{4.12}$$

$$v(\mathbf{x}) = \max_j (v_j(\mathbf{x})) \tag{4.13}$$

---

[13] Why?

[14] A similar discussion appears on page 63.

[15] This is based on Eq. 4.1. Readers should figure out the constraint violation after we have already transformed the equality constraints into inequality ones.

[16] What is the meaning of Eq. 4.13?

Furthermore, for the overall constraint violation Eq. 4.12, we can assign weights, denoted as *penalty coefficients*, ω_i, for each constraint violation to represent our preference or implement the scaling. Thus the *penalty function* is formulated as follows:

$$\text{penalty}(\mathbf{x}) = \sum_{j=1}^{k} \omega_j v_j(\mathbf{x}) \tag{4.14}$$

Then the fitness value for a solution **x** can be described as follows:

$$f'(\mathbf{x}) = f(\mathbf{x}) + \text{penalty}(\mathbf{x}) \tag{4.15}$$

From Eq. 4.11 we know that the constraint violations are not less than zero, so Eq. 4.15 increases the fitness values of the infeasible solutions according to the extent of their constraint violation and keep the fitness values of the feasible solutions the same as their objective values.[17]

Equation 4.15 penalizes the infeasible solutions by adding their fitness values using plus operators and do not utilize the number of constraints being violated. Methods considering this information will be introduced later.

Thus we have already transferred the COP, Eq. 4.1, into an unconstrained optimization problem, Eq. 4.15. All the techniques discussed in Chaps. 2 and 3 can be used.

It is not difficult to understand that the feasible optimal solution of Eq. 4.1 is the global optimal solution of Eq. 4.15. So we can find the feasible optimal solution of Eq. 4.1 by optimizing Eq. 4.15.

4.3.1 Static Penalty Function

For constraint j, if the penalty coefficient ω_j is predefined by the user and does not change during the evolving process, Eq. 4.14 is called a *static penalty function*.

The main advantage of the static penalty function is its simplicity of implementation. But this method suffers from drawbacks similar to those of parameter tuning, which is discussed in Sect. 3.5.1, and fixed weights sum, which will be introduced in Sect. 6.2.1.1. We can summarize the shortcomings of this method as follows:

- By assigning penalty coefficient ω_j with a preference, we tell the EAs which constraint is more important or more difficult or with a larger violation. But this information is generally hard to acquire. Sometimes the difficulty of providing adequate penalty coefficients is severer than that of finding a feasible optimal solution to the original problem.
- Too small penalty coefficients, which might ultimately bring about infeasible solutions, and too large penalty coefficients, which might ultimately cause a local

[17] We discuss minimum problems in this chapter. So smaller fitness values means fitter individuals.

optimal solution, are both inappropriate. But we do not know how large is large enough.
- Different penalty coefficient vectors might provide different solution landscapes of Eq. 4.15, which might induce different treatment of EAs. But we do not know the relationship between the penalty coefficient vector and the shape of the solution landscape beforehand.

An extreme example of a static penalty function is the *death penalty*, which rejects infeasible solutions and reuses the variation operators and initialization process to generate new individuals again. For simple COPs, the death penalty is easy but requires more and more computational costs with the increase in the complexity of constraints.

4.3.2 Dynamic Penalty Function

Similar to the deterministic parameter control discussed in Sect. 3.5.1, we can let the penalty coefficient increase with the evolving process. So in the early stages, a small penalty might promote the exploration of both feasible and infeasible regions; and in the late states, a large penalty might guide the population toward a feasible region. There are many types of deterministic penalty coefficient control schemes. We just introduce one proposed by Joines and Houck [3]. The penalized fitness function can be illustrated as follows:

$$f^{'}(\mathbf{x}) = f(\mathbf{x}) + (c \times gen)^{\alpha} \sum_{j=1}^{k} v_j^{\beta}(\mathbf{x}) \tag{4.16}$$

where α, β, and c are control parameters predefined by users, $\alpha = \beta = 2$ and $c = 0.5$ are suggested values, and *gen* is the generation number that increases with the evolving process. So Eq. 4.16 is called a *dynamic static penalty*.

Dynamic penalty functions, Eq. 4.16, require fewer parameters than static penalty functions and can increase the selective pressure with the evolving process, which is exactly what we want. But the problem is that the proper way of increasing the selective pressure might be problem dependent. Also, if the global optimal solution without constraints is far from the feasible optimal solution, Eq. 4.15 might take the search in the wrong direction.

4.3.3 Adaptive Penalty Function

If we can gather information from the current population and use it in penalty coefficient control, fewer parameters are necessary and the penalty function might thereby acquire a problem-independent ability. This is an *adaptive penalty function*. We will discuss several examples with the intent of spurring readers' creativity.

Gen and Cheng suggested an adaptive multiplication penalty function for maximum optimization problem as follows [4]:

$$f'(\mathbf{x}) = f(\mathbf{x}) \times \left[1 - \frac{1}{k} \sum_{j=1}^{k} \left(\frac{v_j(\mathbf{x})}{v_j^{\max}} \right)^{\alpha} \right] \tag{4.17}$$

where α is the control parameter, $v_j(\mathbf{x})$ is the violation of constraint j defined by Eq. 4.11, and $v_j^{\max}$ is the maximum violation of constraint j in the current population, i.e., $v_j^{\max} = \max\left(\varepsilon, \max_{\mathbf{x}} (v_j(\mathbf{x}))\right)$, where ε is to avoid dividing by zero when all individuals are feasible.

In Eq. 4.17, the objective value of an individual with a larger average constraint violation will be decreased further. The strength of penalty depends on the current maximum constraint violation. So we group it as an adaptive penalty function.

Hadj–Alouane and Bean suggested another type of adaptive penalty function in 1997 [5]. It uses the same penalty coefficient, Ω, for all constraints in one generation, that is,

$$f'(\mathbf{x}) = f(\mathbf{x}) + \Omega v(\mathbf{x}) \tag{4.18}$$

where $v(\mathbf{x})$ is defined by Eq. 4.12.

The penalty coefficient Ω changes according to whether the individual with the smallest objective value is in F or not, i.e.,

$$\Omega(gen+1) = \begin{cases} \frac{\Omega(gen)}{\beta_1}, & \text{if } \mathbf{b}^i \in F \text{ for all } gen - \tau + 1 \le i \le gen \\ \beta_2 \Omega(gen), & \text{if } \mathbf{b}^i \notin F \text{ for all } gen - \tau + 1 \le i \le gen \\ \Omega(gen), & \text{otherwise} \end{cases} \tag{4.19}$$

where $\beta_1 > \beta_2 > 1$ and τ are control parameters predefined by users, gen is the current generation number, and $\mathbf{b}^i$ is the best individual, with the smallest objective value, at generation i.

If in the previous successive τ generations the individuals with the smallest objective value are all feasible, we need to decrease the penalty coefficient so as to promote exploration in the infeasible region. If in the previous successive τ generations the individuals with the smallest objective value are all infeasible, we need to increase the penalty coefficient so as to push the population toward F. Otherwise, the penalty coefficient will be kept unchanged.

A similar idea was adopted by Hinterding for adaptive equality control handling, Eq. 4.2, [6].[18] For every τ generations, we would like to change the tolerance in the following way.

[18] Readers should review the "1/5 success rule" in Sect. 3.5.1.

if ($gen \bmod \tau = 0$) then

$$\delta(gen) = \begin{cases} \frac{\delta(gen-\tau)}{c}, & p_f > 25\% \\ \delta(gen-\tau)c, & p_f < 15\% \\ \delta(gen-\tau), & 15\% \leq p_f \leq 25\% \end{cases} \quad (4.20)$$

else

$$\delta(gen) = \delta(gen-1)$$

end

where p_f is the feasible percentage of the current population and $c = 1.3$ is the suggested control parameter. If the current feasible percentage is large, i.e., $p_f > 25\%$, we need to decrease the tolerance δ so as to find a more precise solution. If the current feasible percentage is small, i.e., $p_f < 15\%$, we need to increase the tolerance δ to promote the exploration of F. Otherwise, δ will be kept unchanged. After δ is adaptively changed, all individuals in the current generation need to be re-evaluated with respect to their constraint violation.

4.3.3.1 Adaptive Segregational Constraint Handling Evolutionary Algorithm

Hamida and Schoenauer adopted and improved the above penalty coefficient and tolerance adaptive control method and suggested the *adaptive segregational constraint handling evolutionary algorithm* (ASCHEA) [7]. Since then ASCHEA has gradually becoming one of the classical COEAs and most current suggested COEAs need to compare their numerical results on benchmark problems with ASCHEA.

ASCHEA is a $(\mu + \lambda)$-ES, which was introduced in Sect. 2.3.1. The main characteristics of ASCHEA can be summarized as follows:

1. **Adaptive penalty coefficient control.** ASCHEA is similar to Eq. 4.19, but treats penalty coefficients separately.
2. **Dynamic and adaptive tolerance control.** ASCHEA develops the idea of tolerance control, like Eq. 4.20, and considers the situation more delicately.
3. **Segregational selection.** ASCHEA has special considerations on parental selection with preference for feasible individuals.
4. **Mating restriction.** ASCHEA embodies the idea of exploring F in a negative assortative mating way.

The penalty coefficients for constraint j are adaptively controlled as follows:

$$\begin{aligned}&\text{if } \left(p_f(j) > p_{target}\right) \quad \omega_j(g+1) = \frac{\omega_j(g)}{fact}\\&\text{otherwise} \qquad\qquad \omega_j(g+1) = \omega_j(g)\, fact\end{aligned} \tag{4.21}$$

where $p_f(j)$ is the feasible percentage of constraint j in the current population, p_{target} is the ideal feasible percentage, $\omega_j(g)$ and $\omega_j(g+1)$ are, respectively, the penalty coefficients of the current and next generation for constraint j, and $fact > 1$ is a user-defined parameter.[19] The analysis of Eq. 4.21 is similar to that of Eq. 4.19.

Hamida and Schoenauer suggested two kinds of tolerance control for equality constraints: dynamic adjustment (DA) and adaptive adjustment (AA). DA is similar to Eq. 4.20:

$$\begin{aligned}&\text{if } \left(p_f(j) > p_{reduct}\right) \quad \delta_j(g+1) = \frac{\delta_j(g)}{fact_\delta}\\&\text{otherwise} \qquad\qquad \delta_j(g+1) = \delta_j(g)\, fact_\delta\end{aligned} \tag{4.22}$$

where $p_f(j)$ is the feasible percentage of equality constraint j in the current population, p_{reduct} is the ideal feasible percentage for equality constraints, $\delta_j(g)$ and $\delta_j(g+1)$ are, respectively, the tolerances of the current and next generation for equality constraint j, and $fact_\delta > 1$ is a user-defined parameter.[20] The analysis of Eq. 4.22 is similar to that of Eq. 4.20.

DA only considers the feasibility number for equality constraints in the current population and neglects the extent of violation. So the reduction of δ_j might be too fast, with large $fact_\delta$, so that it will contain no feasible individuals after Eq. 4.22.[21] AA ensures that it will always retain $p_{equality}$ proportional the feasible individuals for equality constraint j if its current feasible percentage $p_f(j) > p_{reduct}$.[22]

To implement this, we need to refine the relaxation of equality constraint, Eq. 4.2, as follows:

$$\delta_j^- \le h_j(\mathbf{x}) \le \delta_j^+, \quad j = q+1, \cdots, k \tag{4.23}$$

where δ_j^- and δ_j^+ are the negative and positive tolerance for equality constraint j, respectively.

Then for equality constraint j, if its feasible percentage $p_f(j) > p_{reduct}$, then we know not only the number of individuals on both sides of the relaxation but also the extent of violation on equality constraint j for each of these individuals. Suppose there are n_F^+ and n_F^- individuals on the positive and negative relaxation sides of equality constraint j, respectively, we can rank them according to the extent of violation on j in ascending order and then take the violation of the $p_{equality} \times n_F^+$th and $p_{equality} \times n_F^-$th individual as the tolerance in the next generation.[23] A graphical

---

[19] Hamida and Schoenauer suggested $p_{target} = 0.5$ and $fact = 1.1$.

[20] Hamida and Schoenauer suggested $p_{reduct} = 0.6$ and $fact_\delta = 1.01$ for DA.

[21] This explains the suggestion of such a smaller value, $fact_\delta = 1.01$, by Hamida and Schoenauer.

[22] DA and AA are both adaptive controls of the tolerance value. But the "adaptive" represented by the first letter of "AA" is referred to as the adaptive control of $fact_\delta$ in DA.

[23] Hamida and Schoenauer suggested $p_{reduct} = 0.7$ and $p_{equality} = 0.3$.

illustration for determining the positive tolerance in the next generation is in Fig. 4.6, where v_i is the ith violation of the equality constraint in ascending order.

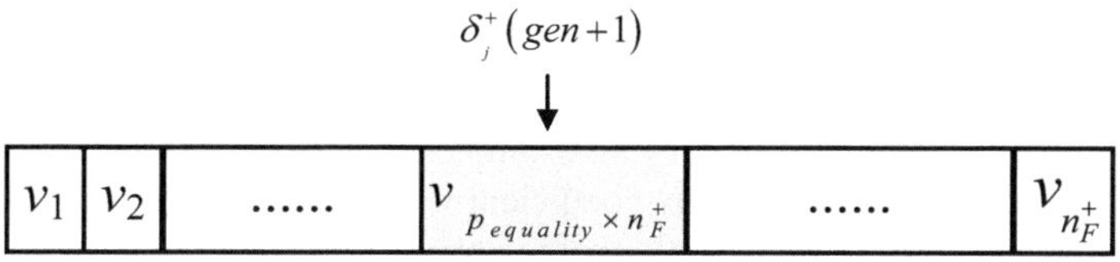

Fig. 4.6 Using the violation of $p_{equality} \times n_F^+$th individual in the positive relaxation region to determine the positive tolerance for constraint j in the next generation

If the COP has only one equality constraint, AA could always maintain a certain number of feasible individuals. But as AA treats the equality constraints separately, in the worst scenario, there might be no feasible individuals after AA in COPs with multiple equality constraints.

The segregation selection in ASCHEA segregates the survivor selection process, replacement, into two parts: feasibility-based selection and penalty-function-based selection. First, $p_{select} \times \mu$ feasible individuals are selected from $\mu + \lambda$ individuals without replacement based on their objective values. If the number of feasible individuals is less than $p_{select} \times \mu$, just select all the feasible individuals without replacement. Then a penalized fitness value is used for selection until a total of μ individuals are picked.[24] In this way, apart from the penalty function, ASCHEA further prefers feasible individuals.

Some COPs have their feasible optimal solutions on the boundary of feasible region F, so in ASCHEA, if one feasible individual is selected to mate, its mating counterpart must be an infeasible individual so as to promote the exploration of the feasible boundary. But Hamida and Schoenauer also suggested that this mating restriction could only be used when too few feasible individuals are in the current population to allow the exploration of the feasible region.

These techniques – adaptive control of the penalty coefficients and the equality tolerances, preference for feasible individuals, and mating restriction promoting the exploration of feasible boundaries – are well balanced in ASCHEA. Hamida and Schoenauer used $(100+300)$-ES to solve benchmark problems and achieve satisfactory results.

[24] Hamida and Schoenauer suggested $p_{select} = 0.3$.

4.3.4 Self-adaptive Penalty Function

Basically, self-adaptive control for penalty coefficients is not possible for COEAs with proportional selection.[25] But Eiben *et al.* suggested an interesting idea to overcome the difficulty in solving CSP [8].

The key point of their self-adaptive penalty function is tournament selection. We take binary tournament selection as an example. For two randomly selected individuals with a self-adaptive penalty coefficient vector, i.e., $(\mathbf{x}^1, \boldsymbol{\omega}^1)$ and $(\mathbf{x}^2, \boldsymbol{\omega}^2)$,[26] we can find the maximum penalty coefficient of constraint j in the two values $\omega_j^{\max} = \max(\omega_j^1, \omega_j^2)$. Then the penalty function for both individuals is as follows:

$$\text{penalty}(\mathbf{x}) = \sum_{j=1}^{k} \omega_j^{\max} v_j(\mathbf{x}) \tag{4.24}$$

where $v_j(\mathbf{x})$ is the violation of constraint j defined by Eq. 4.11.

In this way, the penalty coefficient could not determine the fitness value of its corresponding individual, which avoids the "cheating" of the individual by making its penalty coefficients all zero.

4.4 Separation of Constraint Violation and Objective Value

The main concern in designing COEAs is how to handle infeasible individuals that have small objective values.

We have discussed the first two classes of COEAs using the taxonomy introduced in Sect. 4.1.2. The first class has a special encoding and decoding procedure and variation operators to ensure that the search is only in F. In this way, the above question has been evaded. The representative algorithm is HM, which is introduced in Sect. 4.2.2.

The second class uses penalty coefficients to combine the objective value and the constraint violation and then change COPs into unconstrained problems. Depending on the assignment of the penalty coefficients, the infeasible individuals with small objective values might have advantages compared with feasible individuals with large objective values. The representative algorithm is ASCHEA, which is introduced in Sect. 4.3.3.1.

The third class, which might be the cutting edge of COEAs, separates the objective value and the constraint violation. In the parental selection or replacement procedure, when comparisons are necessary between every two individuals, special rules are suggested to maintain the balance between exploring F and utilizing the information within the infeasible individuals.

---

[25] What will happen if we encode the penalty coefficients into chromosome directly?

[26] $(\mathbf{x}^1, \boldsymbol{\omega}^1) = (x_1^1, \cdots, x_n^1, \omega_1^1, \cdots, \omega_k^1)$ and $(\mathbf{x}^2, \boldsymbol{\omega}^2) = (x_1^2, \cdots, x_n^2, \omega_1^2, \cdots, \omega_k^2)$.

The most straightforward comparison method was suggested by Deb in 2000 [9]. It can be illustrated simply with the following three rules.

- Between two feasible individuals, the one with the better objective value wins.
- Between one feasible individual and one infeasible individual, the feasible one wins.
- Between two infeasible individuals, the one with the smaller overall constraint violation, Eq. 4.12 or Eq. 4.13, wins.

An example of using the above rules was proposed by Kukkonen and Lampinen. They embedded the above ideas into DE for COPs in 2006 [10].

4.4.1 Constrained Optimization Evolutionary Algorithms Based on Rank

4.4.1.1 Stochastic Ranking

Surry and Radcliffe suggested the important idea of using a probability to determine the method of comparing two individuals, by the objective value or the constraint violation [11]. Runarsson and Yao developed this idea into the *stochastic ranking* (SR) procedure and implemented it in (μ,λ)-ES [12]. SR is simple yet powerful, so it has actually become one of the benchmark COEAs.[27]

Bubble sort is a simple ranking method for a list of numbers. It works by repeatedly passing through the list to be ranked, comparing two adjacent items at a time, and swapping them if they are in the wrong order. This procedure is repeated until no swaps are needed, which indicates that the list is sorted. The algorithm gets its name from the way bubbles rise to the surface of water.

For example, we would like to rank the numbers $(1.3, 6.1, 0.5, 1.2, 3.3)$ in ascending order using bubble sort. The ranking procedure is illustrated by Fig. 4.7.

In bubble sort, large elements at the beginning of the list do not pose a problem as they are quickly swapped down, e.g., 6.1 in Fig. 4.7; but small elements toward the end move to the beginning extremely slowly.[28] Bubble sort has a worst-case complexity of $O(n^2)$, where n is the number of items being sorted. So generally, it is a rather slow ranking method. The reason Runarsson and Yao adopted bubble sort might be its convenience in implementation.

In SR-based (μ,λ)-ES, SR is carried out after λ new individuals have been generated by μ parents.[29]

In SR, from the beginning to the end of the population, pairs of individuals are compared and might be swapped using the rules discussed later until there is no swap in one passing through the population. So it is a bubble-sort-like procedure.

[27] The source code of SR can be downloaded at `http://www3.hi.is/~tpr/index.php?page=software/sres/sres`.

[28] How many steps are required if 0.5 is at the end of the initial list?

[29] Refer to Sect. 2.3.1 for details of ES.

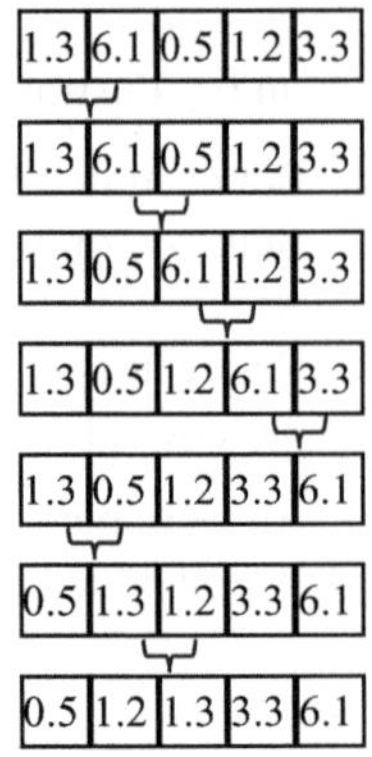

Fig. 4.7 Bubble sort example. The *curly brackets* indicate the pair being compared

After SR, the first μ new individuals in the ranking is selected as the parents for the next generation. The only parameter required by SR is p_o, i.e., the probability of comparison using the objective value. The rules for comparing two individuals, j and $(j+1)$, in the bubble sort, illustrated by Fig. 4.7, is as follows:

```
if (v(j) = v(j+1) = 0) or (rand < p_o) then
    if f(j) > f(j+1) then
        swap(j, j+1)
    end if
else
    if v(j) > v(j+1) then
        swap(j, j+1)
    end if
end if
```

where $v(j)$ and $v(j+1)$ are an overall constraint violation of individuals j and $(j+1)$, respectively, $rand \sim U(0,1)$, $\text{swap}(j, j+1)$ is the function exchange of the location of individuals j and $(j+1)$ in the list.

The rationale of the above SR procedure is that the comparison is carried out on the objective value with probability p_o or the individuals being compared are both feasible; the comparison is carried out on the overall constraint violation otherwise. After SR, the feasible individual with a small objective value, the infeasible individual with a small constraint violation, and the infeasible individual with a small objective value all have the probability of being at the front of the final list, i.e., the best part of the λ new individuals.

If $p_o = 0$, the above procedure is just Deb's comparison idea discussed in Sect. 4.4.[30] If $p_o = 1$, the above procedure only considers the objective value. p_o should be less than 0.5 to capture the bias in favor of feasible solutions.

[30] Why?

Runarsson and Yao used $p_o = 0.45$ in $(30, 200)$-ES to solve 13 benchmark COPs and got excellent results. They further used the DE idea in the mutation procedure of ES in 2005 [13], i.e., the first $\mu - 1$ new individuals are generated with a DE mutation operator as follows:

$$\mathbf{x}_i^{'} = \mathbf{x}_i + \gamma(\mathbf{x}_1 - \mathbf{x}_{i+1}) \tag{4.25}$$

where $\mathbf{x}_i$ and $\mathbf{x}_i^{'}$ are the old and new individuals before and after the DE mutation operator, respectively, $\mathbf{x}_1$ is the first and best individual after SR, $\mathbf{x}_{i+1}$ is the next worse individual $\mathbf{x}_i$, and γ is a parameter predefined by users.[31] Equation 4.25 means that the mutant of individual i starts at $\mathbf{x}_i$ and changes with the direction from the less worse individual to the best individual, i.e., improving direction.

4.4.1.2 Dynamic Stochastic Ranking in Multimember Differential Evolution

In SR, Runarsson and Yao only use fixed $p_o = 0.45$ in all numerical experiments. Zhang *et al.* suggested a dynamic way to linearly control the change in p_o to shift the focus from both the feasible and infeasible region searches to gradually finding a feasible optimal solution [14]. They use multimember differential evolution to solve COPs, where M new individuals are generated for each old individual and the first one after the dynamic stochastic ranking on these $(M+1)$ individuals is the individual for the next generation.

The solution process of the multimember DE based on a dynamic stochastic ranking can be illustrated as follows:

Multimember Differential Evolution Based on Dynamic Stochastic Ranking

Phase 1: Initialization.

Step 1.1: Assign the parameters for multimember DE, such as Np, F, Cr, M, stop criteria (such as *maxgen*), etc. The definition of Np, F, Cr was introduced in Sect. 2.4.2.2.

Step 1.2: Generate Np uniformly distributed individuals randomly to form the initial population and evaluate their fitness values. $gen = 0$.

Phase 2: Main loop. Repeat the following steps until stop criteria are satisfied (such as $gen > maxgen$).

Step 2.1: $gen = gen + 1$. Do the following substeps from the first individual to the last one. $i = 1$

Substep 2.1.1: Get M mutants of individual i using Eq. 2.20 M times with different $\mathbf{r}_1 \neq \mathbf{r}_2 \neq \mathbf{r}_3 \neq \mathbf{x}_i$.

[31] Runarsson and Yao suggested that $\gamma = 0.85$.

Substep 2.1.2: Get M offspring of individual i using Eq. 2.21 M times for M mutants. These M offspring constitute the offspring group for individual i.

Substep 2.1.3: $i = i+1$. Goto substep 2.1.1 if $i \leq Np$, otherwise goto step 2.2.

Step 2.2: Evaluate the objective values and the constraint violations for every offspring in the offspring group for each individual.

Step 2.3: Dynamic control of p_o.

Step 2.4: For each individual and its M offspring, use SR procedure to rank these $(M+1)$ individuals and take the first one as the individual in the next generation. Thus the new population is generated.

Phase 3: Submitting the final Np individuals as the results of EA.

In the above solution process, the dynamic control of p_o in step 2.3 is as follows:

$$p_o = 0.45\left(1 - \frac{gen}{maxgen}\right) \tag{4.26}$$

where *gen* is the current generation number and *maxgen* is the maximum generation number. In the early stage, a large p_o value promotes a search in both feasible and infeasible regions, while in the late stage, a small p_o value focuses on the exploration and exploitation in F.

Zhang *et al.* suggested that $M = 5$ and claimed that the multimember DE with dynamic SR got "significantly better" results than current available methods.

4.4.1.3 Constrained Optimization Evolutionary Algorithm Based on Multiple Ranks

SR considers the rank of objective value and constraint violation in a stochastic way. Ho and Shimizu use three deterministic ranks to guide the search for a feasible optimal solution using (μ,λ)-ES [15].

After λ new individuals have been generated in ES, we can get three ranks considering three properties, i.e., objective value, overall constraint violation, and number of constraints violated, in ascending order. For individual i, we use $R_f(i)$, $R_s(i)$, and $R_v(i)$ to represent its order in these three ranks, respectively.

The order of individual i in any rank is equal to $1+d$, where d is the number of individuals that are better than i in this property. For example, if there are five feasible individuals in the current population, the order of any individual with only one constraint violated in the rank of R_v is 6. It is easy to understand that all the feasible individuals have the order $R_s = 1$ and $R_v = 1$.

In this way, objective value, overall constraint violation, and the number of constraints violated are considered simultaneously and and scaled on the same magnitude.

Then the fitness value of the λ new individuals can be calculated as follows:

if (All λ individuals are infeasible) then

$$f'(\mathbf{x}) = R_s(\mathbf{x}) + R_v(\mathbf{x}) \tag{4.27}$$

else

$$f'(\mathbf{x}) = R_f(\mathbf{x}) + R_s(\mathbf{x}) + R_v(\mathbf{x}) \tag{4.28}$$

end

where $R_f(\mathbf{x})$, $R_s(\mathbf{x})$, and $R_v(\mathbf{x})$ are orders of individual $\mathbf{x}$ in different properties and $f'(\mathbf{x})$ is the fitness value of individual $\mathbf{x}$.

The above procedure requires no predefined parameters, which makes the algorithms easy to implement and robust for COPs.

Ho and Shimizu also suggested adaptive equality constraint tolerance handling techniques. Interested readers are referred to [15].

4.4.2 Simple Multimembered Evolution Strategy

Since 2003, Mezura-Montes and Coello Coello have published a series of papers applying $(\mu+1)$-ES and $(1+\lambda)$-ES in COPs [16, 17]. These algorithms either represent premature convergence in some benchmark COPs or lack the explorative power to allow them to sample large solution landscapes. So they use $(\mu+\lambda)$-ES to solve COPs [18], denoted as simple multimembered evolution strategy (SMES).

As was discussed in the above sections, the crucial part of COEAs is the ability to maintain the diversity of the feasible and infeasible individuals.[32] Mezura-Montes and Coello Coello suggested a novel diversity preservation mechanism for COEAs in the replacement process of $(\mu+\lambda)$-ES, whose solution process was introduced in Sect. 2.3.1.

The main characteristics of SMES are as follows:

1. **Diversity mechanism in replacement.** In step 2.3 of $(\mu+\lambda)$-ES, μ individuals for the next generation are selected from the union of μ parents and λ offspring using Deb's comparison rule introduced in Sect. 4.4 with probability 0.97. For the other 0.03 probability, individuals for the next generation are selected as the "best infeasible individuals" from μ parents and λ offspring with the same chance.[33]

[32] We leave the comparison of these techniques in several classical COEAs as an exercise.

[33] The "best infeasible individuals" are those with the smallest objective value and smallest constraint violation.

2. **Combined recombination.** In step 2.1 of $(\mu + \lambda)$-ES, crossover operators are selected from discrete crossover, introduced in Sect. 3.2.1, and intermediate crossover, introduced in Sect. 2.3.1, with 0.5 probability.
3. **Smaller initial step size.** Schwefel gave suggestions on the initial step size, σ, for mutation in ES [19] and Runarsson and Yao used it directly in RS-based (μ, λ)-ES [12]. Mezura-Montes and Coello Coello suggested using a smaller initial step size, 40% of the suggested value, to promote a finer search.
4. **Dynamic control of the tolerance.** Mezura-Montes and Coello Coello adopt the dynamic control of the tolerance in ASCHEA with $fact_\delta = 1.00195$ in Eq. 4.22.

Mezura-Montes and Coello Coello compared the contribution of the first three characteristics listed above for the performance of SMES in benchmark problems and found that combined recombination is the most important part in these three improvements.

They also compare $(100+300)$-ES with a GA with $popsize = 200$ on benchmark COPs and came to the conclusion that ES is good at COPs for its good search engine, i.e., the self-adaptive mutation. Their numerical comparison, published in 2008, also discuss this issue [20].

To summarize the above two factors, they claimed that the selection of the search engine is more critical than the selection of the constraint handling mechanism for COEAs.

4.4.3 α Constrained Method

Takahama and Sakai suggested the concept of *satisfaction level*, which is similar to the concept of membership function in fuzzy sets, to transform COPs into unconstrained problems without penalty coefficients [21].

For inequality constraints $g_i(\mathbf{x})$ in Eq. 4.1, their satisfaction levels are defined as follows:

$$\mu_{g_i}(\mathbf{x}) = \begin{cases} 1, & g_i(\mathbf{x}) \leq 0 \\ 1 - \frac{g_i(\mathbf{x})}{b_i}, & 0 \leq g_i(\mathbf{x}) \leq b_i \\ 0, & \text{otherwise} \end{cases} \tag{4.29}$$

where b_i is the user-defined threshold.

For equality constraints $h_j(\mathbf{x})$ in Eq. 4.1, their satisfaction levels are defined as follows:

$$\mu_{h_j}(\mathbf{x}) = \begin{cases} 1 - \frac{|h_j(\mathbf{x})|}{b_j}, & |h_j(\mathbf{x})| \leqslant b_j \\ 0, & \text{otherwise} \end{cases} \tag{4.30}$$

where b_j is the user-defined threshold.[34]

[34] Takahama and Sakai suggested that $b_i = b_j = 1000$.

The satisfaction level for inequality and equality constraints are illustrated in Fig. 4.8.[35]

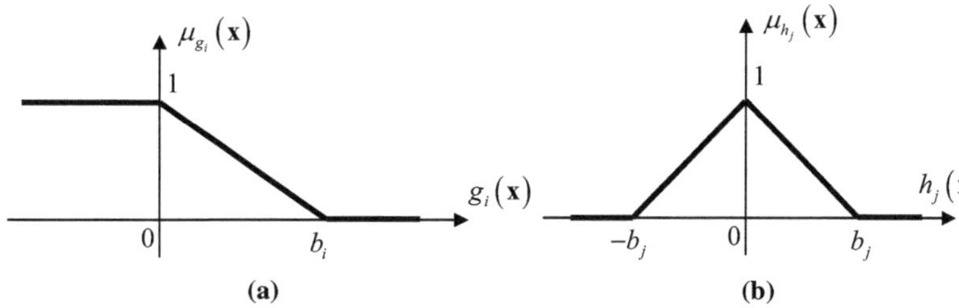

Fig. 4.8 Satisfaction level for inequality constraints and equality constraints: (a) inequality constraints, and (b) equality constraints

Then the overall satisfaction level can be defined as follows:

$$\mu(\mathbf{x}) = \min_{i,j} \left\{ \mu_{g_i}(\mathbf{x}), \mu_{h_j}(\mathbf{x}) \right\} \tag{4.31}$$

Equation 4.31 means that the worst performance in k satisfaction levels determines the overall performance, which is similar to the idea of Eq. 4.13.

Before comparing two individuals, we need to define a level α. If the overall satisfaction levels of two individuals are both above that level or their overall satisfaction levels are the same, we neglect their difference in constraint violation and only compare their objective values. Otherwise, the one with the higher overall satisfaction level wins the competition. This is the idea of *α level comparison*. For a given comparison pair, individuals 1 and 2, 1 is better than 2 at the α level if the following conditions are satisfied:

$$(f_1, \mu_1) \leq_\alpha (f_2, \mu_2) \Leftrightarrow \begin{cases} f_1 \leq f_2, & \mu_1, \mu_2 \geq \alpha \\ f_1 \leq f_2, & \mu_1 = \mu_2 \\ \mu_1 > \mu_2, & \text{otherwise} \end{cases} \tag{4.32}$$

α level comparison neglects the overall satisfaction level above α, i.e., the difference in constraint violation below a certain threshold. In the case of $\alpha = 1$, Eq. 4.32 equals to Deb's comparison rule introduced in Sect. 4.4, which is similar to the case of $p_o = 0$ in SR. If $\alpha = 0$, the Eq. 4.32 only considers the objective value, which is similar to the case of $p_o = 1$ in SR.

After applying the comparison rule between two individuals, the original COP, Eq. 4.1 can be transformed into Eq. 4.33 with a given α level:

$$(P_{\leq\alpha}) \quad \min_{\leq\alpha} f(\mathbf{x}) \tag{4.33}$$

[35] Both Fig. 4.8a and b can be regarded as the membership function of fuzzy sets. That is why we say that satisfaction level is similar to a fuzzy set.

where $\min_{\leq_\alpha}$ denotes minimization based on an α level comparison $\leq_\alpha$.

Takahama and Sakai proved that if Eq. 4.1 has feasible optimal solution $\mathbf{x}^*$ and problem Eq. 4.33 has optimal solution $\hat{\mathbf{x}}_n$ with any $\{\alpha_n\} \sim [0,1]$, a strictly increasing nonnegative sequence $\{\alpha_n\}$ will cause the sequence $\{\hat{\mathbf{x}}_n\}$ to converge to $\mathbf{x}^*$. So we can optimize Eq. 4.33 with an α increase scheme as follows:

$$\alpha(gen) = (1-\beta)\,\alpha(gen-1) + \beta \tag{4.34}$$

where $0 < \beta < 1$ is a control parameter. It is easy to get the general expression of $\alpha(gen)$ as follows:

$$\alpha(gen) = (1-\beta)^{gen}\,\alpha(0) + \beta\left[1 + (1-\beta) + \cdots + (1-\beta)^{gen-1}\right] \tag{4.35}$$

If $gen \to \infty$, the above expression will converge to 1. Takahama and Sakai suggested the pragmatic α dynamic control method, which can be illustrated as follows:

```
if (0 < gen ≤ maxgen/2) and ((gen mod T_α) = 0) then
    α(gen) = (1 − β) α(gen − 1) + β
else if (0 < gen ≤ maxgen/2) and ((gen mod T_α) ≠ 0) then
    α(gen) = α(gen − 1)
else if gen > maxgen/2
    α = 1
end if
```

where T_α is a control parameter. The above procedure increases α for every T_α generations in the first half of the total generations and keeps it constant at 1 in the second half of the total generations.[36] The initial α is suggested as the average of the best satisfaction level and the mean of all satisfaction levels in the initial population, i.e.,

$$\alpha(0) = \frac{1}{2}\left(\max_i \mu\left(\mathbf{x}^i\right) + \frac{\sum_{i=1}^{popsize} \mu\left(\mathbf{x}^i\right)}{popsize}\right) \tag{4.36}$$

After the above transformation, we can solve Eq. 4.33 with the α increase scheme discussed above. Finally, the optimal solution of Eq. 4.33 with $\alpha = 1$ will converge to the feasible optimal solution of Eq. 4.1.

Takahama and Sakai further utilized multiple simplex methods with mutation on the worst solution for solving COPs by the α constrained method. For the implementation detail, readers are referred to [21] and Sect. 2.4.1.1.

---

[36] Takahama and Sakai suggested that $\beta = 0.03$ and $T_\alpha = 50$.

4.5 Performance Evaluation of Constrained Optimization Evolutionary Algorithms

4.5.1 Benchmark Problems

Several factors might affect the difficulty of a COP. Such factors are listed as follows:

- The ratio of solution numbers in the feasible region F to that of the search space S. The analytic expression of such a ratio might be difficult. But we can randomly generate solutions in S and determine whether they are feasible or not.[37] If the sampling points are large enough, such as 10^6, we can get a reliable ratio.
- The dimensionality of variables, i.e., n in Eq. 4.1. Large n will significantly increase the difficulty of COPs.
- The number of nonlinear equality constraints and its ratio to the total number of constraints. Linear equality constraints can be eliminated by GENOCOP but nonlinear equality constraints are difficult even though they could be relaxed with Eq. 4.2.
- The number of inequality constraints that are active in the feasible optimal solution. Being active means that the feasible optimal solution is at the boundary of that inequality constraint that requires a search from both feasible and infeasible sides.

The benchmark COPs can be classified roughly into the following three groups.

1. Problems suggested by different researchers. Koziel and Michalewicz suggested 12 benchmark COPs in 1999 [2]. Then Runarsson and Yao developed them into 13 benchmark COPs in 2000 [12]. These 13 benchmark COPs are actually the standard for comparing and evaluating COEAs. In 2006, the IEEE Congress on Evolutionary Computation held a special session on constrained real parameter optimization. Twenty four benchmark COPs were suggested there.[38]
2. Problem generators suggested by different researchers. Michalewicz *et al.* suggested a test-case generator (TCG) for COPs in 2000 [22]. Schmidt and Michalewicz developed it into TCG-2 in 2000 [23]. In 2003, Craenen *et al.* suggested a systematic way to generate random binary constraint satisfaction problems [24].[39]
3. Real-world problems. In 2004 Mezura-Montes suggested five engineering constrained optimization design problems [25].

The most recently reported best feasible solutions or the analytical optimal feasible solutions for the first two groups are available.

---

[37] Just like the ordinary way in the Monte Carlo method.

[38] `http://www3.ntu.edu.sg/home/EPNSugan/`.

[39] C++ classes for generating the test sets of binary constraint satisfaction problems can be downloaded from `http://freshmeat.net/projects/randomcsp`.

4.5.2 Performance Indices

Several *performance indices* have been suggested by COEA researchers [12, 25, 26]. Here we simply assemble and reformulate them as follows.

For every benchmark COP whose feasible optimal solution is $\mathbf{x}^*$, we need to run the COEA on it many times, e.g., 25, and the maximum number of the objective function evaluation (Max_{NOFE}) needs to be predefined, e.g., 500000, for each run. We define a *feasible run* as a run during which at least one feasible solution is found within Max_{NOFE}[40] and a *successful run* as a run during which a feasible solution $\mathbf{x}$ is found satisfying $|f(\mathbf{x}) - f(\mathbf{x}^*)| < 0.0001$. All of the following PIs are for each benchmark COP.

1. Report the *feasible rate* as the number of feasible runs over the total runs (25). The larger the better.
2. Report the average number of objective function evaluations (NOFE) for finding the first feasible solution. The smaller the better.
3. Report the *successful rate* as the number of successful runs over the total runs (25). The larger the better.
4. Report the average NOFE in successful runs. The smaller the better.
5. Report the *success performance* as ratio of the average number of the objective function evaluation (NOFE) in successful runs over the successful rate. The smaller the better.
6. Report the best, median(mean), worst feasible solutions and the standard deviation of the feasible solutions for the 25 runs.

4.6 Summary

Constrained optimization problems are realistic models coming from the real world. OR has done much research on this topic. If the selection procedure and the variation operators are designed elaborately, EAs can evolve in different feasible regions, find new feasible regions, and approach the feasible optimal solution from different directions, which might be very useful for COPs whose feasible regions are disconnected or whose feasible optimal solution is at the edge of the feasible region.

GENOCOP and HM, discussed in Sect. 4.2, use special decoding approaches and variation operators to limit the search within feasible region F. In the case of convex F and a large ratio of solution numbers of F to that of S, this "feasibility maintenance" algorithm might search effectively.

The penalty function method is an all-purpose method for COPs but the assignment of the penalty coefficients requires different techniques. Deterministic, adaptive, and self-adaptive controls for penalty coefficients were discussed with many examples in Sect. 4.3. Numerical experimental results illustrated that a "properly"

[40] The tolerance for equality constraints is $\delta = 0.0001$.

chosen penalty function might accelerate the search and contribute high-quality solutions [27].

The main part of this chapter is Sect. 4.4. The essences of three famous algorithms – SR, SMES, and α constrained method – are introduced separately and many variation algorithms derived from them are discussed. They all use uncertainty to determine the winner of the competition between feasible and infeasible individuals. It is promising to change, with time or with feedback, the uncertainty that controls the comparison rules.

Several comments can be made from the introduction of this chapter as follows:

- None of the famous COEAs introduced in this chapter has a complicated constraint handling mechanism. So there might be large opportunities for combining traditional constraint handling techniques with EAs.
- The majority of the prevailing COEAs use ES, which might be the result of ES's fine search ability using the self-adaptive control of the step size σ.
- The idea of directional search, e.g., scatter search, simplex search, and differential evolution, might be adopted by COEAs to improve their search ability.

After reading this chapter, you should understand the difficulty of COPs and why COEAs might have an advantage over COPs, be familiar with the basic idea of several state-of-the-art MOEAs, and know how to compare COEAs with standard PIs.

In all, designing a COEA is the art of tradeoff between feasible and infeasible individuals.

Suggestions for Further Reading

Gen and Cheng's book [28], Eiben and Smith's textbook [29], and Bäck *et al.*'s book [30] all have special chapters on constrained optimization.

Michalewicz and Schoenauer published a survey in 1996 [31]. Eiben's 2001 survey [32] and Coello Coello's 2002 survey [33] are also the most consulted ones.

In the field of penalty function, there are many successful cases. Some guidelines are given for designing penalty functions by Richardson *et al.* [34]. An intensive runtime analysis is given by Zhou and He for a better understanding of penalty functions [27]. A numerical comparison of five penalty-based constraint handling techniques was published by Miettinen *et al.* in 2003 [35]. Readers fascinated by the idea of designing adaptive penalty functions are referred to three other examples by Smith and Tate [36],[41] Powell and Skolnick [38], and Farmani and Wright [39].

Readers interested in SMES are referred to Mezura-Montes's Ph.D. thesis, published in 2004 [25], and a numerical comparison of different ESs in solving COPs published in 2008 [20].

---

[41] It was used by Tasgetiren and Suganthan in 2006 [37].

Paredis used competitive coevolution in 1994 [40], and Coello Coello used cooperative coevolution [41] separately to solve COPs in the framework of COEAs.

The must-read papers of this chapter are [12] for SR, [21] for the α constrained method, [25] for a detailed discussion of ES-based COEAs, and [33] for a survey.

Exercises and Potential Research Projects

4.1. How can a whole arithmetic crossover maintain the feasibility of the offspring in GENOCOP?

4.2. How can the location of $\mathbf{r}^0$ affect the optimization results of HM? Can you offer some advice for selecting the reference point?

4.3. In HM search, all points in the convex cube $[-1,1]^n$ are treated identically, which means that all points in F are treated identically. Can you suggest a way to give preference to the boundary points?

4.4. Implement the DA and AA of ASCHEA in your programming environment. Considering the discussion about them in the textbook, can you suggest other improvements? Report your basic idea, your implementation details, your numerical experimental results on the benchmark problems in Appendix, and the statistical conclusions using methods introduced in Sect. 3.6.

4.5. Implement the stochastic ranking in your programming environment or learn to use the source code provided by Runarsson and Yao. Study the sensitivity of p_o in SR with benchmark COPs. Report your numerical experimental results on the benchmark problems in Appendix and the statistical conclusions using methods introduced in Sect. 3.6.

4.6. Can you adopt other ranking methods apart from the bubble sort in SR and implement it within your own COEA? Report your basic idea, your implementation details, your numerical experimental results on the benchmark problems in Appendix, and the statistical conclusions using methods introduced in Sect. 3.6.

4.7. Summarize and compare the feasible and infeasible individual diversity preservation mechanisms in ASCHEA, SR, SMES, and the α constrained method.

4.8. Summarize and compare the equality constraint handling mechanisms in GENOCOP, HM, ASCHEA, SR, SMES, and the α constrained method. Please refer to original paper for details.

4.9. Mezura-Montes and Coello Coello claimed that the selection of the search engine is more critical than the selection of the constraint handling mechanism for COEAs [20]. Read this paper carefully and do your own numerical comparisons on ES and GA for COPs. Report your numerical experimental results on the benchmark problems in Appendix, and the statistical conclusions using methods introduced in Sect. 3.6. What is your conclusion?

References

1. Michalewicz Z (1998) Genetic algorithms + data structures = evolution programs. Springer, Berlin Heidelberg New York
2. Koziel S, Michalewicz Z (1999) Evolutionary algorithms, homomorphous mappings, and constrained parameter optimization. Evol Comput 7(1):19–44
3. Joines J, Houck C (1994) On the use of non-stationary penalty functions to solve nonlinear constrained optimization problems with GA's. In: Proceedings of the IEEE conference on evolutionary computation, pp 579–584
4. Gen M, Cheng R (1996) A survey of penalty techniques in genetic algorithms. In: Proceedings of the IEEE international conference on evolutionary computation, pp 804–809
5. Hadj-Alouane AB, Bean JC (1997) A genetic algorithm for the multiple-choice integer program. Oper Res 45(1):92–101
6. Hinterding R (2001) Constrained parameter optimisation: equality constraints. In: Proceedings of the IEEE congress on evolutionary computation, 1:687–692
7. Hamida SB, Schoenauer M (2002) ASCHEA: new results using adaptive segregational constraint handling. In: Proceedings of the IEEE congress on evolutionary computation, 1: 884–889
8. Eiben A, Jansen B, Michalewicz Z *et al* (2000) Solving CSPs using self-adaptive constraint weights: how to prevent EAs from cheating. In: Proceedings of genetic and evolutionary computation conference, pp 128–134
9. Deb K (2000) An efficient constraint handling method for genetic algorithms. Comput Methods Appl Mechan Eng 186:311–338
10. Kukkonen S, Lampinen J (2006) Constrained real-parameter optimization with generalized differential evolution. In: Proceedings of the IEEE congress on evolutionary computation, pp 207–214
11. Surry PD, Radcliffe NJ (1997) The COMOGA method: constrained optimisation by multiobjective genetic algorithms. Control Cybern 26(3):391–412
12. Runarsson T, Yao X (2000) Stochastic ranking for constrained evolutionary optimization. IEEE Trans Evol Comput 4(3):284–294
13. Runarsson T, Yao X (2005) Search biases in constrained evolutionary optimization. IEEE Trans Sys Man Cybern, C 35(2):233–243
14. Zhang M, Luo W, Wang X (2008) Differential evolution with dynamic stochastic selection for constrained optimization. Inf Sci 178(15):3043–3074
15. Ho PY, Shimizu K (2007) Evolutionary constrained optimization using an addition of ranking method and a percentage-based tolerance value adjustment scheme. Inf Sci 177(14):2985–3004
16. Mezura-Montes E, Coello Coello CA (2003) A simple evolution strategy to solve constrained optimization problems. In: Proceedings of the genetic and evolutionary computation Conference, p 198
17. Mezura-Montes E, Coello Coello CA (2003) Adding a diversity mechanism to a simple evolution strategy to solve constrained optimization problems. In: Prceedings of the IEEE congress on evolutionary computation, pp 6–13
18. Mezura-Montes E, Coello Coello CA (2005) A simple multimembered evolution strategy to solve constrained optimization problems. IEEE Trans Evol Comput 9(1):1–17
19. Schwefel H (1995) Evolution and optimum seeking. Wiley-Interscience, New York
20. Mezura-Montes E, Coello Coello CA (2008) An empirical study about the usefulness of evolution strategies to solve constrained optimization problems. Int J Gen Syst 37(4):443–473
21. Takahama T, Sakai S (2005) Constrained optimization by applying the α constrained method to the nonlinear simplex method with mutations. IEEE Trans Evol Comput 9(5):437–451
22. Michalewicz Z, Deb K, Schmidt M *et al* (2000) Test-case generator for nonlinear continuous parameter optimization techniques. IEEE Trans Evol Comput 4(3):197–215
23. Schmidt M, Michalewicz Z (2000) Test-case generator TCG-2 for nonlinear parameter optimisation. In: Proceedings of the international conference on parallel problem solving from nature, pp 539–548

24. Craenen B, Eiben A, van Hemert J (2003) Comparing evolutionary algorithms on binary constraint satisfaction problems. IEEE Trans Evol Comput 7(5):424–444
25. Montes EM (2004) Alternative techniques to handle constraints in evolutionary optimization. Ph.D. thesis, CINVESTAV-IPN
26. Liang JJ, Runarsson TP, Mezura-Montes E *et al* (2006) Problem definitions and evaluation criteria for the cec 2006 special session on constrained real-parameter optimization. Tech. rep., Nanyang Technological University, University of Iceland, CINVESTAV-IPN, France Télécom, Indian Institute of Technology, Singapore
27. Zhou Y, He J (2007) A runtime analysis of evolutionary algorithms for constrained optimization problems. IEEE Trans Evol Comput 11(5):608–619
28. Gen M, Cheng R (1997) Genetic algorithms and engineering design. Wiley-Interscience, New York
29. Eiben AE, Smith JE (2003) Introduction to evolutionary computing. Springer, Berlin Heidelberg New York
30. Bäck T, Fogel D, Michalewicz Z (2000) Evolutionary computation 2: advanced algorithms and operations. Taylor & Francis, London, UK
31. Michalewicz Z, Schoenauer M (1996) Evolutionary algorithms for constrained parameter optimization problems. Evol Comput 4:1–32
32. Eiben AE (2001) Evolutionary algorithms and constraint satisfaction: definitions, survey, methodology, and research directions. In Kallel L, Naudts B, Rogers A (eds) Theoretical aspects of evolutionary computing. Springer, Berlin Heidelberg New York, pp 13–30.
33. Coello Coello CA (2002) Theoretical and numerical constraint-handling techniques used with evolutionary algorithms a survey of the state of the art. Comput Methods Appl Mechan Eng 191(11):1245–1287
34. Richardson JT, Palmer MR, Liepins GE *et al* (1989) Some guidelines for genetic algorithms with penalty functions. In: Proceedings of the 3rd international conference on genetic algorithms, pp 191–197
35. Miettinen K, Mäkelä MM, Toivanen J (2003) Numerical comparison of some penalty-based constraint handling techniques in genetic algorithms. J Glob Optim 27(4):427–446
36. Smith AE, Tate DM (1993) Genetic optimization using a penalty function. In: Proceedings of the 5th international conference on genetic algorithms, pp 499–505
37. Tasgetiren M, Suganthan P (2006) A multi-populated differential evolution algorithm for solving constrained optimization problem. In: Priceedings of the IEEE congress on evolutionary computation, pp 33–40
38. Powell D, Skolnick MM (1993) Using genetic algorithms in engineering design optimization with non-linear constraints. In: Proceedings of the 5th international conference on genetic algorithms, pp 424–431
39. Farmani R, Wright J (2003) Self-adaptive fitness formulation for constrained optimization. IEEE Trans Evol Comput 7(5):445–455
40. Paredis J (1994) Co-evolutionary constraint satisfaction. In: Proceedings of the international conference on parallel problem solving from nature, pp 46–55
41. Coello Coello CA (2000) Use of a self-adaptive penalty approach for engineering optimization problems. Comput Ind 41(2):113–127

Chapter 5
Multimodal Optimization

Abstract Sometimes you run a EA for a problem several times. The algorithm might provide different solutions with similar qualities. You may feel uncomfortable with this. We will show you in this chapter that there really exist problems with several high-quality solutions and we want to find all of them in a single run of an EA. Techniques in this chapter could also help to adjust the tradeoff between selective pressure and population diversity, which is an eternal subject in designing and analyzing EAs.

5.1 Problems We Face

We introduced the concept of *multimodality* in Chap. 1, where it is used to describe the situation where there are several local optimal solutions in a solution space and we want the algorithm to find the only global optimal solution as soon as possible. But here we start from a different viewpoint.

5.1.1 Multimodal Problems

In this chapter, without loss of generality, we will only discuss the unconstrained maximum problem whose objective values are all above zero:[1]

$$\begin{array}{c} \max f(\mathbf{x}) \geq 0 \\ \mathbf{x} \in \mathbb{R}^n \end{array} \tag{5.1}$$

[1] Fitness transferral, which changes other problems into the required form, was discussed in Sect. 3.3.3.

The concept of *multimodal problems* has different meanings. The general explanation has been stated above. Sometimes there are other situations that make us concern with finding multiple solutions.

- There are several global optimal solutions and we want to find them all, which is illustrated by Fig. 5.1a.
- There are global optimal solutions and local optimal solutions (not too many). We are interested in finding all of them, which is illustrated by Fig 5.1b.

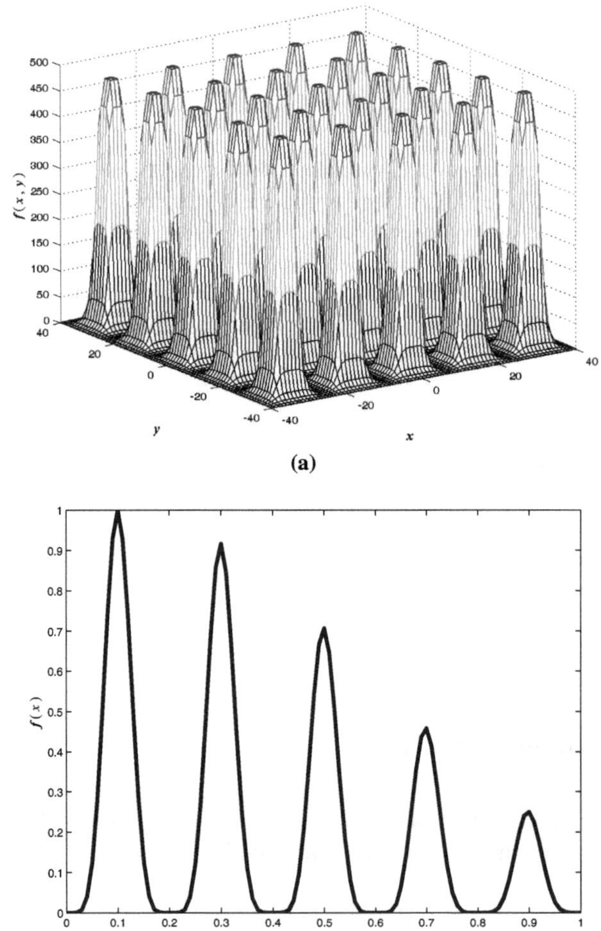

(b)

Fig. 5.1 Multimodal optimization examples: (a) example 1, and (b) example 2

If we use the EAs introduced in Chaps. 2 and 3 to solve the problems illustrated by Fig. 5.1a directly, they might find multiple global solutions after some generations but will lose most of them and converge all the individuals into one peak due to genetic drift. So we need some special techniques.

The reason for finding multiple global optimal solutions and other interested local optimal solutions can be summarized as follows:

- In real-world applications, there will be some factors that are hard to model mathematically, such as manufacturing degree of difficulty, maintenance degree of difficulty, reliability, etc. Finding multiple solutions with similar quality provides the decision maker with multiple options to be further determined by other vague factors.
- Multiple solutions with similar quality is important for finding a robust solution and useful for the sensitivity analysis of a problem.
- As has been discussed, counteracting the effect of genetic drift requires an elaborate tradeoff between selective pressure and population diversity. Multimodal EAs provide useful techniques for such a task, which might be used elsewhere, such as multiobjective optimization.
- If the search ability of the algorithm cannot guarantee that the global optimal solution will be found, then the ability to find multiple solutions of similarly high quality increases the possibility of finding the global optimal solution [1].

In all, *multimodal EAs* need to identify and maintain multiple optimal solutions (global/local) within multimodal domains. Basically, this ability has little relationship to variation operators, so we will only discuss techniques in steps 2.1 and 2.4 of the general solution process of EA introduced in Sect. 3.4.1.

5.1.2 Niche, Species, and Speciation

Analogy is a useful way to innovate. In multimodal EAs, researchers use the concepts of *niche*, *species*, and *speciation*, which come from natural ecosystems, to describe the feature of a problem and an algorithm.

Niche refers to the conditions of the environment within which a particular type of living thing can live successfully.[2] A species is a class of plants or animals whose members have the same main characteristics and are able to breed with each other.[3] Finally, speciation is the formation of new and distinct species in the course of evolution [4].

We use niche as a metaphor for the partial solution landscape in which only one peak resides, species as a metaphor for the subpopulation maintained in a specific peak/niche, and speciation as a metaphor for the formation process of such a

---

[2] From the Oxford Advanced Learner's Dictionary.

[3] From the Collins Cobuild Dictionary.

[4] From Concise Oxford English Dictionary.

species.[5] EAs using the concepts of niche, species, and speciation to solve multimodal problems are called *niche EAs*.

The common characteristics of one species in multimodal EAs is that individuals of this species reside in one niche (surrounding one peak), which is often interpreted as their being close to each other. So we need a distance metric for individuals. The Hamming distance for binary code and the Euclidean distance for real value code are often used, and the latter is getting more popular.

In the latter sections of this chapter, we will ignore the difference between a niche and the attraction basin of a peak.

Every niche has its center, which might be the peak or the centroid of the niche. For reasons of simplicity, we assume every niche has a radius σ_i, which means that if the distance of individual j to niche center i, $d(i, j)$, is larger than σ_i, then individual j does not belong to niche i. Furthermore, we sometimes consider that the radii of all peaks are the same or assume that setting the same radii for all peaks will not influence the identification of multiple peaks. If so, we say that this problem is *distinguishable*.

Figure 5.2 illustrates the number of papers indexed by the SCI on EA-based multimodal optimization.[6]

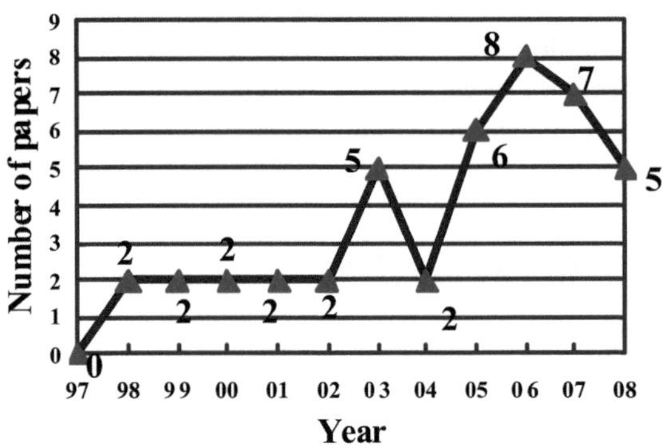

Fig. 5.2 Number of papers indexed by SCI on EA-based multimodal optimization

Even though the number is not impressive, the techniques suggested in niche and speciation have a strong influence on preserving population diversity.

[5] The concepts of niche, species, and speciation were used initially to adjust population diversity and now often refer to multimodal EAs.

[6] TS = ((“multimodal optimization”) AND (“genetic algorithm” OR “genetic algorithms” OR “evolutionary computation” OR “evolutionary computing” OR “evolutionary algorithms” OR “evolutionary intelligence”)). The SCI index “TS” is for the search topic in the title, the keywords, and the abstract.

5.2 Sequential Niche

We will first introduce a very interesting algorithm called a *sequential niche*, suggested by Beasley *et al.* in 1993 [2]. It is not an EA but a multimodal optimizing infrastructure.

In a sequential niche, after finding a global optimal solution using any search algorithm, such as EAs, we can modify the solution landscape so that the attraction basin of the found peak will be punished to ensure that the next run of the search algorithm will find another peak. Continuing the above process will find multiple peaks sequentially.

The key to sequential niche is the modification of the solution landscape. Generally, punishment for the attraction basin of the found peak(s) means decreasing the objective value of these points, which can illustrated as follows:

$$G(\mathbf{x},\mathbf{r}) = \begin{cases} \frac{d(\mathbf{x},\mathbf{r})}{\sigma} & , d(\mathbf{x},\mathbf{r}) < \sigma \\ 1 & , \text{otherwise} \end{cases} \tag{5.2}$$

where σ is the peak radius, $\mathbf{r}$ is the location of the found peak, $\mathbf{x}$ is the location of one point, and $d(\mathbf{x},\mathbf{r})$ is the distance between them. Equation 5.2 means that points closer to $\mathbf{r}$ have a smaller $G(\mathbf{x},\mathbf{r})$ and $G(\mathbf{r},\mathbf{r}) = 0$, which represents the strongest punishment.

Then the modified solution landscape could be illustrated as

$$f'(\mathbf{x}) = f(\mathbf{x}) \times \prod_{i=1}^{k} G(\mathbf{x},\mathbf{r}_i) \tag{5.3}$$

where $f(\mathbf{x})$ is the raw objective value of point $\mathbf{x}$, $f'(\mathbf{x})$ is the modified objective value, and k is the number of found peaks. Equation 5.3 means that after finding a peak, we decrease the objective value around it and then use the search algorithm again to find another peak.

If we properly set $\sigma = 0.1$ for Fig. 5.1b after $x = 0.1$ has been found using Eq. 5.3, the solution landscape is illustrated by Fig. 5.3a. Then the second run of the search method might find $x = 0.3$.

But generally we have no idea what the real peak radius is. A larger-than-necessary peak radius might diminish real peaks, which is illustrated by Fig. 5.3b with $\sigma = 0.2$, and a smaller-than-necessary peak radius might generate fake peaks, which is illustrated by Fig. 5.3c with $\sigma = 0.01$. Both conditions introduce difficulties into the following search on the modified solution landscape.

There are other ways to define the punishment function apart from Eq. 5.2, but they all suffer from the sensitive parameter σ. Another drawback of sequential niche is its sequential approach, i.e., one run for one peak, which is ineffective for real-time applications.

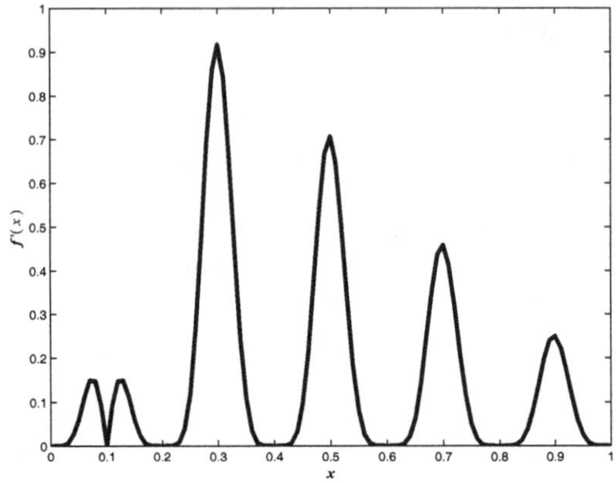

(a)

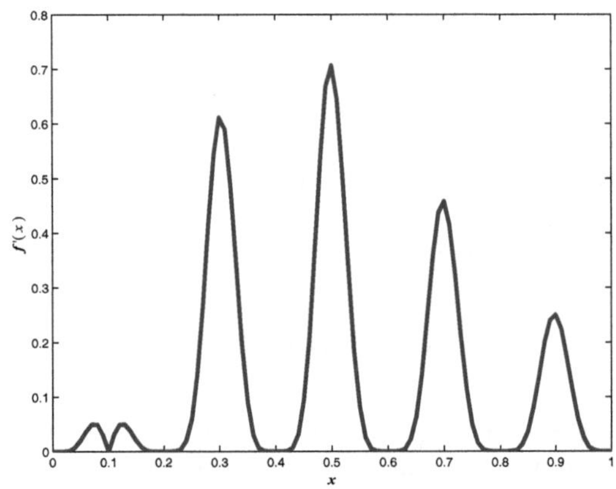

(b)

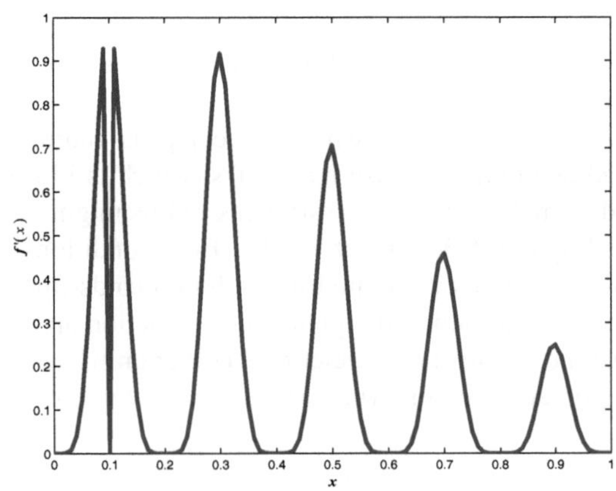

(c)

Fig. 5.3 Settings of peak radius in sequential niche for Fig. 5.1b: (a) $\sigma = 0.1$, (b) $\sigma = 0.2$, and (c) $\sigma = 0.01$

5.3 Fitness Sharing

5.3.1 Standard Fitness Sharing

The concept of *fitness sharing* was introduced by Goldberg and Richardson in 1987 [3]. They think of the height of a peak as the resources in the niche. Individuals of the same species residing in that niche share the resources by decreasing their fitness values according to the crowdedness of the species. Standard fitness sharing is carried out before the parental selection process, i.e., by adding the following steps to the general solution process of EA on page 77.

Standard Fitness Sharing

Step 2.01: Calculate the distance d_{ij} between every two individuals i and j.

Step 2.02: Calculate the *sharing function* value between every two individuals i and j as

$$sh(d_{ij}) = \begin{cases} 1 - \left(\frac{d_{ij}}{\sigma}\right)^{\alpha} & d_{ij} < \sigma \\ 0 & \text{otherwise} \end{cases} \tag{5.4}$$

where σ is the niche radius, α is a control parameter and often set as $\alpha = 1$, and $sh()$ defines the similarity between i and j within the range $[0,1]$. Readers are suggested to draw the sharing function when $\alpha = 1$ and analyze the reason we name it as the similarity metric.

Step 2.03: Calculate the *niche count* for every individuals i as

$$m_i = \sum_{j=1}^{popsize} sh(d_{ij}) \tag{5.5}$$

Step 2.04: Calculate the *shared fitness value* for every individual i as

$$f_i' = \frac{f_i}{m_i} \tag{5.6}$$

where f_i is the raw fitness of i.

According to Eq. 5.6, fitness sharing can also be regarded as a kind of fitness scaling technique, whose intention is also to adjust the selective pressure and the population diversity.

Suppose the five peaks have heights of 1, 0.9, 0.7, 0.48, and 0.26, respectively, in Fig. 5.1b; there is an equilibrium point in the evolving process, i.e., exactly 26 individuals reside in the lowest peak, ..., and exactly 100 individuals reside in the highest peak. The shared fitness value of every individual is exactly 1, which makes them have the same probability of being selected and cancels the effect of genetic drift. Thus this equilibrium is stable and these five peaks could be *maintained* if there were no selective bias and the variation operators did not break it.[7]

So a selection process with no selective bias, such as SUS, is often used in niche EAs. To avoid variation operators breaking the equilibrium, positive assortative mating or mating restriction, introduced on page 63, is often used. In Goldberg and Richardson's design, only individuals within a distance of σ have the opportunity to cross over. p_{m} is often set to be very small (or even zero) with the same intention.

According to the above discussions, standard fitness sharing decreases the selective pressure and promotes population diversity, which is necessary for niche EAs. By contrast, there are several situations in which we need to scale the raw fitness value before fitness sharing to balance the selective pressure between scaling and sharing.

- There are some global peaks and many local peaks with heights similar to that of the global peaks. We only want to find these global peaks.
- There is a strong bias in the solution landscape [4]. Suppose in Fig. 5.1b we add 10 to all objective values, which makes the function value change between $[10, 11]$. There are three individuals in the population. Two are at the global peak with a height of 11 and one individual has the minimum value 10 and is far from global optimal solutions. The shared fitness value for the two at the global peak is 5.5 and 10 for the third one, which is an unreasonable scheme. If we can use power law or exponential scaling to enlarge the difference, things will be better.[8]

What an amazing technique standard fitness sharing is! It provides the possibility of identifying and maintaining multiple peaks in a single run of EAs, which makes it one of the most often used techniques in niche EAs. But in-depth analysis for standard fitness sharing reveals its drawbacks as follows:

1. The niche radius parameter σ, which is hard to estimate for real-world problems and sensitive to the algorithm's performance, is required.
2. The niche radii are assumed to be the same in the solution landscape, Eq. 5.4, which is unrealistic for real-world problems. Empirical equations for estimating σ under the above two assumptions are reported in [5], but these assumptions are too strong for many applications.
3. Standard fitness sharing does not generate species explicitly. So peaks are found implicitly. To make things worse, individuals might cross over with individuals belonging to other species if we use the mating restriction discussed above,[9] which might generate useless offspring, known as *lethal* offspring.
4. One must calculate the distance for $O(popsize^2)$ times to get the shared fitness value, which is an expensive cost when we want to find many peaks with large $popsize$.

---

[7] Twenty-six individuals for the lowest peak is just an example. The number can be scaled.

[8] Consider $\alpha = 10$ in Eq. 3.42.

[9] Why?

Many researchers have suggested various improvements for standard fitness sharing.

5.3.2 Clearing Procedure

Petrowski suggested a *clearing* procedure in 1996 to overcome drawbacks 2–4 of standard fitness sharing [6].[10] The clearing procedure also requires σ but defines it as the minimum niche radius in the solution landscape. The following steps constitute a clearing procedure, which can be embedded in the general solution process of EA on page 77.

Clearing Procedure

Step 2.01: Sort the population in descending order of the raw fitness values.

Step 2.02: Assign the first individual as the first species center and call it a *winner*.

Step 2.03: Do the following substeps from the second individual to the last one.

Substep 2.03.1: Assign the current individual as the new species center if its distance to all the assigned species centers is larger than σ, and call it a *winner*.

Substep 2.03.2: Assign the current individual to one species if the distance between the individual and the species center is less than σ and the number of individuals in that species is no larger than κ, and also call it a *winner*. Numerical experiments show that κ is a nonsensitive parameter for clearing procedure. Users sometimes set $\kappa = 1$.

Substep 2.03.3: Assign others as *losers*.

Step 2.04: The shared fitness values of winners are the same as their raw fitness values but, zero for losers. Save winners to another memory.

...

Step 2.4: For every saved winner, a competition between it and its closest neighbor in the current population is carried out and the winner will survive.

A clearing procedure is like a winner-takes-all mechanism: winners will not be punished and losers will die.[11] The selective pressure for a clearing procedure is higher than that of standard fitness sharing. So Petrowski does not use a mating restriction

---

[10] Lee *et al.* suggested a similar procedure, restricted competition selection, in 1999 [7].

[11] If we use $m = 1$ for winners and $m = \infty$ for losers in Eq. 5.6, the clearing procedure could be regarded as a kind of fitness sharing technique.

with the intention of promoting exploration. With the help of elitism, step 2.4 above, a clearing procedure can identify and maintain multiple peaks effectively, provided σ is set adequately.

5.3.3 Clustering for Speciation

Speciation is a process of grouping individuals according to their locations. So clustering techniques are suitable for it. The first study on this topic was published by Yin and Germay in 1993 [8]. Many clustering techniques [9–13] and many applications using hierarchical clustering [14] and nonhierarchical clustering [15, 16] with a predefined peak number or niche radius to group individuals have been proposed recently. Here we just use Yin and Germay's original suggestion, adaptive *k-means clustering*, to illustrate how to combine clustering and sharing.

A minimum distance between species centers $\sigma_{\min}$ (coarsening parameter) and a maximum distance between an individual and its species center $\sigma_{\max}$ (refining parameter) are required by Yin and Germay's algorithm. The following steps constitute adaptive k-means clustering with fitness sharing, which could be embedded in the general solution process of EA on page 77.

Adaptive k-means Clustering with Fitness Sharing

Step 2.01: Sort the population in descending order of the raw fitness values.

Step 2.02: Generate a random integer k in the range $[1, popsize]$.

Step 2.03: Put the first k individuals into different species as the *species centers* and ensure the distance between every two species centers is larger than $\sigma_{\min}$. If not, merge the species and take the centroid of the new species as the *species center*.

Step 2.04: Calculate the distances of other $(popsize - k)$ individuals to all species centers. If the smallest distance of one individual is larger than $\sigma_{\max}$, this individual forms a new species and is the center of that species. Otherwise, the individual belongs to the nearest species. After every allocation, the *species center* needs to be recalculated as its centroid. We need to ensure that the distance between every two species centers is larger than $\sigma_{\min}$. If not, merge the species and take the centroid of the new species as the *species center*.

Step 2.05: After all individuals have been allocated, fix the location of all *species centers*, reallocate every individual to its closest species center, and count the number of individuals in each species.

Step 2.06: Calculate the niche count as

$$m_i = n_c - n_c \times \left(\frac{d_{ic}}{2\sigma_{\max}}\right)^{\alpha}, \quad x_i \in C_c \tag{5.7}$$

where n_c is the number of individuals in species c, d_{ic} is the distance between individual i and its corresponding species center, C_c is the individual set in species c, and α is the control parameter, which is often set as 1.

Step 2.07: Calculate the shared fitness using Eq. 5.6.

Yin and Germay also implemented a mating restriction that only allows breeding between members in the same species. For the control parameters, they suggested that $\sigma_{\max} \approx (2 \sim 3) \times \sigma_{\min}$ for their test problems, but a more delicate choice of those parameters for generalized multimodal problems is required.

5.3.4 Dynamic Niche Sharing

Miller and Shaw suggested *dynamic niche sharing* (DNS) in 1996 [17, 18]. DNS requires two parameters: the minimum peak radius σ and the peak number k. The following steps constitute DNS, which can be embedded in the general solution process of EA on page 77.

Dynamic Niche Sharing

Step 2.01: Sort the population in descending order of the raw fitness values.

Step 2.02: Assign the first individual as the first *species center* and form the first species.

Step 2.03: Do the following substeps from the individual to the last individual.

Substep 2.03.1: Form a new species and assign the current individual as the new *species center* if its distance to all the assigned species centers is larger than σ and the formed species number is smaller than k.

Substep 2.03.2: Assign the current individual to the closest species if its smallest distance to all the assigned species centers is smaller than σ.

Substep 2.03.3: Assign the current individual as an *independent individual* if its distance to all the appointed species centers is larger than σ and the formed species number is larger than k.

Step 2.04: For those who belong to one species, its niche count is the number of individuals belonging to that species, call the *occupation number*. For independent individuals, use Eq. 5.5 to calculate its niche count.

Step 2.05: Calculate the shared fitness value for every individual using Eq. 5.6.

In the above step 2.03, we can identify the peaks, as the species center, in every generation dynamically, so this is sometimes referred as *dynamic peak identification* (DPI).

Miller and Shaw also suggested two mating restriction rules for selecting parents. In *dynamic line-breeding*, the first parent is selected through tournament selection based on shared fitness values and the second parent is its closest dynamic species center. Dynamic line-breeding always makes individuals belonging to the same species mate. In *dynamic inbreeding*, the first parent is the same as dynamic line-breeding and the second parent is the one with the highest fitness value and belongs to the same species as the first parent within the MF, a user-defined parameter, randomly selected individuals from the mating pool. If there are no individuals belonging to the same species, the closest one in the mating pool is selected as the second parent to decrease the possibility of lethal offspring.[12] Dynamic inbreeding relaxes dynamic line-breeding by enabling mating between different members and species, which might not be the centers of the species.

DNS has the following characteristics.

- It has an explicit and clear speciation process, which makes further improvement easy.
- It assigns the best individual as the peak center, which makes the speciation unsymmetrical.
- It has stronger punishment power comparing to standard fitness sharing,[13] which makes it faster than standard fitness sharing.
- It requires two kinds of information on the solution landscape, which is harder to provide comparing to standard fitness sharing.

Setting *popsize* and σ for Fitness Sharing

As was discussed above, σ is a sensitive and critical parameter for fitness sharing. Cioppa *et al.* suggested a way to estimate *popsize* and σ [19]. Their method uses a procedure, inspired by DPI above, to identify the niche number, $\nu(g, popsize, \sigma)$, in generation g. Then the effects of different σ and *popsize* on the mean, $\mu(popsize, \sigma)$, and the standard deviation, $\delta(popsize, \sigma)$, of $\nu(g, popsize, \sigma)$ over generations are used to estimate the proper *popsize* and σ.

12 Line-breeding and inbreeding are genetic terms that describe the mating process of relatives breeding with one another and related individuals breeding with each other, respectively, which is the nature of positive assortative mating. The nature of negative assortative mating is called *outcrossing*, which describes the mating process of two completely unrelated individuals.

13 Why?

To begin with, they suggest a *dynamic niche identification* process, which is carried out after we get a niche count for every individual in standard fitness sharing, to identify the niche maintained in the current generation with a predefined parameter σ, illustrated as follows.[14]

Dynamic Niche Identification

Phase 1: Sorting the population in descending order of the niche counts. The niche count is calculated by Eq. 5.5. Cioppa *et al.* use the centroid rather than its fittest individual to represent a niche.

Phase 2: Generating a dynamic niche set. Unmark all individuals, empty the dynamic niche set, and do the following steps from the first individual $i = 1$ to the last individual $i = popsize$.

Step 2.1: If current individual i has not been marked, calculate its distance to the following individuals $j = (i+1)$–$popsize$. If any following individual j is close to i, i.e., $d(i,j) < \sigma$, and it is not marked, mark it as a member of the niche represented by individual i and increase the number of individuals in that niche by one.

Step 2.2: If the number of individuals in the niche represented by individual i is no less than two, set individual i as the *niche center* and *species master* and insert it into the dynamic niche set.

Step 2.3: $i = i+1$ and go to step 2.1.

Phase 3: Submitting the dynamic niche set.

With the help of dynamic niche identification, we can find the niche number $\nu(g, popsize, \sigma)$ in every generation g of standard fitness sharing with parameters $popsize$ and σ. If we make T runs for a set of parameters $popsize$ and σ, there is a matrix with T rows and $maxgen$ columns whose elements are the niche numbers in every generation and run with parameters $popsize$ and σ. Then $\nu(g, popsize, \sigma)$ can be calculated by averaging the niche numbers in the T runs at generation g. Finally, $\mu(popsize, \sigma)$ and $\delta(popsize, \sigma)$ can be calculated using $\nu(g, popsize, \sigma)$.

In order to estimate the proper $popsize$ and σ for a problem, we need to analyze the effect of different $popsize$ and σ on $\mu(popsize, \sigma)$ and $\delta(popsize, \sigma)$. Cioppa *et al.* provided a thorough analysis and their key results are listed as follows:

- Provided that $popsize$ has been fixed, with an increase of σ from underestimation to proper estimation and overestimation, μ decreases and δ has a minimum, which is close to zero, at proper estimation, $\sigma = \sigma^*$, and will finally drop to zero with very large σ.
- Provided that σ has been fixed, with an increase of $popsize$ from underestimation to proper estimation and overestimation, $\delta(popsize, \sigma^*)$ will become increasingly smaller.

---

[14] Readers should compare dynamic peak identification and dynamic niche identification.

Now we can start running standard fitness sharing with $\sigma_{\min}$ and $popsize_{\min}$, which are user-defined parameters. We first fix $popsize = popsize_{\min}$ and increase σ from $\sigma_{\min}$ to find the minimum of δ, which is close to zero. We record σ^* and $popsize^*$ as the current σ and $popsize$, respectively, set the upper bound $\sigma_{\max}$ as the current σ, and increase $popsize$. Then we increase σ again from $\sigma_{\min}$ to find the minimum of δ. The process will continue until we cannot find σ^* in $[\sigma_{\min}, \sigma_{\max}]$. Then we take the previous σ^* and $popsize^*$ as the proper estimation for σ and $popsize$. In this way, we can find the smallest σ^* and its corresponding $popsize^*$, which is quite useful for multimodal problems with different but distinguishable niche radii .

The above process could be regarded as a metasearch process that uses standard fitness sharing as the local searcher. So this process can find the proper parameters for standard fitness sharing at the cost of a rather time-consuming performance evaluation for different $popsize$ and σ.[15]

Dynamic Fitness Sharing

In 2007, Cioppa *et al.* suggested *dynamic fitness sharing* (DFS) to further improve DNS with only σ [20].

They use a procedure called *dynamic species identification* to identify the *species master* in the current generation. Dynamic species identification is quite similar to dynamic niche identification introduced above. The only difference is that the sorting in Phase 1 is done according to the raw fitness instead of the niche count. After dynamic species identification process, we get a dynamic species set with ν species, say $S^1, \cdots, S^\nu$, the element of which contains at least two individuals.

So the population P could be divided into $(\nu+1)$ groups. The $(\nu+1)$th group, S^*, contains isolated individuals that do not belong to any species and do not contain any other isolated individuals.

Then the niche count for individuals belonging to one species is calculated as follows:

$$m_i = \sum_{j \in S^k} sh(d_{ij}) \tag{5.8}$$

where individual i belongs to species S^k and $sh()$ is defined by Eq. 5.4. The niche count for dynamic species identification considers the contribution of distance and only considers individuals within the same species, which makes it different from DNS and standard fitness sharing. But the niche count for isolated individuals is set at 1 so as to improve the chances for those individuals to generate a new species. Then we can use Eq. 5.6 to calculate the shared fitness value and go to an EA selection process. Elitism is also adopted in DFS to copy the species masters found at each generation into the next population.

---

15 No free lunch again!

With the help of the dynamic species identification, the redesigned niche count function, and elitism, Cioppa *et al.* claimed that DFS performs significantly better than the three other fitness sharing EAs.

5.3.5 Coevolutionary Shared Niching

In 1997, Goldberg and Wang incorporated fitness sharing into coevolution, which was discussed in Sect. 3.7.1, and suggested *coevolutionary shared niching* (CSN) [21].

There are two groups of individuals in CSN, i.e., *businessmen* and *customers*. Businessmen are for locating the peaks and customers are ordinary individuals of EAs. The number of businessmen is k, which is the same as, or a bit larger than, the predefined peak number, and the number of customers is $popsize$. The fitness value of a customer is the shared fitness value calculated with Eq. 5.6, and the fitness value of a businessman is the sum of the raw fitness values of the customers belonging to this businessman.

CSN also needs σ to proceed. The following steps constitute CSN, which could be embedded in the general solution process of EA on page 77.

One Generation of Coevolutionary Sharing Niching

Phase 1: Assigning every customer to its closest businessman.

Phase 2: Calculating the customer niche count, which is the number of customers belonging to their corresponding businessman.

Phase 3: Calculating the customer shared fitness value using Eq. 5.6.

Phase 4: Generating a new population of customers with selection (by shared fitness values), crossover, and mutation.

Phase 5: Updating the businessman group with the following mutation process, called *imprint*. *Genomic imprinting* is a genetic phenomenon by which certain genes are expressed depending upon whether they resided in a male or female in the previous generation.

Step 5.1: For every businessman b, randomly select one customer c from its current customer group. Calculate its fitness value as a businessman and distance to other businessmen. If the new one has a higher fitness value and its smallest distance to other businessmen is larger than σ, it replaces the old businessman.

Step 5.2: If the mutant does not satisfy both of the above conditions, pick another customer until the mutation has been carried out n_{limit} times.

When the stop criteria are satisfied, provide businessmen as the found peaks.

5.4 Crowding

Standard crowding was first introduced by De Jong in 1975 as a way of adjusting population diversity [22]. With the application of crowding to multimodal problems, standard crowding has been seldom used.

Apart from fitness sharing, crowding has a different perspective. We think that the solution landscape is crowded with individuals. So the newly generated individual need to compete with the old ones in order to replace a worse old individual close to it.

According to the above considerations, crowding is often implemented in the steady state EA form and the main concern is with the replacement process.

5.4.1 Deterministic Crowding

Deterministic crowding (DC) was suggested by Mahfoud in 1995 [23]. It uses parents to compete with offspring in a deterministic way and does not require any predefined parameters.

DC is carried out in the replacement process, i.e., replacing Phase 2 in the general solution process of EA on page 77 with the following steps.

Deterministic Crowding

Phase 2: Main loop. Repeat the following steps until stop criteria are satisfied.

Step 2.1: Randomly select two parents p_1 and p_2 with replacement from the current population.

Step 2.2: Generate two offspring c_1 and c_2 using the variation operators introduced in Sect. 3.2.

Step 2.3: Evaluate the fitness value of offspring, $f(c_1)$ and $f(c_2)$, and calculate their distances to parents, i.e., $d(p_1,c_1)$, $d(p_1,c_2)$, $d(p_2,c_1)$, and $d(p_2,c_2)$.

Step 2.4: Identify a close competition pair. If $[d(p_1,c_1)+d(p_2,c_2)] \leq [d(p_1,c_2)+d(p_2,c_1)]$, the competition is between $p_1 \Leftrightarrow c_1$ and $p_2 \Leftrightarrow c_2$. Otherwise, the competition is between $p_1 \Leftrightarrow c_2$ and $p_2 \Leftrightarrow c_1$.

Step 2.5: Determine the winner. Individuals with higher fitness values win the competition and will stay in the population. Losers are discarded.

With the competition between parents and offspring, DC maintains the population diversity effectively. Thus it increases the possibility of identifying and maintaining multiple peaks.

Different variation operators may contribute different locations of the offspring. Thus it is possible that the offspring will be far from their parents. So genetic drift might be possible in DC, which explains why sometimes it discovers niches very fast but only maintains them for a few generations [24]. Gomez suggested an improvement for DC in 2004 [25]. His method makes only slight changes to the above step 2.5. If the winner of the competition is offspring, it can replace the corresponding parent only if the distance between them are smaller than a predefined parameter δ. Otherwise, parents will stay in the population.

DC's competition rule is deterministic, which promotes convergence but might suffer from premature. Mengshoel suggested *probabilistic crowding* in 1999 to introduce down-hill possibility in competition [26]. In the above step 2.5, a offspring c competes with parent p. Offspring c has the probability $p = \frac{f(c)}{f(c)+f(p)}$ to replace parent p. In 2008, Mengshoel and Goldberg further developed probability crowding and suggested a framework of *local tournament algorithms* [27].

5.4.2 Restricted Tournament Selection

With the same objective of avoiding excessive distance between the competitors of DC, Harik suggested *restricted tournament selection* (RTS) in 1995 [28]. The difference between RTS and DC shows up in steps 2.3–2.5, which is illustrated as follows:

Restricted Tournament Selection

Step 2.3: Evaluate the fitness values of offspring $f(c_1)$ and $f(c_2)$. Randomly select w (*windowsize*) individuals from the current population with replacement as the *comparison set*.

Step 2.4: Without loss of generality, suppose d_1 and d_2 are the closest individuals in the comparison set to c_1 and c_2, respectively.

Step 2.5: Determine the winner. If $f(c_1) > f(d_1)$, replace d_1 with c_1, otherwise keep d_1 and discard c_1. The same operation is carried out between d_2 and c_2.

It is necessary to mention that Gomez's suggestion and RTS improve DC in finding and maintaining multiple peaks in a single run provided that parameters, δ and w, have been predefined correctly.

Cedeño suggested *multiniche crowding* (MNC) in 1995 with a similar idea to RTS [29]. MNC has more control parameters and gives a better performance with adequate parameters.

5.4.3 Species Conserving Genetic Algorithm

Li *et al.* suggested a *species conserving genetic algorithm* (SCGA) in 2002 [1]. SCGA uses σ to determine the peak in a niche, denoted as *species seeds*, which is similar to the dynamic niche identification and dynamic species identification discussed in Sect. 5.3.4. SCGA does not use specie seeds to calculate the niche count and punish the overcrowded species, but saves them until the replacement process to replace the close and worse new individuals.

The procedure for determining the species seeds is carried out before the selection process as follows:

Determining Species Seeds

Phase 1: Sorting the population in descending order of the raw fitness values.

Phase 2: Generating a species seed set (SSS). Empty SSS and do the following steps from the first individual $i = 1$ to the last individual $i = popsize$.

Step 2.1: If the distances of the current individual i to all SSS elements are larger than σ, insert it into SSS. $i = i + 1$.

Step 2.2: If the smallest distances of the current individual i to all SSS elements is smaller than σ, $i = i + 1$.

Phase 3: Submitting SSS.

Then EAs execute selection (with raw fitness value), crossover, and mutation to generate a new population. Step 2.4 of the general solution process of EA introduced on page 77 needs to be changed as follows:

Elitism

Step 2.4: Elitism.

Substep 2.4.1: Assign every new individual to its closest species seed in SSS if the distance between it and its closest species seed is smaller than σ. There might be some new individuals that do not belong to any SSS element and there might be some SSS elements that do not contain any new individuals.

Substep 2.4.2: If each SSS element i contains new individuals, select the worst one (with the smallest raw fitness value) j. If $f(i) > f(j)$, replace j with i.

Substep 2.4.3: If SSS element i contains no new individuals, select the worst individual j in the new individuals that do not belong to any SSS element. Replace j with i.

SCGA uses σ to find SSS and it uses elitism in the replacement process but does not change the raw fitness, so we introduce it here as a crowding method. In 2003, Li *et al.* published another paper where they compared SCGA with CSN [30], which is introduced in Sect. 5.3.5.

5.5 Performance Indices for Multimodal Optimization

We want niche EAs to find multiple peaks in a single run. If such peaks have different heights, special considerations are necessary in evaluating the performance of niche EAs.

Suppose we know the real location and the height of the peaks of the benchmark problem.[16] After niche EAs stop, we can identify the peaks found by these EAs.[17] If the location and height of one peak found by a niche EA are within the threshold of the real peak, we claim that this peak has been maintained by the niche EA. For real peak $\mathbf{x}$ with height $h_{\mathbf{x}}$, we often define $0.9 \times h_{\mathbf{x}}$ or $0.8 \times h_{\mathbf{x}}$ as the height threshold and $d(\mathbf{x}, \mathbf{y}) < \varepsilon$ as the distance threshold, where ε is the user-allowed discrepancy.

The most often used *performance indices* for multimodal optimization are listed as follows.

Effective Number of the Peaks Maintained

We run the niche EAs several times and use the average and the standard deviation of the effective number of peaks maintained (ENPM) to illustrate the niche EAs' ability. A larger ENPM value indicates a better ability to identify and maintain multiple peaks.

Maximum Peak Ratio

ENPM does not consider the influence of peak heights. Thus in 1996, Miller and Shaw proposed the maximum peak ratio (MPR) [18].

Suppose a problem has k peaks whose heights are $h_1, \cdots, h_k$. Also suppose that the niche EA finds m peaks whose heights are $h_1^{'}, \cdots, h_m^{'}$.[18] MPR is defined as follows:

---

[16] If the explicit form of the optimal solution is not available, the best results of the multiple runs of different niche EAs might be used as the "real" location and height of the peaks.

[17] Dynamic species identification, introduced in Sect. 5.3.4, might be an option.

[18] These peaks need to satisfy the follow conditions: they are within the radius of one real peak; their heights are 80% or higher than that of the corresponding real peaks; they have the largest fitness values in the species.

$$MPR = \frac{\sum_{i=1}^{m} h_i^{'}}{\sum_{j=1}^{k} h_j} \tag{5.9}$$

If no real peak has been found, the corresponding part in the numerator is zero.

As can be seen from Eq. 5.9, MPR grants higher peaks with more preference. The largest value of MPR is 1, which means all the peaks have been identified and maintained correctly. A larger MPR value means a better convergence to peaks.

Chi-square-like Performance Criterion

ENPM and MPR do not consider the distribution of individuals in the last generation.[19] Deb and Goldberg suggested a Chi-square-like (CSL) performance criterion in 1989 to deal with this [5].

Suppose, after a "perfect" run of our niche EA, every individual at the end of the evolution converges to one peak. We can consider a probabilistic event for peak i as: an individual is on peak i. It is a Bernoulli experiment. It's natural and straightforward to suppose that the probability p_i of event A is

$$p_i = \frac{h_i}{\sum_{l=1}^{k} h_l} \tag{5.10}$$

where h_l is the height of peak l and k is the relevant peak number.

Then we can define a Bernoulli distribution random number y_j for individual j in the final population. $y_j = 1$ means individual j is on peak i with probability p_i, and $y_j = 0$ means it is not on peak i with probability $1 - p_i$. It is easy to determine that the mean and the standard deviation of this Bernoulli distribution are p_i and $p_i(1 - p_i)$, respectively.

For the final population, there are *popsize* Bernoulli distribution random numbers for peak i. According to the central limit theorem, if *popsize* is large enough,

$$\frac{y_1 + \cdots + y_{popsize} - popsize \times p_i}{\sqrt{popsize \times p_i \times (1 - p_i)}} \sim N(0,1) \tag{5.11}$$

We can then define the number of individuals on peak i as $x_i = y_1 + \cdots + y_{popsize}$. And by defining $\mu_i = popsize \times p_i$ and $\sigma_i^2 = popsize \times p_i \times (1 - p_i)$, we can reformulate Eq. 5.11 as follows:

$$\frac{x_i - \mu_i}{\sigma_i} \sim N(0,1) \tag{5.12}$$

[19] We want to populate our individuals with the proportion to peaks' heights, which was discussed in Sect. 5.3.1.

So for k peaks in the solution landscape, we have k standard normally distributed random numbers $x_1, \cdots, x_k$. According to the definition of chi-square distribution,

$$\sum_{i=1}^{k} \frac{(x_i - \mu_i)^2}{\sigma_i^2} \sim X^2(k) \tag{5.13}$$

where $X^2(k)$ is the chi-square distribution with (k) degrees of freedom.

To further emphasize the final individuals that do not belong to any peak, Deb and Goldberg consider them as the $(k+1)$th group. It can be represented as a random number $x_{k+1} = popsize - \sum_{i=1}^{k} x_i$ whose mean and variance can be calculated easily as $\mu_{k+1} = 0$ and $\sigma_{k+1}^2 = \sum_{i=1}^{k} \sigma_i^2$, respectively.

So CSL can be illustrated as follows:

$$CSL = \sqrt{\sum_{i=1}^{k+1} \frac{(x_i - \mu_i)^2}{\sigma_i^2}} \tag{5.14}$$

where x_i is the real number of individuals on peak i by the end of a niche EA, μ_i and σ_i can be calculated from previous discussions with the known height of every peak of the benchmark problem.

By comparing Eqs. 5.14 and 5.13 we can see that CSL borrows the form of chi-square distribution and can be used to evaluate the distribution of a population, i.e., if the number of individuals on every peak equals the mean of that peak, the CSL value is zero. A smaller CSL value means a better individual distribution.

5.6 Application Example

The distributed multipump Raman amplifier (DMRA) can offer ultra-wide and ultraflat gain bandwidth, improve noise characteristics, and realize ultralong-haul systems reach. Additionally, DMRA can mitigate fiber nonlinear effects compared to erbium-doped fiber amplifiers. So the design of the DMRA is quite attractive. In 2004, Liu and Lee suggested a clustering-based niche EA to provide multiple DMRAs with equal quality in a single run [31].

Liu and Lee's algorithm needs the peak number k and smallest peak radius σ as the control parameters. The solution process can be illustrated as follows:

Phase 1: Initialization.

Step 1.1: Assign the parameters for the niche EA, such as k, σ, and other control parameters for SGA.

Step 1.2: Generate *popsize* uniformly distributed individuals randomly to form the initial population and evaluate fitness values. $gen = 0$.

Phase 2: Main loop. Repeat the following steps until the stop criteria are satisfied.

Step 2.1: Sort the population in descending order of the raw fitness value.

Step 2.2: Find k species centers.

Substep 2.2.1: Assign the first individual as the first confirmed species center, and mark it.

Substep 2.2.2: Select k individuals orderly from the population that satisfy the following two conditions: the individuals are not marked; the distances from the individual to all of the confirmed species centers are larger than σ.

Substep 2.2.3: Calculate the sum of distances between each of k individuals and all the confirmed species centers, assign the individual with the largest sum of distances as the next confirmed species center, and mark it.

Substep 2.2.4: Repeat substeps 2.2.2 and 2.2.3 until k species centers are confirmed.

Step 2.3: Allocate other individuals to their nearest species center.

Substep 2.3.1: For each unmarked individual, calculate its distances to all the confirmed species centers.

Substep 2.3.2: Select the shortest distance and allocate the individuals to their corresponding species.

Substep 2.3.3: Repeat substeps 2.3.1 and 2.3.2 until all *popsize* individuals are allocated.

Step 2.4: Calculate the niche count for each individual, which is equal to the individual number of the species it belongs to. Calculate its shared fitness value using Eq. 5.6. Select M individuals to be saved as the *elitism set* according to the shared fitness values.

Step 2.4: Selection (using shared fitness value), crossover, and mutation are carried out to generate a new population. $g = g + 1$.

Step 2.5: Replace M worst individuals (according to the raw fitness value) of the new population with individuals in the elitism set.

Phase 3: Submitting the final k species centers in the final generation as the results of the niche EA.

The optimal results show that the broadest bandwidth of a DMRA can be 100.08 nm with just five pumps under the design conditions of an ON-OFF gain of 10 dB, a relative gain flatness of 0.1, and a fiber length of 80 km.

Liu and Lee assume that $k = 6$, i.e., they want to find approximately six peaks in the solution landscape, and their niche EA provides four design schemes with the broadest bandwidths being 100.08 nm and two design schemes with bandwidths of 94.76 nm and 92.28 nm, respectively.

5.7 Summary

Multimodal problems have real-world applications. Niche EAs can provide both multiple high-quality solutions for multimodal problems and techniques for improving population diversity.

Fitness sharing downscales the fitness values of overcrowded individuals so as to maintain an equilibrium in which individuals are distributed proportionally to the peak heights. These techniques contain useful information for adjusting the selective pressure in a selection process according to different requirements.

Crowding has a competition mechanism between the offspring and their close parents (or other close individuals) within the framework of a steady state EA. These techniques also contain useful information for adjusting the selective pressure in a replacement process according to different requirements.[20] Apart from the above classification, we can give the taxonomy for niche EAs in a broader view as illustrated in Fig. 5.4.

Niche EAs can be classified into three categories. Sequential niche tries to solve multimodal problems by modifying the objective function in a serial way. Whole population niche tries to improve population diversity in a global way, and subpopulation niche uses rarely communicated subpopulations to improve population diversity in a local way.

Furthermore, we can classify fitness sharing according to the required information before calculation. Standard fitness sharing, adaptive k-means clustering with fitness sharing, and clearing procedure belong to the "niche radii" group. The "niche number" group includes the algorithms reported in [8, 15]. DNS, DFS, and CSN belong to the "niche radii and number" group. The "blind" group does not require any predefined information. Interested readers are referred to [32, 33].

Deterministic crowding involves competition with parents but restricted tournament selection, multiniche crowding, and species conserving genetic algorithm have competition with others.

In the *island* model, the population is divided into multiple subpopulations, which evolve independently for a fixed number of generations. After several generations the procedure of migration arises, in which a portion of each subpopulation migrates to other subpopulations [34].

The *Tagging* algorithm gives individuals different tags by which we can divide them into several groups. Only parents with the same tags are able to breed. Spears claimed that tagging was shown to be a rather efficient implementation [35].

The *Multinational* model considers the population as the world of "nations," "governments," and "politicians," and also uses the migration of individuals and the merging of subpopulations in the evolving process [36].

The effective understanding of this chapter means in-depth comprehension of related concepts, e.g., niche, species, and speciation, grasping the essence of fitness sharing and crowding, and being able to adopt adequate selective pressure adjust-

[20] Readers are referred to Sect. 3.4.1 for further information.

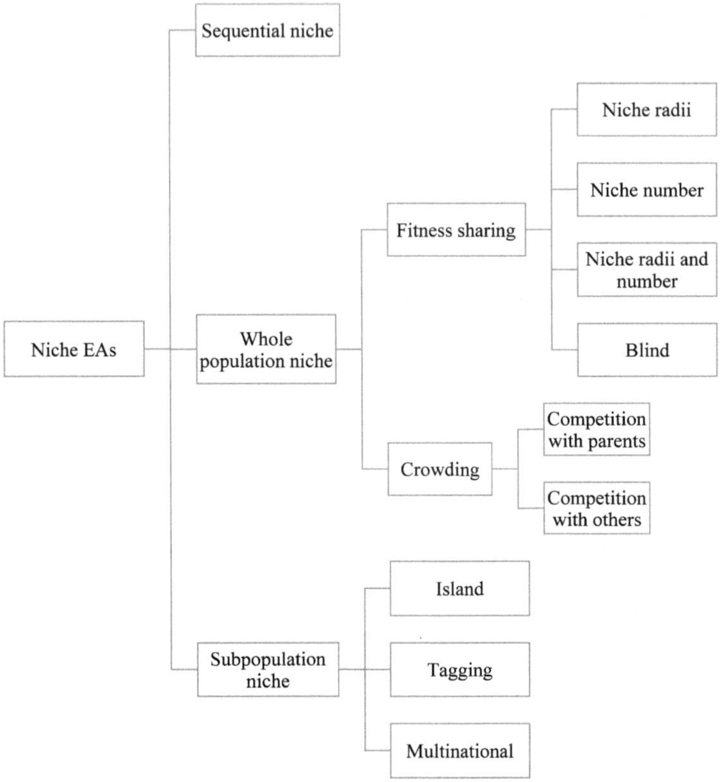

Fig. 5.4 The taxonomy of niche EAs

ment techniques to maintain population diversity, which might contribute to finding multiple high-quality solutions in a single run.

In all, designing an EA for multimodal optimization is the art of the tradeoff between converging to peaks and distributing individuals in different peaks.

Suggestions for Further Reading

There are two intensive surveys and comparisons for niche EAs published in 1998 and 2006 respectively. Interested readers are referred to [37, 38].

Shir and Bäck use CMA to solve multimodal problems by ES and also self-adaptively controlled peak radius σ. Readers interested in (derandomized) ES are referred to Shir's Ph.D. thesis, published in 2008 [32], and a paper published in 2009 [39].

In 2005, Wei and Zhao combined a clearing procedure (Sect. 5.3.2) and Nelder–Mead's simplex method (Sect. 2.4.1) to find the only global optimal solution [40]. Although finding the only global optimal solution is not the task of this chapter, we still suggest it as an application of the clearing procedure and want our readers to understand that innovative ideas from other fields may help you in solving your problem, which is why we explain so many intuitive ideas in this book.

The must-read papers of this chapter are [37] for a general survey, [3] for standard fitness sharing, [23] for deterministic crowding, [18] for DNS, and [20] for an elaborate way to estimate the key control parameters σ and *popsize*.

Exercises and Potential Research Projects

5.1. Why does standard fitness sharing technique not suffer from genetic drift after all relevant peaks have already been found?

5.2. Sareni and Krahenbuhl have reported that tournament selection is not as good as SUS in standard fitness sharing with a mating restriction [37]. Use the benchmark problems in Appendix and PI in Sect. 5.5 to compare SUS and unbiased tournament selection, introduced in Sect. 3.3.5. Use the techniques introduced in Sect. 3.6 to do a fair comparison and draw statistical conclusions according to your numerical experiments.

5.3. How many distance calculations are necessary for clearing, adaptive k-means clustering with fitness sharing, DNS, and CSN discussed in Sect. 5.3?

5.4. Produce a table to compare the differences between various sharing techniques, e.g., standard fitness sharing, clearing, adaptive k-means clustering with fitness sharing, DNS, and CSN discussed in Sect. 5.3. The content of comparison might include the information about the solution landscape necessary to proceed, whether or not there exists explicit speciation, how to obtain the niche count, whether the species/niche center is the peak or the centroid, and the time of the distance calculation.

5.5. Implement standard fitness sharing and deterministic crowding in your programming environment. Use the benchmark problems in Appendix and PI in Sect. 5.5 to compare these two classical niche EAs. Use the techniques introduced in Sect. 3.6 to do a fair comparison and draw statistical conclusions according to your numerical experiments.

5.6. Suggest a way to expand DE into multimodal optimization. Use the benchmark problems in Appendix and PI in Sect. 5.5 to compare your novel algorithm with standard fitness sharing. Use the techniques introduced in Sect. 3.6 to do a fair comparison and draw statistical conclusions according to your numerical experiments.

5.7. What is the difference between dynamic niche identification, dynamic species identification, and determining the species seeds? The first two procedures were introduced in Sect. 5.3.4 and the last one in Sect. 5.4.3.

References

1. Li J, Balazs ME, Parks GT *et al* (2002) A species conserving genetic algorithm for multimodal function optimization. Evol Comput 10(3):207–234
2. Beasley D, Bull D, Martin R (1993) A sequential niche technique for multimodal function optimization. Evol Comput 1(2):101–125
3. Goldberg DE, Richardson J (1987) Genetic algorithms with sharing for multimodal function optimization. In: Proceedings of the 2nd international conference on genetic algorithms and their applications, pp 41–49
4. Ursem R (2001) When sharing fails. In: Proceedings of the IEEE congress on evolutionary computation, pp 873–879
5. Deb K, Goldberg DE (1989) An investigation of niche and species formation in genetic function optimization. In: Proceedings of international conference on genetic algorithms, pp 42–50
6. Petrowski A (1996) A clearing procedure as a niching method for genetic algorithms. In: Proceedings of the IEEE conference on evolutionary computation, pp 798–803
7. Lee C, Cho D, Jung H (1999) Niching genetic algorithm with restricted competition selection for multimodal function optimization. IEEE Trans Magnet 35(3):1722–1725
8. Yin X, Germay N (1993) A fast genetic algorithm with sharing scheme using cluster analysis methods in multimodal function optimization. In: Proceedings of international conference on artificial neural nets and genetic algorithms, pp 450–457
9. Duda RO, Hart PE, Stork DG (2000) Pattern classification, 2nd edn. Wiley-Interscience, New York
10. Bishop CM (2007) Pattern recognition and machine learning. Springer, Berlin Heidelberg New York
11. Xu R, Wunsch D (2008) Clustering. Wiley-IEEE, New York
12. Alpaydin E (2004) Introduction to machine learning. MIT Press, Cambridge, MA
13. Xu R, Wunsch D (2005) Survey of clustering algorithms. IEEE Trans Neural Netw 16(3):645–678
14. Shan S, Xinjie Y (2006) Multi-peak function optimization using a hierarchical clustering based genetic algorithm. In: Proceedings of the sixth international conference on intelligent systems design and applications, pp 425–428
15. Yu X (2005) A novel clustering fitness sharing genetic algorithm. In: Proceedings of the international conference on natural computation, pp 1072–1079
16. Ando S, Sakuma J, Kobayashi S (2005) Adaptive isolation model using data clustering for multimodal function optimization. In: Proceedings of the ACM conference on genetic and evolutionary computation, pp 1417–1424
17. Miller BL, Hansen N, Shaw MJ (1995) Genetic algorithms with dynamic niche sharing for multimodal function optimization. Tech. Rep. 95010, University of Illinois at Urbana-Champaign
18. Miller B, Shaw M (1996) Genetic algorithms with dynamic niche sharing for multimodal function optimization. In: Proceedings of IEEE conference on evolutionary computation, pp 786–791
19. Cioppa AD, Stefano CD, Marcelli A (2004) On the role of population size and niche radius in fitness sharing. IEEE Trans Evol Comput 8(6):580–592
20. Cioppa AD, Stefano CD, Marcelli A (2007) Where are the niches? dynamic fitness sharing. IEEE Trans Evol Comput 11(4):453–465
21. Goldberg D, Wang L (1997) Adaptive niching via coevolutionary sharing. In: Miettinen K (ed) Genetic algorithms and evolution strategies in engineering and computer Science. Wiley, New York, pp 21–38
22. Jong KAD (1975) An analysis of the behavior of a class of genetic adaptive systems. Ph.D. thesis, University of Michigan
23. Mahfoud S (1995) Niching methods for genetic algorithms. Ph.D. thesis, University of Illinois at Urbana-Champaign

24. Mahfoud SW (1995) A comparison of parallel and sequential niching methods. In: Proceedings of the international conference on genetic algorithms, pp 136–143
25. Gomez J (2004) Self adaptation of operator rates for multimodal optimization. In: Proceedings of the IEEE conference on evolutionary computation, pp 1720–1726
26. Mengshoel OJ (1999) Efficient Bayesian network inference: genetic algorithms, stochastic local search, and abstraction. Ph.D. thesis, University of Illinois at Urbana-Champaign
27. Mengshoel OJ, Goldberg DE (2008) The crowding approach to niching in genetic algorithms. Evol Comput 16(3):315–354
28. Harik G (1995) Finding multimodal solutions using restricted tournament selection. In: Proceedings of the international conference on genetic algorithms, pp 24–31
29. Cedeño W (1995) The multi-niche crowding genetic algorithm: analysis and applications. Ph.D. thesis, Universiteit California at Davis
30. Li J, Balazs ME, Parks GT *et al* (2003) Erratum: a species conserving genetic algorithm for multimodal function optimization. Evol Comput 11(1):107–109
31. Liu X, Lee B (2004) Optimal design of fiber raman amplifier based on hybrid genetic algorithm. IEEE Photon Technol Lett 16(2):428–430
32. Shir OM (2008) Niching in derandomized evolution strategies and its applications in quantum control. Ph.D. thesis, Leiden University
33. Yu X (2005) Fitness sharing genetic algorithm with self-adaptive annealing peaks radii control method. In: Proceedings of the international conference on natural computation, pp 1064–1071
34. Martin W, Lienig J, Cohoon J (1997) Island migration models: evolutionary algorithms based on punctuated equilibria. In: Bäck T, Fogel DB, Michalewicz Z (eds) Handbook of evolutionary computation. Oxford University Press, Oxford, UK, C6.3:1–16
35. Spears WM (1994) Simple subpopulation schemes. In: Proceedings of the 3rd annual conference on evolutionary programming, pp 296–307
36. Ursem RK (1999) Multinational evolutionary algorithms. In: Proceedings of the IEEE congress on evolutionary computation, pp 1633–1640.
37. Sareni B, Krahenbuhl L (1998) Fitness sharing and niching methods revisited. IEEE Trans Evol Comput 2(3):97–106
38. Singh G, Deb K (2006) Comparison of multi-modal optimization algorithms based on evolutionary algorithms. In: Proceedings of the ACM annual conference on genetic and evolutionary computation, pp 1305–1312
39. Shir OM, Bäck T (2009) Niching with derandomized evolution strategies in artificial and real-world landscapes. Nat Comput 8(1):171–196
40. Wei L, Zhao M (2005) A niche hybrid genetic algorithm for global optimization of continuous multimodal functions. Appl Math Comput 160(3):649–661

Chapter 6
Multiobjective Optimization

Abstract EAs are developed to solve real-world problems, such as designing and scheduling. In real conditions, there are many requirements to fulfill. In previous chapters, we sometimes wanted to model them into constraints because it is hard to compare two objectives simultaneously.[1] Pareto gave us the idea of dominance, so we can divide the relationship between two vectors into three categories: one is better than the other, the converse is true, or they are incomparable or incommensurable. For problems with multiple objectives, there exist many "good" points that are no worse than any other point in the objective space. EAs contain a group of individuals. So if we can distribute the individuals evenly on these "good" points, it will be very helpful for designers and decision makers. This chapter discusses how to use EAs to solve such problems. It is a fascinating and hot research area. You will experience the shining wisdom of other researchers, which will deepen considerably your understanding of EAs.

6.1 Introduction

6.1.1 Problems We Face

As was introduced in Chap. 1, there are many realistic applications that call for multiple objectives. Let us start with a fictitious product you are designing. Every product has many features to describe how good it is. Among them, you mainly care about failure rate and cost. As a designer, you want the cost to be as low as possible, which makes your product more competitive in the harsh market. At the same time, you must consider the failure rate seriously so as to diminish the return repair rate.

Unfortunately, these two considerations are in conflict. Improving the failure rate means utilizing more reliability approaches, which will definitely increase the cost. The dilemma can be illustrated by Fig. 6.1.

[1] Consider how to compare two vectors in 2-D space.

According to previous discussions, square points a_1 and a_2 are two *incommensurable* designs, sometimes referred as tradeoff designs, and circle point a_3 is worse than square point a_2.

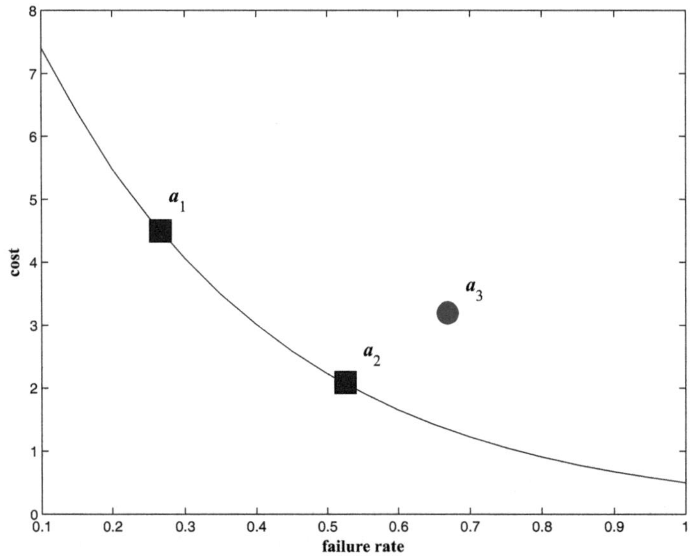

Fig. 6.1 Failure rate vs. cost of a fictitious product

Vilfredo Pareto, an Italian economist, gave the definitions of the relationships between these designs, which will be discussed in the next subsection.

6.1.2 Terminologies

Let us first formulate the problem and then give the definitions of the terms used in this chapter.

Consider the following optimization problem:[2]

$$\begin{aligned} \min\ & \{z_1 = f_1(\mathbf{x}), z_2 = f_2(\mathbf{x}), \cdots, z_m = f_m(\mathbf{x})\} \\ \text{s.t.}\ & g_i(\mathbf{x}) \leqslant 0, \quad i = 1, \cdots, q \\ & h_j(\mathbf{x}) = 0, \quad j = q+1, \cdots, k \\ & \mathbf{x} \in \mathbb{R}^n \end{aligned} \tag{6.1}$$

where $\mathbf{x}$ is the decision variable, f_i is objective i, g_i is inequality constraint i, and h_j is equality constraint j.

[2] Unless specifically noted, minimum objectives are discussed in this chapter.

There are m objectives in Eq. 6.1, which makes the model a *multiobjective optimization problem* (MOP).[3]

Two spaces need to be defined. *Decision space* (or *parameter space*) is defined as follows:

$$S = \left\{ \mathbf{x} \in \mathbb{R}^n \left| g_i(\mathbf{x}) \leq 0, \quad i = 1, \cdots, q \cap h_j(\mathbf{x}) = 0, \quad j = q+1, \cdots, k \right. \right\} \tag{6.2}$$

where $\mathbf{x} \in S$ is called a *decision vector*(or *parameter vector*).

Criterion space(or *objective space, fitness space*) is defined by Eq. 6.3. $\mathbf{z} \in Z$ is called an *objective vector*.

$$Z = \ \{ \mathbf{z} \in \mathbb{R}^m \left| z_1 = f_1(\mathbf{x}), z_2 = f_2(\mathbf{x}), \cdots, z_m = f_m(\mathbf{x}) \right. \} \tag{6.3}$$

The most important term given by Pareto is "*dominance*." Dominance can be defined in either objective space or parameter space.

In objective space, "point i dominates point j" means

$$\begin{aligned} z_l^i \leq z_l^j, \quad \forall l \in \{1, 2, \cdots, m\} \\ z_k^i < z_k^j, \quad \exists k \in \{1, 2, \cdots, m\} \end{aligned} \tag{6.4}$$

where z_l^i is the lth objective value of point i.

In parameter space, "point i dominates point j" means

$$\begin{aligned} f\left(\mathbf{x}^i\right)_l \leq f\left(\mathbf{x}^j\right)_l, \quad \forall l \in \{1, 2, \cdots, m\} \\ f\left(\mathbf{x}^i\right)_k < f\left(\mathbf{x}^j\right)_k, \quad \exists k \in \{1, 2, \cdots, m\} \end{aligned} \tag{6.5}$$

where $f\left(\mathbf{x}^i\right)_l$ is objective l's value of point i.

We use $i \prec j$ to demonstrate that point i dominates j.[4] If $z_l^i = z_l^j, \forall l \in \{1, 2, \cdots, m\}$, we say i is *equivalent* to j, expressed as $i = j$. If i either dominates or is equivalent to j, we say i *covers* j. If $z_l^i \leq z_l^j, \forall l \in \{1, 2, \cdots, m\}$, we say i *weakly dominates* j, expressed as $i \preceq j$. If $z_l^i < z_l^j, \forall l \in \{1, 2, \cdots, m\}$, we say i *strongly dominates* j, expressed as $i \prec\prec j$.[5]

In the objective space, we are most caring for those points which are not dominated by any other point. Here comes the definition of *nondominated solution* (or *Pareto optimal solution*). A point $\mathbf{z}^0 \in Z$ in objective space is a nondominated solution if and only if there is not any point $\mathbf{z} \in Z$ which dominates $\mathbf{z}^0$. In other words, any improvement in one objective of a nondominated solution will cause deteriora-

---

[3] Other terms include *multicriteria optimization problem, vector optimization problem, multiattribute optimization*, and *multicriteria decision making*.

[4] $i \prec j$ only means that i is "better" than j. For maximum problems, Eqs. 6.4 and 6.5 have different forms. But we still use $i \prec j$.

[5] We strongly encourage readers to determine the areas that are dominated, weakly dominated, and strongly dominated by a_2 in Fig. 6.1. Further, readers are urged to draw the covering relationship between these three sets.

tion in at least one other objective. All the nondominated points consist the *Pareto front*,[6] often expressed as PF*.

Analogously, a point $\mathbf{x}^0 \in S$ in parameter space is an *efficient solution* (or *non-inferior solution*) if and only if there is no point $\mathbf{x} \in S$ that dominates $\mathbf{x}^0$. All the effective solutions consist of the *Pareto set*,[7] often expressed as P*. PF* is the map of P*.

Here is a simple example. For the optimization model expressed by Eq. 6.6,

$$
\begin{aligned}
\min\ & z_1(x_1,x_2) = x_1 - 3x_2 \\
\min\ & z_2(x_1,x_2) = 3x_1 + x_2 \\
\text{s.t. } & g_1(x_1,x_2) = x_1 + 2x_2 - 2 \leqslant 0 \\
& g_2(x_1,x_2) = 2x_1 + x_2 - 2 \leq 0 \\
& x_1, x_2 \geq 0
\end{aligned}
\tag{6.6}
$$

Figure 6.2 illustrates the above definitions in which the wide line is the P* and PF* of the model, respectively.

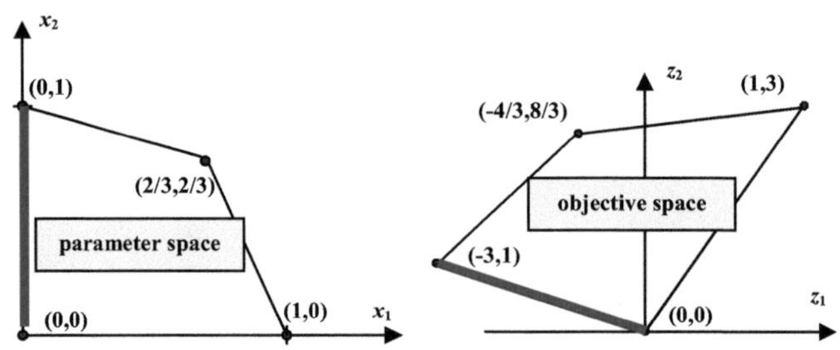

Fig. 6.2 Illustration of Pareto solution set and Pareto front

6.1.3 Why Are Evolutionary Algorithms Good at Multiobjective Optimization Problems?

Apart from single objective optimization problems,[8] MOPs generally do not have a single optimal solution (cf. Fig. 6.2), which means there are many solutions with incommensurable quality for designers and decision makers. This characteristic of MOPs is a two-edged sword. In one respect, there might be some features that are hard to mathematically model. So designers welcome multiple solutions with the same good quality (or incommensurable quality) so that they can select from them

[6] Other names include *Pareto frontier* and *Pareto optimal front*.

[7] Other names include *Pareto solution set* and *Pareto optimal set*.

[8] Here we neglect the multimodal problem discussed in Chap. 5.

according to their subjective preferences. But on the other hand, special concerns are necessary to handle the solutions, especially when the number of nondominated solutions in |PF∗|[9] is of substantial size. Considering both aspects, we generally welcome moderate information about the objective space. So an algorithm is considered to be efficient if it provides an adequate number of nondominated solutions in a single run.

EAs designed for MOPs are called *multiobjective EAs* (MOEAs), and this kind of optimization is called *evolutionary multiobjective optimization* (EMO).

What is the most distinguishing feature of EAs? A group of individuals evolve together to search the solution landscape in a parallel way. So if we can control the diversity of the population by preventing all individuals from converging to one point and encouraging them to reside in different points on the Pareto front, EAs could provide designers with at most *popsize* nondominated solutions in a single run. That would be fantastic!

Apart from the above reason, by generating many nondominated solutions and researching the location of these solutions, we could gain insights into a problem's solution landscape, which could facilitate the design of the problem-specific operators in MOEAs. As has been discussed before, problem-specific operators will dramatically alleviate search difficulties.

Figure 6.3 illustrates the number of papers indexed by the SCI on MOEAs and EMO.[10] By watching the increased frequency of papers on the topic and comparing to other SCI-indexed graphs in this textbook, it is easy to conclude that MOEAs have conspicuously attracted research interests.

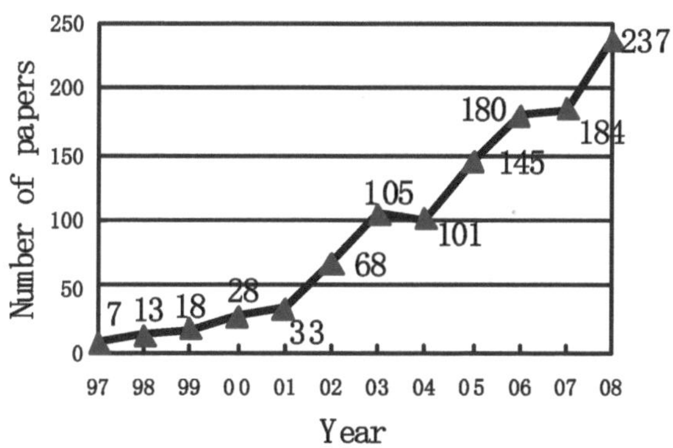

Fig. 6.3 Number of papers indexed by SCI on MOEAs

Let us review the MOEAs that are of interest to us.

[9] The size of PF*, i.e., how many points there are in the PF*.

[10] TS = ((“multiobjective optimization” OR “multi-objective optimization”) AND (“genetic algorithm” OR “genetic algorithms” OR “evolutionary computation” OR “evolutionary computing” OR “evolutionary algorithms” OR “evolutionary intelligence”)). The SCI index “TS” is for the search topic in the title, the keywords, and the abstract.

6.2 Preference-based Approaches

Approaches to deal with multiple objectives can be divided roughly into two parts: *preference-based approaches* and *construction approaches*. If some preference information is provided before optimization (such as which objective decision maker cares more), preference-based approaches can embed the user requirement into the algorithm and generate high-quality solutions quickly. But under blind circumstances, constructing approaches can provide multiple nondominated solutions for users to make decisions.

Generally, preference-based approaches transform MOPs into single-objective problems in different ways. So in this section, we only discuss the transformation procedure. After that, single-objective problems can be optimized by any algorithm (such as GA).

6.2.1 Weight Sum Method

If the decision maker can provide the relative importance for each objective, MOPs will be easier. Let us consider the design problem of Fig. 6.1. If the designer thinks the importance of the cost is double that of the failure rate, she can express this idea mathematically in the following way:

$$\text{Satisfaction} = \frac{2}{3}\text{cost} + \frac{1}{3}\text{failure rate} \tag{6.7}$$

In this way, if there exists an algorithm that can optimize Eq. 6.7, which is a single-objective problem, then the design problem has been solved. This is the initial idea of the *weight sum method*, and sometimes formulations like Eq. 6.7 are called the *aggregate fitness function*.

6.2.1.1 Fixed Weight Sum Method

Let us first introduce the weight sum method formally. If the designer can provide m relative importance weights for m objectives, the MOP (Eq. 6.1) can be transformed as

$$\begin{aligned} \min \quad & \sum_{i=1}^{m} \omega_i z_i = \sum_{i=1}^{m} \omega_i f_i(\mathbf{x}) \\ \text{s.t.} \quad & g_i(\mathbf{x}) \le 0, \quad i = 1, \cdots, q \\ & h_j(\mathbf{x}) = 0, \quad j = q+1, \cdots, k \\ & \mathbf{x} \in \mathbb{R}^n \end{aligned} \tag{6.8}$$

where $\omega_i \in [0,1]$ is the weight of objective i assigned by the user in advance.[11] That is the model of the *fixed weight sum method*.

6.2.1.2 Random Weight Sum Method

But actually the fixed weight sum method requires too much from the decision maker, who generally cannot provide the overall relative importance weight for all objectives.[12] In addition, only one nondominated solution can be found by successfully optimizing Eq. 6.8,[13] which makes the second purpose of researching MOPs hard to fulfill.

This is where the *random weight sum method* comes in. Suppose we have no idea of any preference and we also want to use the weight sum method to explore the Pareto front. We could do the following transformation:

$$\omega_k = \frac{r_k}{\sum_{j=1}^{m} r_j}, \quad k = 1, 2, \cdots, m \tag{6.9}$$

where $r_j \sim U(0,1)$ is the jth uniformly distributed random number. The MOP could be transformed into a single-objective problem by substituting Eq. 6.9 into Eq. 6.8.

6.2.1.3 Adaptive Weight Sum Method

Please stop for a few minutes before continuing with this subsection by considering why we need other methods after the random weight sum method has been introduced.

The main reason is scale. Consider this situation. There are two objectives. Objective 1 is for cost and its value is in the interval $[10^3 \sim 10^6]$. Objective 2 is for the failure rate and its value is in the interval $[0.01 \sim 0.85]$. If we use Eqs. 6.8 and 6.9 to do the optimization, what is the expected result? The main focus is on objective 1 and there is little concern for objective 2, which makes the exploration process for objective 2 incomplete.

How can we improve the exploration ability of the random weight sum method? What about assigning the weight according to the objective function value interval? How do we define the objective function value interval before the optimization process? It is difficult.[14] But EAs are population-based algorithms that contain a group

---

[11] The user of the fixed weight sum method needs to assign normalized m weights that satisfy $\sum_{i=1}^{m} \omega_i = 1$. Readers are encouraged to think why there is an additional requirement.

[12] Readers are urged to reread Sect. 3.5.1 for parameter tuning and Sect. 4.3 for the static penalty function.

[13] Why is the optimal solution of Eq. 6.8 an efficient solution?

[14] If we know the range for all objective values, we can normalize them to the range $[0,1]$ and assign weights on objectives for the fixed or random weight sum method.

of individuals. The population of the current generation represents some kind of sampling over the solution space. So we can get the approximate objective function value interval according to the current population. This is where the definition of the *ideal point* comes in.

For m dimensions of objective space, if we can find a point with the smallest value in all dimensions, it is absolutely the only nondominated point. But these ideal things always fail in real situations.[15] Thus we can only define the virtual ideal points. In a group of points, the virtual point $\mathbf{z}^+ = (z_1^+, z_2^+, \cdots, z_m^+)$ constructed by $z_k^+ = \min\{f_k(\mathbf{x})|\mathbf{x} \in S\}, k = 1, 2, \cdots, m$ is called the *ideal point* or *positive ideal point*. Conversely, in a group of points, the virtual point $\mathbf{z}^- = (z_1^-, z_2^-, \cdots, z_m^-)$ constructed by $z_k^- = \max\{f_k(\mathbf{x})|\mathbf{x} \in S\}, k = 1, 2, \cdots, m$ is called the *negative ideal point*.[16]

In every generation, there is one positive ideal point $\mathbf{z}^+$ and one negative ideal point $\mathbf{z}^-$. Considering the aforementioned scale problem, the following normalized weight sum scheme is natural:

$$z = \sum_{k=1}^{m} \frac{z_k - z_k^+}{z_k^- - z_k^+} \tag{6.10}$$

where z_k is the kth objective value of $\mathbf{z}$. Using Eq. 6.10 as the optimization objective, we obtain the *adaptive weight sum method*.

If there really exists a positive ideal point, its normalized objective is 0. Similarly, the normalized objective for a negative ideal point is m. In this way, every individual can be mapped into the interval $[0, m]$. The smaller z is, the better $\mathbf{z}$ is.

6.2.2 Compromise Method

Let us consider a new idea. What about evaluating how far an individual is from the positive ideal point? This would be interesting.

To accomplish this, we need to define a *regret function* ($r(\mathbf{z})$) to describe how bad the current individual is. The better (closer to the positive ideal point), the smaller (regret function value).

$$r(\mathbf{z}) = \left\|\mathbf{z} - \mathbf{z}^+\right\| = \sqrt{\sum_{k=1}^{m} \left(z_k - z_k^+\right)^2} \tag{6.11}$$

Then the designer can normalize the expression in the following way:

$$eval(\mathbf{z}) = \frac{r_{\max} - r(\mathbf{z}) + \gamma}{r_{\max} - r_{\min} + \gamma} \tag{6.12}$$

---

[15] If a MOP has a real ideal point, in the general meaning, the model for the MOP is not adequate because there are no conflicts between objectives, which is the essence of MOP modeling.

[16] Remember we deal with minimum problems in this chapter. So here positive means "good direction" and negative means "bad direction."

where $eval(\mathbf{z})$ is the fitness value of $\mathbf{z}$, $r(\mathbf{z})$ is the regret function value of $\mathbf{z}$, $r_{\max}$ and $r_{\min}$ are the maximum and minimum regret function values in the current population, respectively, and γ is a small number to avoid the divide-by-zero error. Equation 6.12 is the fitness function of the *compromise method*. The fitness of the best individual is 1, and that of the worst individual is close to 0, which changes the model into a maximum problem.

Even though there seems to be no need for a preference to be provided by the decision maker, readers are encouraged to consider why the adaptive weight sum method and compromise method are classified as preference-based approaches.

6.2.3 Goal Programming Method

Here we consider another situation. The designer cannot provide the relative importance for every objective. But she can provide the goal for every objective, which could transform the MOP into a single-objective problem in a different way.

Let us revisit the "failure rate vs. cost" problem. If the designer realizes she needs to make a tradeoff between two objectives, she may think that it is good to make the cost close to 4.0 (the unit is unrelated) and the failure rate close to 0.3, i.e., she sets the "goal" for programming. Not the smaller the better, but the closer to the designated goal, the better. In this way, the designer provides the preference information too.

If the goal of the objective k is f_k^*, we can define the positive and negative deviations as follows:

$$\delta_k^+ = \begin{cases} f_k - f_k^*, & f_k \geqslant f_k^* \\ 0, & f_k < f_k^* \end{cases} \tag{6.13}$$

$$\delta_k^- = \begin{cases} 0, & f_k \geqslant f_k^* \\ f_k^* - f_k, & f_k < f_k^* \end{cases} \tag{6.14}$$

Then the MOP described by Eq. 6.1 can be transformed into the following single-objective problem:[17]

$$\begin{aligned} \min \quad & \sum_{k=1}^{m} \left(\delta_k^+ + \delta_k^-\right) \\ \text{s.t.} \quad & g_i(\mathbf{x}) \leqslant 0, \quad i = 1, \cdots, q \\ & h_j(\mathbf{x}) = 0, \quad j = q+1, \cdots, k \\ & \mathbf{x} \in \mathbb{R}^n \end{aligned} \tag{6.15}$$

Optimizing Eq. 6.15 properly could result in locating a nondominated point in Pareto front, which reflects the designer's preference.

[17] Proofs to the correction of the transformation can be found in specific books on goal programming.

6.3 Vector-evaluated Genetic Algorithm

The initial idea of applying EAs to MOP came from Schaffer's *vector-evaluated genetic algorithm* (VEGA) suggested in 1985 [1].

VEGA is straightforward: dividing the population into m subpopulations, each of which evolves toward a single objective. The whole algorithm can be illustrated by Fig. 6.4.

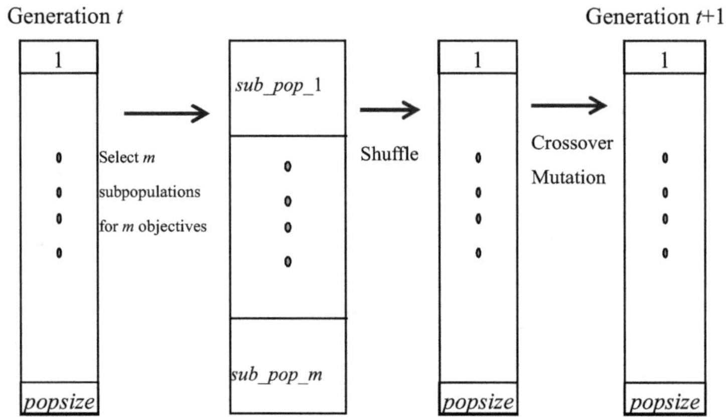

Fig. 6.4 VEGA

VEGA seems amazing at first glance. It evolves the objectives in a parallel way. The selection process in Fig. 6.4 could be any of those introduced in Chaps. 2 and 3 because only one objective needs to be considered. The shuffle procedure is quite necessary when crossover and mutation are carried out from individual 1 to *popsize*. In this way, individuals from different subpopulations are to exchange information, which promotes the exploration of the objective space.

Unfortunately, the result of VEGA is not good. Let us take a simple two-objective problem as an example. The model is illustrated by Eq. 6.16:

$$\begin{aligned} &\min\ z_1 = x^2 \\ &\min\ z_2 = (x-2)^2 \\ &\text{s.t.}\ \ x \in R \end{aligned} \tag{6.16}$$

Running VEGA with the following parameters results in the final population distribution in Fig. 6.5, $maxgen = 500$, $popsize = 100$, $p_c = 1$, $p_m = 0$.

Why did VEGA fail in such a simple MOP? Readers are encouraged to consider this question deeply before jumping to the next section.

Although VEGA is not good at directly distributing individuals along PF*, it can be utilized in other ways. The following sections will discuss this issue .

6.4 Considerations for Designing Multiobjective Evolutionary Algorithms

A good MOEA for MOPs should at least satisfy the following three requirements.

1. **Convergence.** MOEAs should have a convergence mechanism so that they can find the PF* as soon as possible. In single-objective problems, the search direction is clear. But in MOPs, a good algorithm should search the objective space in a different way.
2. **Distribution.** MOEAs should try to distribute their individuals as evenly as possible along PF* so as to provide more nondominated solutions. In single-objective problems, the selective pressure and the genetic drift will ultimately converge the population into a single optimal solution. But how can a MOEA maintain the distribution?
3. **Elitism.** The importance of elitism was discussed in Chap. 3. In a MOP environment, which individuals need to be archived? How many of them should be archived? Is there an insert/replacement mechanism? All these questions need to be pondered before designing and analyzing a MOEA.

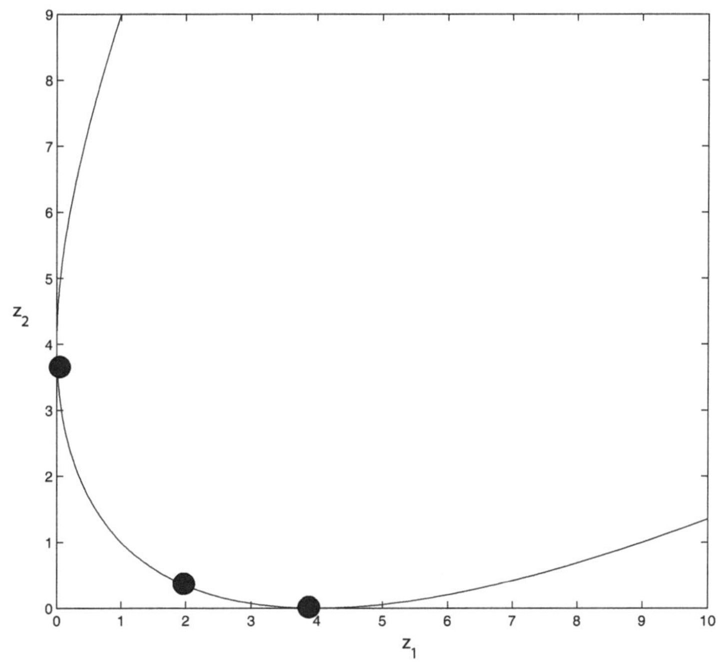

Fig. 6.5 Final population of VEGA for model Eq. 6.16

In Sect. 6.1, we showed that generally there are many nondominated points in the objective space in MOPs. So the elitism mechanism for MOEAs should contain a set of nondominated individuals found thus far, often called the *archive* in MOEAs. The variation operators, i.e., crossover and mutation, of MOEAs are the same as single optimization EAs. The main difference is the selection and replacement process for both the population and the archive.

6.4.1 Quality

For the aforementioned reasons, different measurement approaches for convergence should be designed for MOPs.

Let us review the selection process of SGA first. If we do not transform the multiple objective values into one, we cannot use proportional selection. But ranking and tournament selection can also be used if we define the rank and the comparison rule according to MOP characteristics. We can categorize those approaches roughly into two groups.

6.4.1.1 Quality Measure Approaches Considering Global Information

Goldberg suggested a way to rank a population: the *Pareto rank method* [2], which can be implemented easily.

In Pareto rank, all individuals need to be compared with others using a Pareto dominance concept to determine the nondominated solutions in the current population. Those individuals are given rank 1. Then remove rank 1 individuals from the population and determine the nondominated solutions in the remaining individuals and give them rank 2. Do the procedure until all individuals have been assigned a rank number.[18] Figure 6.6 illustrates the results of many different quality measure approaches introduced in this chapter using a two-objective problem. The first number represents the point's Pareto rank.

Let us discuss the two points with Pareto rank 2 in Fig. 6.6. Do we want to distinguish them further? The point above is dominated by four rank 1 individuals, but the point below is only dominated by one rank 1 individual. Even though these two points cannot be compared in the Pareto dominance concept, we could assign different quality signs for them. This is the Fonseca and Fleming's idea [3]: *MOP rank*.

The implementation is not sophisticated. All individuals in the current populations should be compared with others using the Pareto dominance concept. So for every individual, we know how many individuals in the current population dominate it. The fewer, the better. Then we assign the nondominated individuals rank 1 and

[18] How does one evaluate the Pareto dominance relationship between two individuals, how does one get the nondominated solutions in the current population, and how does one get the rank for every individual in a programming environment?

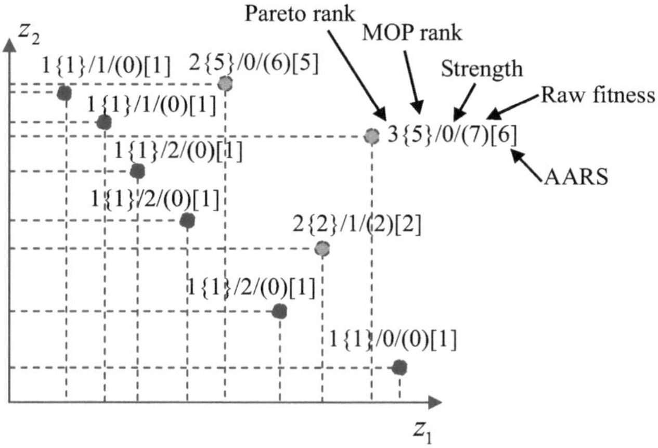

Fig. 6.6 Results of different rank-based quality measure approaches

other individuals' rank equals 1 plus the number of individuals who dominated it. In Fig. 6.6, the number in { } illustrates the Fonseca and Fleming's rank result.[19]

Many other quality measure approaches will be introduced in later sections. We will refer to Fig. 6.6 again and again to augment your understanding of the quality of individuals.

6.4.1.2 Quality Measure Approaches Considering Local Information

In Chap. 3, we introduced tournament selection. As the binary tournament selection, two individuals are randomly selected from the current population, the better individual (according to the fitness value) wins the tournament and is taken as a selection result. Then those two individuals are put back (with replacement). The procedure continues until *popsize* individuals have been selected.

But in MOP, how do we compare the two randomly selected individuals? It is quite easy to determine the tournament winner if one dominates another, but what if neither dominates the other? More considerations are necessary. Horn *et al.* suggested the *Pareto competition* method to do the selection in a tournament like way [4].

Let us take a binary tournament selection of a Pareto competition as an example. Instead of comparing the two selected individuals directly, the Pareto competition method evaluates each selected individual with a comparison set, which was selected randomly from the current population with *a priori* size. There are two situations:

---

[19] Consider the situation where PF* is a straight line between $(0,1)$ and $(1,0)$ in a normalized two-objective situation and the individuals are distributed evenly above the PF*. Is there any preference in MOP rank? If so, which individuals have advantages?

1. One is dominated by the comparison set and the other dominates the comparison set. Obviously the latter defeats the former.
2. Both or neither of them is dominated by the comparison set. Then they are of the same quality. The one who resides in the uncrowded space wins the competition.

As to the evaluation of crowdedness, there are many ways to do this. The niche count described in Chap. 5 is one option, which was adopted by Horn. Situation 2 is illustrated by Fig. 6.7. The circle represents the species surrounding the individual. The more individuals there are of the comparison set in the circle, the more crowded the competing individual is.[20]

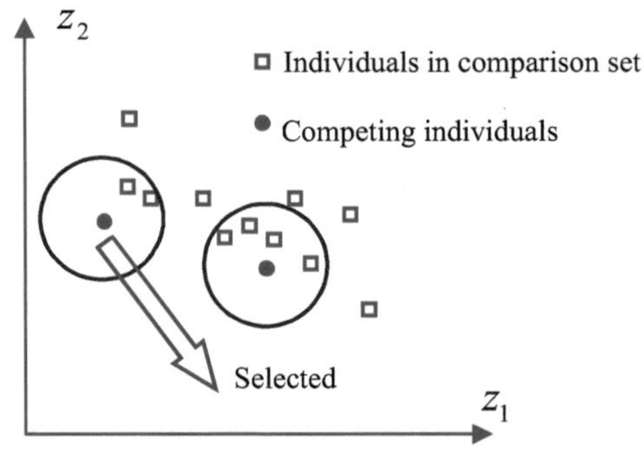

Fig. 6.7 Quality measure result by Horn *et al.*'s Pareto competition method

6.4.2 Distribution

In the selection and the replacement process of MOEAs, the convergence toward PF* is, of course, the first priority. Then the distribution might be considered in the following conditions.

- Select an individual from the archive, which contains the nondominated individuals thus far. A less crowded nondominated individual should gain some advantage.
- Select an individual from those belonging to the same rank, i.e., a Pareto dominance attribute that will be discussed later, of the population. The least crowded one in that rank is preferred.
- Update the archive with a newly generated individual that is nondominated by individuals in the archive. Because there is a content limitation for the archive, the most crowded one in the archive might be replaced by newly generated nondominated individuals with less extent of crowdedness.

[20] According to the discussion in Chap. 5, generally we need a niche radius parameter to get the niche count, which requires more computation resources and limits the application of Pareto competition.

As for the distribution of the population, the Pareto competition method provides one possibility. In addition, there are also two ways to measure the distribution during evolution.

6.4.2.1 Distribution Measure Approaches Using Individual Numbers

One way to measure the distribution (or the crowdedness) of an individual is to divide the objective space into many *hyperboxes* or *hypercubes* and count the number of individuals in the same hyperbox. Corne *et al.* suggested this approach in 2000 to evaluate the extent of proximity in the archive [5]. In two-objective condition, the hyperbox is degenerated into a rectangle (Fig. 6.8). The numbers, called by Corne and Knowles the *squeeze factor*, represent how many nondominated individuals in the archive are in one hyperbox. The smaller the number is, the more uncrowded the hyperbox is. If we want to select one nondominated individual from the archive, we can use the binary tournament selection. The one with the smallest squeeze factor wins the tournament. If a newly generated individual is nondominated by the individuals in the archive, we can calculate the squeeze factor for the new one. If it is smaller than the largest one in the archive, the new individual replaces the most crowded one in the archive. In this way, the algorithm can "squeeze" the overcrowded individuals from the archive.

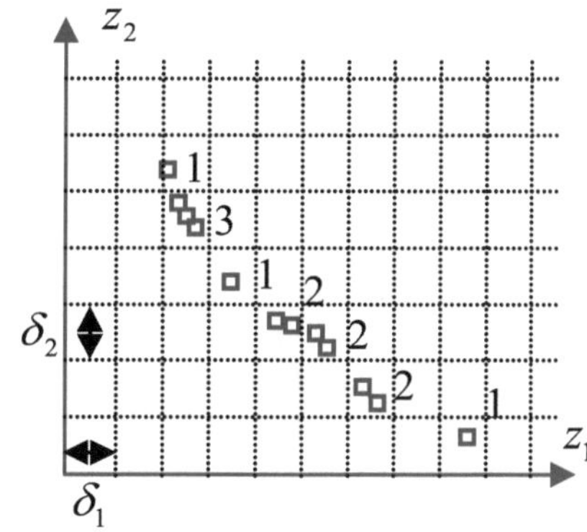

Fig. 6.8 Distribution measure result by Corne and Knowles's hyperbox method

Knowles and Corne suggested an adaptive way to control δ_1 and δ_2 [6]. It is also necessary to mention that δ_1 and δ_2 need not be the same and unchanged.

6.4.2.2 Distribution Measure Approaches Using Distance

In 1996, Osyczka and Kundu use the smallest distance between one individual and the archive to evaluate the extent of proximity of a new individual [7]. After getting the nondominated solutions in the current population, if we need to evaluate the distribution of the new individuals during the evolving procedure, one simple way is to calculate the smallest *modified distance* to the nondominated solutions (Eq. 6.17).[21]

$$d(\mathbf{x}) = \min_j \sqrt{\sum_{k=1}^{m} \left(\frac{f_k^j - \varphi_k(\mathbf{x})}{f_k^j} \right)^2} \tag{6.17}$$

where f_k^j is the kth objective value of the nondominated solution j, and $\varphi_k(\mathbf{x})$ is the kth objective value of individual $\mathbf{x}$.[22] The one with the larger $d(\mathbf{x})$ is preferred.

In 2002, Deb *et al.* suggested a way to measure the distribution without any parameters: *crowding distance* method [8].

When comparing the crowdedness in the same Pareto rank, we give the sequence for the individuals in that rank in ascending order according to the fitness of objective k and let $f_k^{[j]}$ represent the fitness value of individual j in the sequence. Then the crowdedness of individual j in dimension k in that rank can be expressed by Eq. 6.18:

$$c_k^{[j]} = \frac{f_k^{[j+1]} - f_k^{[j-1]}}{f_k^{\max} - f_k^{\min}} \tag{6.18}$$

where $f_k^{\max}$ and $f_k^{\min}$ represent the maximum and minimum values in objective k, respectively.[23] Individual i in a Pareto rank has m values for m objectives according to Eq. 6.18. So one can simply add them up to represent the overall crowdedness, crowding distance, of this individual (Eq. 6.19).

$$c^i = \sum_{k=1}^{m} c_k^i \tag{6.19}$$

where c_k^i is calculated by Eq. 6.18 and c^i is the crowding distance of individual i.

Figure 6.9 represents the two-objective condition. All the points in Fig. 6.9 are in the same rank and we have already done the normalization so that the maximum distance in both objectives is 1. The crowding distance of individual i (represented as a square) is exactly half of the normalized perimeter of the rectangle surrounding it. The smaller the value is, the more crowded it is to others.

Do you have any questions on the above discussion? Isn't it an amazing idea? Yes, but more careful considerations is needed. How about the *edge points*, i.e., the top left point and the bottom right point? There is only one individual larger or smaller than the edge point in the specific objective. You may argue that we can just assign the one normalized distance to represent its crowdedness. Yes, but on the other hand, edge points are valuable in exploring the PF* because they have more

---

[21] Why don't they use the smallest distance to all the other individuals in the current population? Why is it called *modified*?

[22] Readers are encouraged to consider why there is an f_k^j in the denominator.

[23] Why is there a $f_k^{\max} - f_k^{\min}$ term in the denominator?

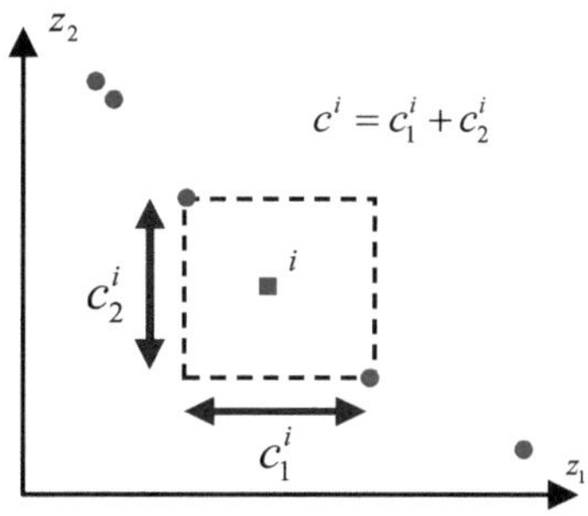

Fig. 6.9 Distribution measure result by Deb *et al.*'s crowding distance method

chances to generate new nondominated solutions by crossover and mutation. So edge points can represent the spread ability of the MOEA, which will be discussed in later sections. Thus generally, nondominated edge points need to be preserved in a special way. Deb *et al.* assigned the infinite crowding distance for edge points.

Up to now, we have introduced all the necessary elements for designing an efficient MOEA.[24] Let us now analyze several famous MOEAs to see how these elements are combined.

6.5 Classical Multiobjective Evolutionary Algorithms

6.5.1 Nondominated Sorting Genetic Algorithm II

Srinivas and Deb suggested their *nondominated sorting genetic algorithm* (NSGA) in 1994 by combining Goldberg's Pareto rank method and fitness sharing. After Pareto ranking, every individual in the same rank r gets the same dummy fitness value f_r $(f_1 > f_2 > \cdots)$ [9]. Then individuals share their fitness with others in the same rank. In this way, separated lower rank (good quality in Pareto dominance) individuals gain a selection advantage, which pushes NSGA toward PF* with good distribution.

Deb *et al.* suggested *NSGA-II* in 2002 to improve its predecessor in three ways: improving the Pareto ranking procedure so it had less time complexity, adding the elitism mechanism, and eliminating the requirement for the niche radius niche by the new crowding distance method [8]. We will not discuss the first improvement

[24] We need to mention that all the techniques discussed in Sect. 6.4 provide effective elements for guiding the population of MOEAs toward PF* evenly. In Sect. 6.7, we will discuss the problem of evaluating the results of a MOEA considering convergence and distribution. They are two different topics.

for page limitation.[25] The crowding distance method was discussed in Sect. 6.4.2. So we will focus on the elitism mechanism.

The importance of elitism in MOEA was discussed in Sect. 6.4. In NSGA-II, the capacity of the archive is *popsize*. The solution process of one generation in NSGA-II is as follows.[26]

One Generation of NSGA-II

Phase 1: Assigning Pareto rank and crowding distance.

Step 1.1: Combine the population $P(t)$ ($popsize$) and the archive $A(t)$ ($popsize$) to get $2 \times popsize$ individuals.

Step 1.2: Assign each individual a Pareto rank.

Step 1.3: Calculate the crowding distance for each individual.

Phase 2: Generating the new archive $A(t+1)$.

Step 2.1: Insert the individuals into $A(t+1)$. The individuals in rank 1 should be inserted first, then rank 2 ... If rank r cannot be fully inserted into $A(t+1)$, then insert individuals in descending order of the crowding distance until $A(t+1)$ is full with $popsize$ individuals.

Phase 3: Generating the new population $P(t+1)$.

Step 3.1: Select from $A(t+1)$ using binary tournament selection to form a mating pool. If two individuals in $A(t+1)$ have different ranks, the one with the lower rank wins the tournament. Or the one with the same rank but larger crowding distance wins the tournament.

Step 3.2: Generate the new population $P(t+1)$ by crossover and mutation from the mating pool. In standard NSGA-II the crossover operator is SBX and the mutation operator is polynomial mutation, which were introduced in Sect. 3.2.2.2 and 3.2.2.3, respectively.

Figure 6.10 illustrates the evolving process in one generation of NSGA-II.[27]

Suppose individual j reside in PF* and far from others. From the solution process and Fig. 6.10 above we can see that individual j will never be lost. It will always be in rank 1 with a large crowding distance, which means it will absolutely reside in $A(t+1)$. How about the crossover and the mutation that will destroy it? It does not matter because in the next generation, $A(t+2)$ will be selected based on the union of $A(t+1)$ and $P(t+1)$ and individual j is in $A(t+1)$. So it is necessary to mention

[25] The first improvement is concerned with how to find a Pareto rank for each individual efficiently. We omit this part to attract readers to theoretical issues at first and encourage them to read [8] for this part before implementation.

[26] The C language implementation of NSGA-II can be downloaded at: `http://www.iitk.ac.in/kangal/codes.shtml`.

[27] In the original paper, Deb *et al.* did not explicitly label the two groups of individuals as population and archive. Here we denote them in this way for pedagogical reasons.

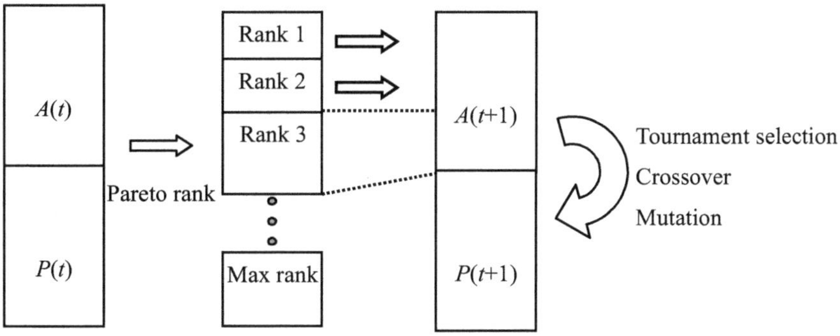

Fig. 6.10 The evolving process in one generation of NSGA-II

that NSGA-II can be regard as a kind of $(popsize + popsize)$-ES with rapid Pareto ranking and crowding distance, even though it is called a GA.[28]

The convergence, distribution, and elitism mechanisms in NSGA-II are Pareto rank and tournament selection, the crowding distance, and the introduction of the archive A, respectively. In this way, NSGA-II establishes an elaborate balance in the three requirements and requires no additional parameters besides SGA.

6.5.2 Strength Pareto Evolutionary Algorithm 2 and Pareto Envelope-based Selection Algorithm

6.5.2.1 Strength Pareto Evolutionary Algorithm 2

Zitzler *et al.* suggested a *strength Pareto evolutionary algorithm* (SPEA) in 1999 and improved it into *SPEA2* in 2001 [10].

Let us suppose we want another efficient MOEA starting from NSGA-II.[29] Tasking another group of individuals, an archive, with storing the Pareto solution up till now is very fascinating. But why does the size of the archive need to be the same as that of the population? In addition, in NSGA-II, rank is considered twice, as is crowding distance (steps 2.1 and 3.1). Can we shorten them into one?

Based on these two questions, the evolving process in one generation of SPEA2 can be illustrated by Fig. 6.11.[30] $A(t)$ represents the archive at generation t with size $\bar{N}$, and $P(t)$ represents the population at generation t with size N.

---

[28] We would like to remind readers again that sometimes an algorithm that combines many advanced searching techniques is hard to classify.

[29] We need to mention that all these discussions are based on a pedagogical purpose, not the same developing procedure as the real development of SPEA2.

[30] A C language implementation of SPEA2 can be downloaded at: `http://www.tik.ee.ethz.ch/sop/pisa/`.

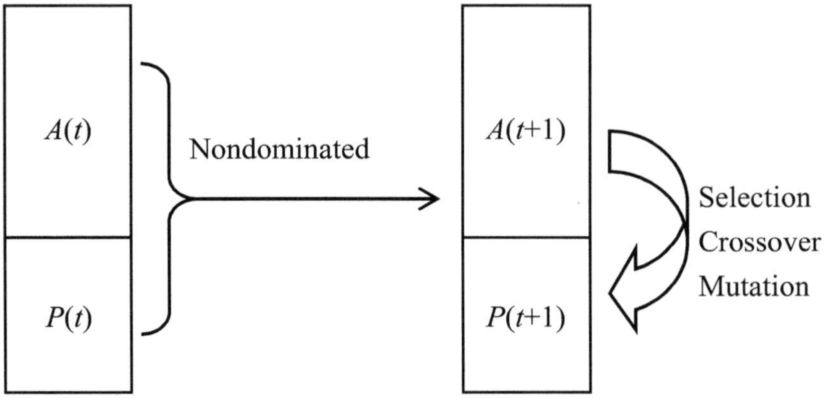

Fig. 6.11 The evolving process in one generation of SPEA2

Two questions might arise in consideration of Fig. 6.11:

1. What is the mechanism of selection? How can we combine the convergence requirement and the distribution requirement into one selection process?
2. What is the mechanism of archive updating? Specifically, what if the number of nondominated solutions in the union of $A(t)$ and $P(t)$ is not equal to $\bar{N}$?

Let us consider the first question first. In SPEA2, every individual has been assigned a number to describe its "*strength*" as follows:

$$S(i) = |j|(j \in P + A) \cap i \prec j| \tag{6.20}$$

where $|\,|$ is the function of cardinality, i.e., the number of elements in one set. According to Eq. 6.20, an individual's strength is the number of individuals it dominates in the union of A and P. The larger the number is, the stronger the individual is. In Fig. 6.6, the number in $//$ illustrates the strength of each individual.

But just comparing the strength of every individual will lead to selection bias.[31] Zitzler *et al.* defined another term, *raw fitness*, to describe how good an individual is of convergence.[32] An individual's raw fitness is defined as follows:

$$R(i) = \sum\nolimits_{j \in P+A,\ j \prec i} S(j) \tag{6.21}$$

[31] Why?

[32] "Raw" here means the fitness needs to be modified later. Another thing that needs to be mentioned is that here fitter does not mean larger. Only with proportional selection is a larger fitness value to be the better one.

If an individual i is the nondominated solution in the union of A and P, it needs to be assigned the best raw fitness (such as zero).[33] In Fig. 6.6, the number in () illustrates the raw fitness of each individual.

As regards distribution, in SPEA2, one must know the distances of every individual to all other individuals. Then the distances of one individual are ranked in ascending order. Zitzler *et al.* defined the term *density* to describe the crowdedness of individual i as follows:[34]

$$D(i) = \frac{1}{\sigma_i^k + 2} \tag{6.22}$$

where σ_i^k is the kth shortest distance to individual i, i.e., the kth number in the distance rank.[35] The farther an individual is away from others, the larger σ_i^k is, and the smaller $D(i)$ is. Zitzler *et al.* suggested that $k = \sqrt{N + \bar{N}}$ might be a good parameter option.

Finally, every individual is granted fitness by Eq. 6.23, which constitutes the basis of the binary tournament selection in Fig. 6.11.

$$F(i) = R(i) + D(i) \tag{6.23}$$

For nondominated individual i, $R(i) = 0$ and $D(i) < 1$. So the MOP has been transformed into a single-objective minimum problem in this way.

Now we consider how to make decisions when the number of nondominated solutions in the union of $A(t)$ and $P(t)$ is not equal to $\bar{N}$. Zitzler *et al.*'s answer was straightforward. Let $B(t)$ represent the nondominated set selected from the union of $A(t)$ and $P(t)$ and $|B(t)|$ represent its cardinality. If $|B| < \bar{N}$, we add more dominated individuals to fill in the gap according to fitness described by Eq. 6.23 until $|B| = \bar{N}$. If $|B| > \bar{N}$, we delete those individuals from $B(t)$ with the smallest distance to others (σ_i^1), then delete individuals with the smallest second distance to others (σ_i^2)... until $|B| = \bar{N}$.

From Fig. 6.6 we can see that for dominated individuals, their raw fitness values are several, dozens, or hundreds, large enough to neglect their density, which is less than 0.5. So SPEA2 directs the search toward the PF*. For nondominated individuals, their raw fitness values are all 0, so sparser individuals, those with smaller densities, gain the advantage.

Suppose individual j resides in PF* and far from others. From Fig. 6.11 we can see that individual j will never be lost. So the elitism mechanism in SPEA2 works fine.

The convergence, distribution, and elitism mechanisms in SPEA2 are raw fitness assignment, density, and the introduction of the archive A, respectively.

---

[33] Consider the situation where PF* is a straight line between $(0,1)$ and $(1,0)$ in a normalized two-objective situation and the individuals are distributed evenly above the PF*. Is there any preference in strength and raw fitness? If so, which individuals have advantages?

[34] "Density" here describes the crowdedness of an individual. From its literal meaning, if we prefer evenly distributed individuals, the density of the individual should always be small.

[35] Try to understand why we do not directly use the smallest distance?

Recently nearly every new published paper on MOEA design and improvement has found it necessary to compare the suggested algorithm at least with NSGA-II and SPEA2 on MOEA benchmark problems, which actually makes these two algorithms the benchmark algorithms in MOEAs.

6.5.2.2 Pareto Envelope-Based Selection Algorithm

Corne *et al.* suggested a *Pareto envelope-based selection algorithm* (PESA) in 2000 [5]; a similar idea is illustrated in Fig. 6.12.[36] Compared to SPEA2, PESA has different considerations in answering the two questions raised on page 212.

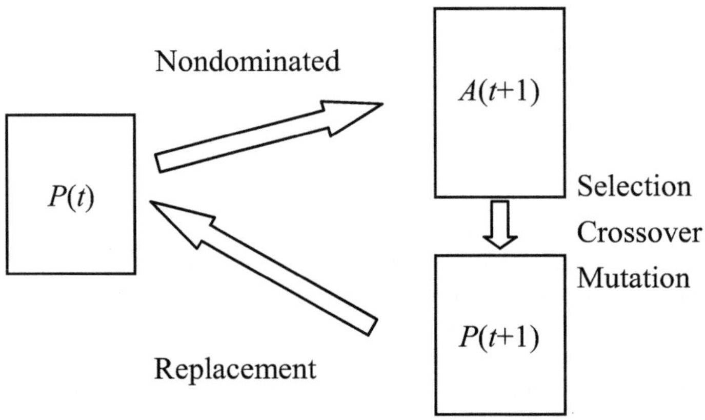

Fig. 6.12 The evolving process in one generation of PESA

Before answering the two questions, we need to distinguish the diversity mechanisms of NSGA-II, SPEA2, and PESA. As was discussed above, NSGA-II uses the crowding distance and SPEA2 the density function. In PESA, the hyperbox method and squeeze factor concept are used (Fig. 6.8).

For the first question, PESA uses binary tournament selection to generate new population from the archive. The archive in PESA *only* contains the nondominated solutions found thus far. They are not comparable with respect to objectives. So the one with the smaller squeeze factor, i.e., the one residing in the less crowded hyperbox, wins the tournament.

For the second question, the archive-updating mechanism, if a new individual is nondominated in both the population and the archive and the archive is full, select the individual in the archive with the largest squeeze factor to be replaced by the new one.

[36] We guess that the meaning of "envelope" in the name of the algorithm is that Corne *et al.* consider the archive as an envelope to store the nondominated individuals found thus far, which is different from SPEA2 and NSGA-II.

In 2001, Corne *et al.* developed PESA by suggesting a region-based selection. Then the algorithm was named *PESA-II* [11]. PESA-II differs only in the selection mechanism. Every hyperbox has its own squeeze factor. The hyperbox with the smallest squeeze factor will be selected first and then a randomly chosen individual is selected. Corne *et al.* claimed that region-based selection could ensure a good distribution along PF*.

6.5.3 Pareto Archived Evolution Strategy

All the above-discussed algorithms are based on GAs, this subsection discusses ES.

The evolution strategy was introduced in Chap. 2. Here we briefly review $(1+1)$-ES. We only have a current individual ($c(t)$) and an updated individual ($u(t)$). $c(t)$ evolves (mutation) into $u(t)$.[37] If $u(t)$ is better than $c(t)$, it replaces $c(t)$ and becomes $c(t+1)$. Else, $c(t)$ remains unchanged and becomes $c(t+1)$.

How can $(1+1)$-ES be developed into a MOP? First, there should be an archive containing the nondominated individuals. And special considerations are necessary in the replacement procedure. Knowles and Corne proposed a *Pareto archived evolution strategy* (PAES) in 2000 to expand ES to solve MOPs [6].

In $(1+1)$-PAES, there are three groups: one current individual $c(t)$, one updated individual $u(t)$, and one archive A containing all the nondominated individuals found thus far. First, $c(t)$ evolves (mutation) into $u(t)$.

For $u(t)$, there are two targets: try to be the current individual and try to be added to the archive.

If $u(t)$ is bad (is dominated by $c(t)$ or A), it has failed and $c(t)$ is used again to generate a new updated individual.

For the good $u(t)$ (not dominated by $c(t)$ and A), if any of the following conditions is satisfied, it enters the archive: (1) $u(t)$ dominates $c(t)$, (2) the archive is not full, (3) there is at least one individual x in A which is dominated by $u(t)$, or (4) $u(t)$ and A are nondominated but there is at least one individual x in A whose crowdedness is larger than that of $u(t)$. In the third condition, $u(t)$ replaces any of its dominated ones and in the fourth condition, it replace the most crowded one.

For the good $u(t)$ (not dominated by $c(t)$ and A), if any of the following conditions is satisfied, it replaces $c(t)$: (1) $u(t)$ dominates $c(t)$ or (2) the crowdedness of $c(t)$ in A is larger than that of $u(t)$. The crowdedness evaluation in PAES is the squeeze factor, introduced by Fig. 6.8.

In all, PAES ensures that the nondominated solutions residing in an uncrowded location will survive.

Starting from $(1+1)$-PAES, Knowles and Corne also discussed $(1+\lambda)$-PAES and $(\mu+\lambda)$-PAES [6].

[37] The evolution from $c(t)$ to $u(t)$ is discussed in Chap. 2.

6.5.4 Micro-GA for Multiobjective Optimization

Goldberg suggested the idea of *micro-GA* in 1989 [12]. Micro-GA has a small population size (e.g., 3) and small maximum generation number (e.g., 4). By the end of the micro-GA, the population reaches the status of *nominal convergence*, and all individuals are identical or similar. Then the best individual in the final generation is used as one of the initial individuals for the next micro-GA, and other initial individuals are generated randomly. Goldberg claimed that a micro-GA evolved in this way might search the solution space more efficiently.

All the MOEAs discussed above need a global dominance relationship (rank or at least the nondominated solutions in the population), which is very time consuming. Coello Coello and Pulido developed the micro-GA into the field of MOP, the *micro-GA for multiobjective optimization*, in 2001 [13]. The initial population needs to be updated by the results of a micro-GA and the archive is necessary to store the nondominated individuals. Coello Coello and Pulido divided the individuals for initialization into two groups: I_r (replaceable potential initial individuals) and I_n (nonreplaceable potential initial individuals) with a predefined percentage. I_r will be replaced by the good micro-GA results and I_n will remain unchanged since it was randomly generated, which ensures diversity.

The solution process of micro-GA for multiobjective optimization is as follows:

One Meta-Generation of Micro-GA for Multiobjective Optimization

Phase 1: Initialization.

Step 1.1: Combine I_r and I_n to form I.

Step 1.2: Select randomly from I to form the initial population for a micro-GA.

Phase 2: Evolving micro-GA until nominal convergence.

Step 2.1: Select by binary tournament selection based on the Pareto dominance relationship.

Step 2.2: Use crossover and mutation to explore the objective space.

Step 2.3: Store one nondominated individual arbitrarily and copy it intact to the following generation.

Phase 3: Updating the archive and I_r in a local way.

Step 3.1: Determine whether the best nondominated individuals from the micro-GA (1 or 2) could be inserted into the archive. If the archive is not full and the individual is not dominated by the archive, it is inserted. If the archive is full and the individual in not dominated by the archive, the individual replaces the one in the archive with a larger crowdedness than it.

Step 3.2: Determine whether the best nondominated individuals from micro-GA (1 or 2) will be inserted into I_r. Compare the nondominated individual with the randomly selected one from I_r by the Pareto dominance relationship. If the former dominates the latter, it will replace the latter.

Phase 4: Updating the archive and I_r in a global way.
Step 4.1: After every several generations, discard the individual in I_r and use the archive to fill I_r.

During the evolving process, the micro-GA will start from points getting closer and closes to PF*, which makes the micro-GA for multiobjective optimization very efficient.

The crowdedness evaluation in the micro-GA for multiobjective optimization is the squeeze factor, introduced by Fig. 6.8.

Up to now, we have discussed several classical MOEAs that satisfy the requirements suggested in Sect. 6.4. They (1) contain a convergence mechanism that allows the algorithm to approach PF*; (2) contain a distribution mechanism that allows individuals to reside in different places on PF*; (3) contain an elitism mechanism that the uncrowded nondominated solutions will never be lost in the evolving process. If you would like to design a new MOEA or improve the classical one, you need to check whether these requirements are satisfied in a balanced way.

All of these algorithms were proposed at the turn of the century, which explains the first high slope in Fig. 6.3 (from 2001 to 2003). Many new algorithms were proposed and comparisons were carried out. After 2004, research interests have been spilt out many areas. How does one provide persuasive comparison results for MOEAs? How does one adapt MOEAs to uncertain environments? How can MOEAs be used to handle constraints? All these questions make researchers strive for a better understanding of MOEAs. We will discuss these topics in the next part of this chapter. Before that, we need to mention some of the latest developments in the design of MOEAs.

6.6 Cutting Edges of Multiobjective Evolutionary Algorithms

In this section, several different research directions will be introduced to excite readers' interest in developing new algorithms, improving current ones, and combining other ideas with MOEAs. All of these techniques are quite useful in the field of algorithm design and analysis. It is necessary to mention again that the selection of the material is based on the authors' interests and the potential of the application.

6.6.1 Expanding Single-objective Evolutionary Algorithms into Multiobjective Optimization Problems

In this subsection, we will introduce three examples of expanding algorithms for single-objective problems, discussed in Chaps. 2 and 3, into MOPs.

6.6.1.1 Embedding Differential Evolution into NSGA-II

In Sect. 2.4.2, we introduced differential evolution, in which directions are used to direct the new individual generation. In standard DE, the direction is randomly generated. During the evolving process, individuals will converge toward the global optimal solution due to the selective pressure, so the randomly generated directions among individuals have some means of improving the quality.

From another aspect, in MOP, we need to quickly make individuals converge toward PF* and distribute them as evenly as possible. Why not use the information of the convergence and distribution to guide the evolving process?

One way to expand DE into MOP is to embed DE into NSGA-II. Iorio and Li proposed their embedding work in 2006 [14]. In Fig. 6.10, standard crossover and mutation are used to generate $P(t+1)$ from $A(t+1)$. Iorio and Li use DE to substitute the new-population-generating process.

Iorio and Li utilized the new-individual-generating formula “current/2” as follows:

$$\mathbf{v}_i = \mathbf{x}_i + K(\mathbf{r}_3 - \mathbf{x}_i) + F(\mathbf{r}_1 - \mathbf{r}_2) \tag{6.24}$$

where $\mathbf{x}_i$ is an individual in $A(t+1)$ and $\mathbf{v}_i$ is the new generated individual in $P(t+1)$. Classically, $\mathbf{r}_1 \neq \mathbf{r}_2 \neq \mathbf{r}_3 \neq \mathbf{x}_i$ are random samplings in $A(t+1)$. Equation 6.24 means that the new individual starts at $\mathbf{x}_i$ and has a direction and length combination of $K(\mathbf{r}_3 - \mathbf{x}_i)$ and $F(\mathbf{r}_1 - \mathbf{r}_2)$.

If we are concerned with convergence, we can assign $\mathbf{r}_3$ as any individual better than $\mathbf{x}_i$ on Pareto dominance, which means it has a lower rank, so that direction $(\mathbf{r}_3 - \mathbf{x}_i)$ might be a Pareto dominance improving direction. This is scheme 1.

If we are concerned with distribution, we can assign $\mathbf{r}_1$ and $\mathbf{r}_2$ to be with same rank. Then $(\mathbf{r}_1 - \mathbf{r}_2)$ might be the disperse direction on the same rank. This is scheme 2.

Or we can consider both aspects, i.e., assign $\mathbf{r}_3$ as any individual better than $\mathbf{x}_i$ on Pareto dominance and $\mathbf{r}_1$ and $\mathbf{r}_2$ to be with same rank. This is scheme 3.

According to the numerical experiments carried out by Iorio and Li, scheme 3 is the best one and scheme 2 has similar performances. Scheme 1 is the worst one but still better than the standard NSGA-II.[38]

It is necessary to point out that the embedding techniques introduced above can also be used with other MOEAs.

6.6.1.2 Multiobjective Scatter Search

Scatter search was introduced in Chap. 2. We demonstrate that scatter search is a kind of direction-based method that utilizes the subtraction of two solutions as the perturbation direction in an evolution episode.

[38] Could you give some intuitive explanation for the results?

The main loop of scatter search is illustrated by Fig. 2.14. Reference set (RS) stores high-quality solutions ($RefSet_1$ for objective value and $RefSet_2$ for crowdedness). With a generation procedure, subsets are generated from RS. After that, a combination procedure is carried out to form new solutions from subsets. Then the new solutions experience the local search by the improvement procedure to become better solutions. There are update rules to determine whether an improved solution could enter an RS.

How does one expand scatter search into MOP?

According to the discussion above, at the very least we need an archive (apart from $RefSet_1$) to store the nondominated individuals found thus far. In addition, several questions need to be answered:

- What is the improvement direction in MOP scatter search? In single-objective situations, it is easy: the lower the objective value, the better.
- How does one update RS? Again, multiple objective values become the focus.
- How does one maintain diversity in an archive? This is the same important question that must be answered by any MOEA designer.

Nebro *et al.* successfully answered the above questions and proposed the *Archive-based hYbrid Scatter Search* (AbYSS) algorithm in 2008 [15]. The main loop of AbYSS is the same as that of the standard scatter search illustrated by Fig. 2.14. We will just present the main differences.

The initialization of $RefSet_1$ is the same as in standard scatter search. After the random initialization and the improvement procedure, AbYSS adopts the fitness of SPEA2 (Eq. 6.23) to determine those who will be in $RefSet_1$.

In the main loop, the improvement procedure is a simple mutation-based EA, like $(1+1)$-ES. After mutation, the parent and the offspring compete according to Pareto dominance. The winner takes the place. If they are nondominated to each other and the mutant is not dominated by the archive (not the $RefSet_1$), the mutant will be inserted into the archive and replace the parent. Otherwise, discard the mutant.

If the improved solution dominates solutions in $RefSet_1$, then all dominated solutions are deleted and the improved one is inserted. If the improved solution is dominated by $RefSet_1$, it is then considered whether it is to be inserted into $RefSet_2$ with the standard procedure (distance to the closest RS solutions).

The archive is updated by the end of every loop. If the archive is not full, then the improved solutions, which are nondominated by the archive, will be inserted into the archive. If the archive is full, Nebro *et al.* use the crowding distance (Eq. 6.19 and Fig. 6.9) to determine which overcrowded archive member should be replaced by the improved one residing in the uncrowded area.

If RS does not change in the updating procedure and the stop criterion has not been satisfied, then the initialization procedure in AbYSS includes some nondominated solutions in the archive based on the standard scatter search reinitialization.

In this way, scatter search has been successfully expanded into the field of MOPs. Nebro *et al.* claimed that AbYSS obtains very competitive results compared with NSGA-II and SPEA2.

6.6.1.3 Cooperative Coevolutionary Multiobjective Optimization

The concept of coevolution was introduced in Sect. 3.7.1. Cooperative coevolution divides the full solution into several parts and promotes the one that has good collaboration relationships with others. Competitive coevolution promotes the one that has good winning percentage with others.

In 2006, Tan *et al.* proposed a cooperative coevolutionary MOEA [16]. The key problem to be resolved in designing a coevolutionary algorithm is the solution division and integration scheme. Other considerations related to MOP also need to be elaborated.

When solving a MOP with n variables, Tan *et al.* divide the population into n subpopulations, each of which contains individuals only representing one variable. They use a binary code scheme for chromosomes.

To evaluate the fitness value for newly generated individual j in subpopulation i, it needs to be combined with other $n-1$ individuals sampled from other $n-1$ subpopulations. The simplest method is to randomly sample the subpopulations, combine them into a full solution, and use its fitness values to represent individual j in subpopulation i. To accelerate the convergence, Tan *et al.* suggested another method. Every subpopulation has a representative individual with the highest fitness values determined previously. The new individual combines with other $n-1$ representative individuals to form a full solution and get its fitness values. After these two methods, the new individual will have two sets of fitness values, each of which contains m fitness values. The point on the objective space generated by the latter approach will represent the new individual unless it is dominated by the point generated by the former approach.

In Tan *et al.*'s cooperative coevolutionary MOEA, the archive is used to store the full nondominated solutions found thus far and the niche-based diversity preserving techniques are adopted. The solution process of Tan *et al.*'s cooperative coevolutionary MOEA is as follows.

Cooperative Coevolutionary MOEA

Phase 1: Initialization.

Step 1.1: Assign the parameters for the cooperative coevolutionary MOEA, such as chromosome length l, p_c, p_m, *popsize* for each subpopulation, clone number c, stop criteria, et.

Step 1.2: Generate n subpopulations, each with *popsize* uniformly distributed individuals, randomly with one variable to form the initial population and evaluate their fitness values in the random sampling way and determine the represent individual for each subpopulation. Initialize the archive as empty. $gen = 0$.

Phase 2: Main loop. Repeat the following steps until the stop criteria are satisfied. In each loop, n subpopulations are evolved in a sequential way, i.e., subpopulation 1 is first evolved and then subpopulation 2 until subpopulation n. $i = 1$.

Step 2.1: Calculate the rank for each individual in subpopulation i. The rank-assigning scheme is the same as the MOP rank introduced in Sect. 6.4.1, i.e., the rank for individual j is one plus the number of solutions in the archive that dominate j.

Step 2.2: Normalize the objective space and calculate the niche count for each individual in subpopulation i. Tan *et al.* suggested an adaptive way to estimate niche radius σ in the evolving process [17].

Step 2.3: Generate a mating pool for subpopulation i with binary tournament selection. The one with the lower rank wins the tournament. If the competitors have same rank, the one with the smaller niche count wins the tournament.

Step 2.4: Generate a new subpopulation using uniform crossover with probability p_c and bit-flip mutation with probability p_m.

Step 2.5: Calculate the fitness values for all new individuals in subpopulation i using the two methods discussed above.

Step 2.6: Update the archive in a way like PESA and PAES. For each individual, its full solution will enter the archive if it is not dominated by the archive. Any solution in the archive dominated by it will be removed. If the capacity of the archive has been reached, the one in the archive with the largest niche count will be removed to make room for the new nondominated solution.

Step 2.7: $i = i + 1$. Go back to step 2.1 until $i > n$.

Step 2.8: After all subpopulations have been evolved in one generation, clone c copies of the solution in the archive with the smallest niche count and put their components into corresponding subpopulations. This clone operator will promote exploration and exploitation over uncrowded regions. $g = g + 1$.

Phase 3: Submitting the archive as the results of an cooperative coevolutionary MOEA.

Tan *et al.* also implemented the above algorithm in a distributed environment [16], and in 2009 they expanded the algorithm into a competitive-cooperative paradigm to solve dynamic MOPs [18].

6.6.2 Archive Maintenance

The main difference between MOEAs and the algorithms introduced in Chap. 3 are selection and archive maintenance. In Sect. 6.5 we used classical MOEAs mainly to introduce the selection methods that promote convergence and diversity, while several advanced techniques about archive maintenance will be discussed here.

Generally we will predefine a capacity limitation for an archive, i.e., the number of nondominated solutions provided at the end of a MOEA.[39] In the following part of this subsection, we maintain the archive $A(t)$ with a capacity limitation of N at generation t. After a new individual c is generated by variation operators, we need to determine $A(t+1)$ using techniques introduced here.[40]

Another thing that needs to be mentioned is that these techniques are generally algorithm independent, i.e., applicable to any MOEA with an archive.

6.6.2.1 Adaptive Archive Maintenance Based on Hyperbox

In Sect. 6.4.2, we introduced the hyperbox concept, suggested by Corne *et al.*, and mentioned that Knowles and Corne suggested an adaptive way to control the size of hyperboxes. They further developed the adaptive control of the hyperbox and in 2003 suggested a technique to archive the evenly distributed nondominated solutions found thus far [20].[41]

Let us review the hyperbox in Fig. 6.8. If the user predefines that the grid numbers along z_1 and z_2 are 9 and 6, respectively, the we can adaptively redraw the hyperbox according to the current nondominated solutions in the archive in Fig. 6.13.

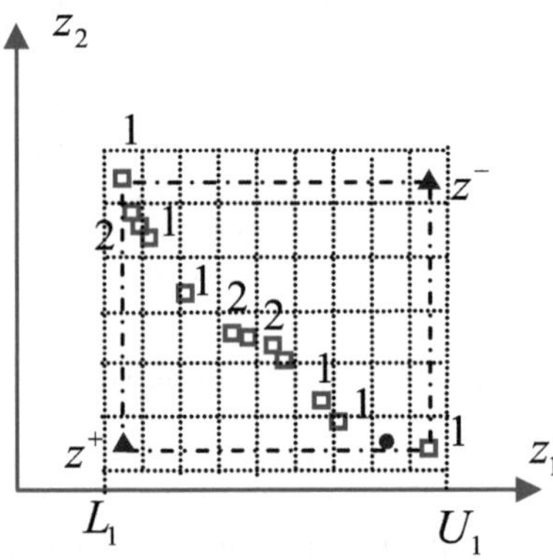

Fig. 6.13 Adaptive squeeze factor determination

We can define the positive and negative ideal points in the archive as $\mathbf{z}^+$ and $\mathbf{z}^-$, respectively, i.e., triangles in Fig. 6.13. For the ith objective, with the predefined

[39] The reason for the limited archive capacity was discussed in Sect. 6.1.3. Interested readers are referred to [19, 20] for more discussions.

[40] Thus we will discuss steady state MOEAs in this subsection, but these techniques can also be applied in generational MOEAs.

[41] This paper is the winner of the IEEE Transactions on Evolutionary Computation Outstanding Paper Award.

grid number div_i, the lower bound, L_i, and the upper bound of the adaptive grid, U_i, are defined as follows:

$$\begin{aligned} L_i &= z_i^+ - \frac{z_i^- - z_i^+}{2div_i} \\ U_i &= z_i^- + \frac{z_i^- - z_i^+}{2div_i} \end{aligned} \tag{6.25}$$

The second parts in the above equations make the edge points at the center of the outer hyperboxes. Then we can get the coordinates for every hyperbox, determine which nondominated solution belongs to which hyperbox, and calculate the squeeze factor for each hyperbox.

Whenever we get a new individual c with variation operators, the archive maintenance method is carried out with the follow logic considerations.

1. If c is dominated by at least one solution in archive $A(t)$, discard it.
2. If c dominates at least one solution in $A(t)$, all the dominated solutions in $A(t)$ are removed and c is inserted into the archive.
3. If the capacity limitation is not reached, $|A(t)| < N$, and c is nondominated with all solutions in $A(t)$, just insert the new one.
4. If $|A(t)| = N$ and c is nondominated with all solutions in $A(t)$, c is accepted with any of the following conditions: (1) it does not reside in the range of the current hyperbox, i.e., it increase the extend of the grid in at least of one objective dimension; (2) it resides in a less crowded hyperbox than some other point in $A(t)$, e.g., the circle in Fig. 6.13. If so, randomly select one solution in the hyperbox(es) with the largest squeeze factor to be replaced by the new one.

After the archive maintenance procedure, the above adaptive procedure is carried out on $A(t+1)$ again to get the new squeeze factor for each hyperbox.

The above adaptive method maintains a good balance in convergence, diversity, edge-point retention, and efficiency in the archive.

6.6.2.2 Archive Maintenance Based on ε-Dominance

In 2002, Laumanns *et al.* suggested a new definition for dominance, ε-*dominance*, to solve the tradeoff between convergence and diversity [19].

They suggested two types of ε-dominance. In objective space, point i ε-dominates point j, expressed as $i \prec_\varepsilon j$, means

$$(1-\varepsilon)z_l^i \leq z_l^j, \quad \forall l \in \{1,2,\cdots,m\} \tag{6.26}$$

where $0 < \varepsilon < 1$. Or

$$z_l^i - \varepsilon \leq z_l^j, \quad \forall l \in \{1,2,\cdots,m\} \tag{6.27}$$

where $\varepsilon > 0$ and z_l^i is the lth objective value of point i.

For a point **i** in objective space, i.e., the square in Fig. 6.14, its ε-dominant domains according to definition Eq. 6.26 are the areas on top and to the right of the dashed line (the lines are included). For a point **j** in objective space, i.e., the circle in Fig. 6.14, its ε-dominant domains according to definition Eq. 6.27 are the areas on top and to the right of the dotted line (the lines are included).

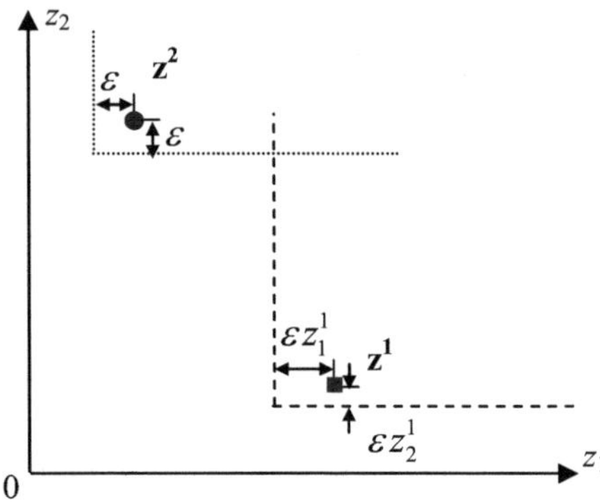

Fig. 6.14 ε-dominance suggested by Laumanns *et al.*

ε-dominance relaxes the concept of weak dominance by making a point dominate more points to the extent of ε. The difference between the two definitions is whether the relaxation is based on absolute value, Eq. 6.27, or relative value, Eq. 6.26.

Let Z be the objective space, a set $PF_\varepsilon \subseteq Z$ is called an *ε-approximate Pareto set* if any point $\mathbf{g} \in Z$ is ε-dominated by at least one point $\mathbf{f} \in PF_\varepsilon$. Those points that belong to PF_ε and are Pareto nondominated solutions of the objective constitute an *ε-Pareto set*, denoted as PF_ε^*. The ε-approximate Pareto set and ε-Pareto set are illustrated by Fig. 6.15.

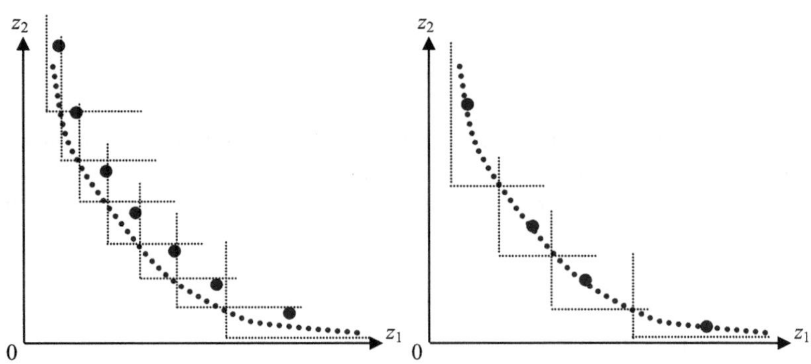

Fig. 6.15 (a) ε-approximate Pareto set, and (b) ε-Pareto set

In Fig. 6.15, we use Eq. 6.26 for the definition of ε-dominance. Thus the relaxation extent of all points is not the same. Figure 6.15a and b have the same PF*. As can be seen from these two figures, with seven solutions in PF_ε and four solutions in PF_ε^*, we can ε-dominate all the solutions in objective space, which gives us an efficient way to pursue convergence and diversity simultaneously.

Based on the above preparations, Laumanns *et al.* suggested two methods for archive maintenance.

Whenever we get a new individual $\mathbf{c}$ with variation operators. The first archive maintenance method converging to an ε-approximate Pareto set is carried out with the follow logic considerations.

1. If $\mathbf{c}$ is ε-dominated by at least one solution in archive $A(t)$, discard it.
2. If $\mathbf{c}$ is not ε-dominated by any solution in archive $A(t)$, insert c into $A(t)$ and remove solutions that are Pareto dominated by c.

Suppose we have scaled the value range for objective i into range $[1, U_i]$, then the cardinality of $A(t)$ can be calculated with predefined parameter ε, illustrated by Fig. 6.16. For the ε-dominance defined by Eq. 6.26, the largest value of ε-dominant 1 is $\frac{1}{1-\varepsilon}$, the largest value of ε-dominant $\frac{1}{1-\varepsilon}$ is $\frac{1}{(1-\varepsilon)^2}$, ... Suppose that the final n_ith point in the above sequence is U_i; we can get $\frac{1}{(1-\varepsilon)_i^n} = U_i$. So with value range $[1, U_i]$ for objective i and ε-dominance defined by Eq. 6.26, the number of grids on objective i is $n_i = \lfloor -\frac{\lg U_i}{\lg(1-\varepsilon)} \rfloor$, where $\lfloor \ \rfloor$ is for rounding toward minus infinity. The number of hyperboxes in the whole value space is $\prod_{i=1}^{m} n_i$. Thus the above archive maintenance method will keep a limited number solutions after convergence, which belong to an ε-approximate Pareto set.

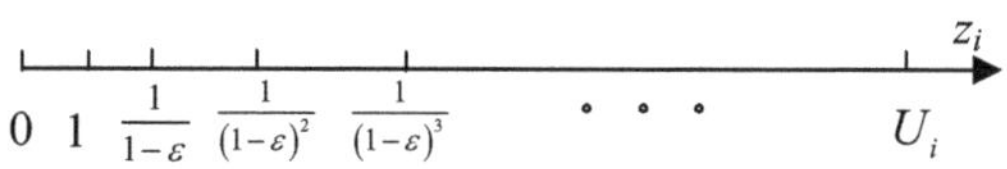

Fig. 6.16 Grid numbers on objective i with ε-dominance

Let us now discuss Fig. 6.15b in more detail. Every solution in an ε-Pareto set has a *representation point*, i.e., the lower left corner of its ε-dominated area. We can use the grid sequence to express the representation point. For the nondominated point $\mathbf{z} = (z_1, \cdots, z_m)$, in objective i, its representation point is the b_ith grid.

$$b_i = \left\lfloor -\frac{\lg z_i}{\lg(1-\varepsilon)} \right\rfloor \tag{6.28}$$

Then the representation point of solution $\mathbf{z}$ can be expressed as $\mathbf{b} = (b_1, \cdots, b_m)$. Thus PF* is occupied by several hyperboxes constructed by the ε-dominance concept.

Then the second archive maintenance method converging to an ε-Pareto set is carried out with the following logic considerations.

1. If the representation point of the newly generated individual $\mathbf{c}$ Pareto dominates the representation point of at least one solution in archive $A(t)$, add $\mathbf{c}$ and remove

those solutions whose representation points are dominated by the representation point of **c**.

2. If the representation point of **c** is not Pareto dominated by any representation point of archive $A(t)$ and **c** shares the hyperbox with the one point **d** in the archive, then **c** can replace **d** only if it Pareto dominates **d**.
3. If the representation point of **c** is not Pareto dominated by any representation point of archive $A(t)$ and there is no solution in the archive sharing the same hyperbox with **c**, add **c**.
4. Otherwise, discard **c**.

In this way, there will always be at most one nondominated solution in one hyperbox, which ensures diversity and convergence simultaneously. For the same reason, the above archive maintenance method will keep a limited number of solutions after convergence that belong to the ε-Pareto set.

There have been extensive comparisons between NSGA-II and SPEA2 on different benchmark problems. One of the conclusions is that, comparing to NSGA-II, "SPEA produced a much better distribution at the expense of a large computational effort" [21]. In order to improve the distribution of NSGA-II, in 2005 Deb *et al.* introduced the steady state into NSGA-II and adopted the ε-dominance concept to control the distribution [21].

Deb *et al.* used Eq. 6.27 for their ε-dominance definition, i.e., objection spaces are divided into hyperboxes evenly.

If users can provide $(\varepsilon_1, \cdots, \varepsilon_m)$ for all objectives, then they express the idea that they are disinterested in the difference between ε_j for the jth dimension. We can divide the objective space into hyperboxes and calculate the index of the representation point by Eq. 6.29:

$$b^i_j = \left\lfloor \frac{z^i_j - f^{\min}_j}{\varepsilon_j} \right\rfloor \tag{6.29}$$

where $\lfloor\,\rfloor$ is for rounding toward minus infinity, z^i_j is the jth fitness value of the ith individual, $f^{\min}_j$ is the jth minimum fitness value in the population, and b^i_j is the index of the grid start from the minimum fitness value along dimension j. In this way, we can construct hyperboxes using $(\varepsilon_1, \cdots, \varepsilon_m)$, as in Fig. 6.17. For a given set of $(\varepsilon_1, \cdots, \varepsilon_m)$ and an archive, hyperboxes are determined. We can use the indices calculated by Eq. 6.29 to express the representation point $\mathbf{b}^i = \left[b^i_1, b^i_2, \cdots, b^i_m\right]$, which resides in the lower left corner of the hyperbox containing $\mathbf{z}^i$. That is all the individuals residing in the same hyperbox have the same representation point.

A steady state GA starts with a population including *popsize* individuals. The selection, crossover, and mutation processes are all individual-based, which means every time a new individual or two, not a new population, is generated, the process will substitute an individual or two in the population if the new one is good enough. The changes in the population are much less dramatic than those of the generational GA. If the initialization process generates widely distributed individuals across the

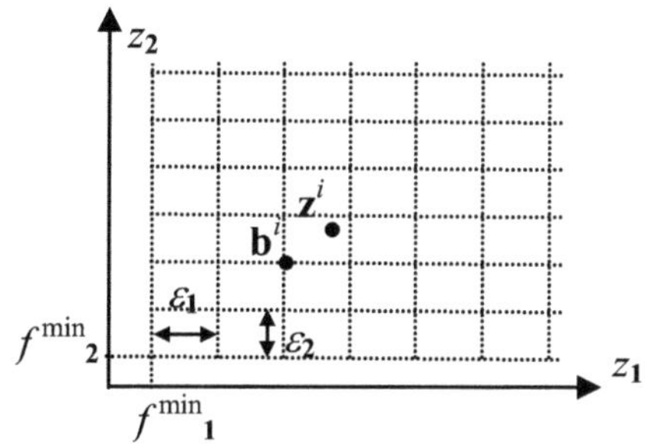

Fig. 6.17 Hyperboxes formulated by $(\varepsilon_1, \cdots, \varepsilon_m)$ provided by users

objective space, it is reasonable to believe that a steady state GA, which replaces the close parents with offspring, could maintain diversity better than a generational one.

Deb *et al.*'s algorithm combines the archive and the steady state GA together and has the main loop step as Fig. 6.18.[42]

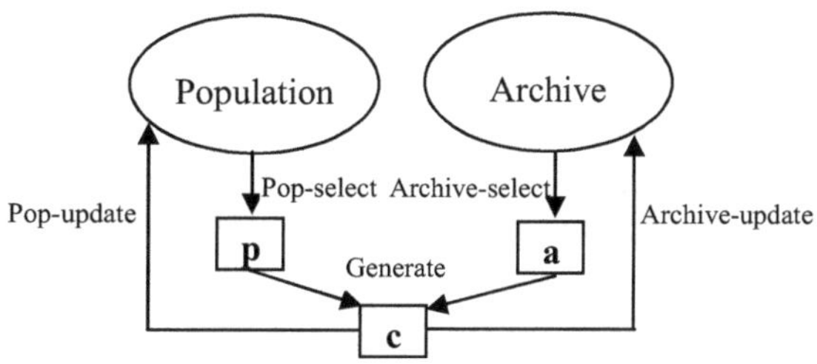

Fig. 6.18 Deb *et al.*'s steady state MOEA

There are five procedures in Fig. 6.18. The "generate" procedure is the same as in ordinary GAs. In the "pop-select" procedure, binary tournament selection based on Pareto dominance is carried out. The winner is selected. If two individuals cannot dominate each other, randomly select one. Because all the individuals thus far in the archive are nondominated solutions, just randomly pick one in the "archive-select" procedure.

After individual **c** is generated by **p** and **a**, it is considered for insertion into both the population and the archive.

The "pop-update" procedure is simple: (a) if **c** dominates at least one individual in the population, **c** replaces any of the dominated ones; (b) if **c** is dominated by at least one individual, discard **c**; (c) otherwise, **c** replaces a random individual in the population.

The "archive-update" procedure is similar to that of the above archive maintenance method converging to an ε-Pareto set. The only difference is in the second logic consideration. If **c** is the winner of the Pareto competition or **c** is closer to the

---

[42] A C language implementation of this algorithm can be downloaded at: `http://www.iitk.ac.in/kangal/codes.shtml`.

representation point of the hyperbox, then **c** replaces **d**. The adoption of distance comparison between nondominated solutions promotes convergence.

Deb *et al.* claimed that the suggested steady state MOEA is a good compromise in terms of convergence, diversity, and computation time.

6.6.3 Rebirth from the Ashes

The main reason for introducing the weighted sum method in Sect. 6.2.1 and VEGA in Sect. 6.3 was to show the historical development process of MOEAs to readers. In addition, recent developments of MOEAs promoted the rebirth of these ancient methods. Generally speaking, new ideas are the results of reforming, reusing, and reuniting old ones.

6.6.3.1 Rebirth of VEGA

We have discussed in depth the tradeoff between convergence and distribution. Different algorithms employ different mechanisms to achieve the best tradeoff. Like Grefenstette's metaevolution discussed in Chap. 3, in 2003 Toffolo and Benini used VEGA to search for two objectives: the best Pareto rank and the best distribution under the scheme of ES [22]. Also in 2003, Lu and Yen implemented the same idea under the scheme of GA [23] and developed it further into a variable *popsize* MOEA [24].

The *rank-density-based genetic algorithm* (RDGA) suggested by Lu and Yen utilizes VEGA (Fig. 6.4) to deal with the tradeoff between convergence and distribution. The main differences between RDGA and VEGA are fitness assignment, crossover operator, and elitism.

RDGA transforms the m-objective problem into a two-objective problem. For each individual, it has two fitness values: rank and density. With respect to both of them, the smaller, the better.

The definition of rank in RDGA of an individual is different from all the aforementioned ranks. The rank for every nondominated individual is 1. Suppose that the dominated individual y is dominated by P individuals $y_1, y_2, \cdots, y_P$, whose rank values are already known as $rank(y_1), \cdots, rank(y_P)$. Then y's rank can be calculated as follows:

$$rank(\mathbf{y}) = 1 + \sum_{j=1}^{P} rank(\mathbf{y}_j) \tag{6.30}$$

Equation 6.30 calculates the rank for each individual from the standard Pareto rank 1 to higher ranks.The entire procedure can be carried out in an automatic accumulated way, hence its name: *automatic accumulated ranking strategy* (AARS).

In Fig. 6.6, the number in [ ] illustrates the AARS rank result.[43]

The density in RDGA is like that of the squeeze factor, illustrated in Fig. 6.8, but in another adaptive way apart from Eq. 6.25. Users need to assign the grid numbers for each objective. Suppose for the ith objective there are K_i grids and the negative and positive ideal values in the current population are z_i^- and z_i^+, respectively, then the width of the box in i's objective is as follows: [44]

$$d_i = \frac{z_i^- - z_i^+}{K_i} \tag{6.31}$$

The negative and positive ideal points in the current population will change during the evolving process, and then the hyperbox size will vary accordingly. With Eq. 6.31 we can determine the coordinates for the center of every hyperbox, thus every individual can be assigned to the closest hyperbox. After that, the numbers of individuals belonging to each hyperbox can be counted easily.

Figure 6.19 illustrates the evolving process in one generation of RDGA.

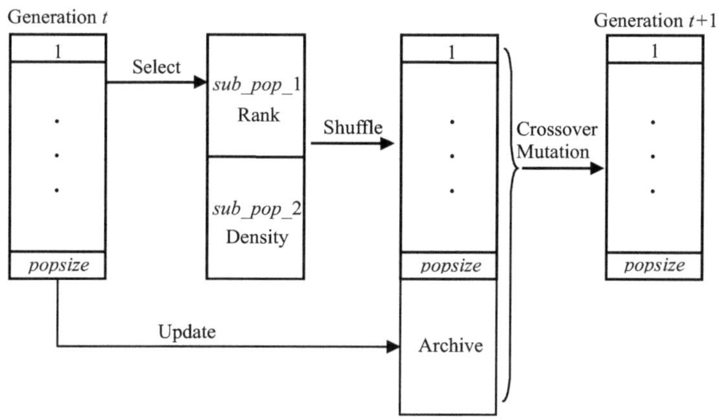

Fig. 6.19 The evolving process in one generation of RDGA

RDGA uses the selection process in the two-objective VEGA, and half of the population is selected according merely to rank and another half merely to density.

In the crossover procedure, RDGA employs a mating restriction rule to promoting the exploration and exploitation toward PF*. Individuals after the selection process reside in the hyperboxes. For each hyperbox, a fixed number of parents are randomly selected to cross over with the best individual (lowest rank) in that hyperbox and the neighboring hyperbox. The offspring can survive only if they dominate their parents or are less crowded than their parents.

---

43 Consider the situation where PF* is a straight line between $(0,1)$ and $(1,0)$ in a normalized two-objective situation and the individuals are distributed evenly above PF*. Is there any preference in AARS? If so, which individuals have advantages?

44 Compare it with the techniques in Sect. 6.6.2.1.

Because RDGA concerns diversity with a probability of 0.5, some bad individuals (dominated by parents but residing in the empty area) may survive, which is not good. So a forbidden region is set up to avoid such a situation.

Another characteristic of RDGA is that individuals in the archive have a probability p_e of being selected as a parent to undergo crossover.

$$p_e = 1 - \left(\frac{|P|}{|P| + |A|}\right)^2 \tag{6.32}$$

where $|P|$ is the size of the population and $|A|$ is the size of the archive. The larger A is, the greater the opportunity of an elitist being selected for crossover.

After new individuals have been generated by crossover and mutation, they might be added to the archive if they dominate some part of the archive.

Finally, it is necessary to mention that the drawback of VEGA's lacking diversity is not important in RDGA because we only want one edge point in the Pareto front of the 2-D problem (rank and density): $rank = 1$ and VEGA is good at maintaining edge points. In this way, Lu and Yen utilized the drawback of VEGA to effectively maintain the balance between quality and diversity. RDGA is competitive with NSGA-II, SPEA2, and PAES according to Lu and Yen's numerical experiments.

6.6.3.2 Rebirth of Weight Sum Method

The weight sum method was introduced in Sect. 6.2.1. Its main advantage over dominance-based MOEAs is fast convergence. As for convergence, Miettinen has proven that the optimal solution of the normalized weight sum of the m objectives $\min \sum_{i=1}^{m} \omega_i f_i(\mathbf{x})$ is always Pareto optimal if $\omega_1, \cdots, \omega_m$ are all positive or the solution is unique [25]. So we can safely use the weight sum method to find nondominated solutions. But there are several drawbacks obstructing the application of the weight sum method in MOPs:

1. If the objective values are disparately scaled, i.e., whose value scales have large differences, special normalization techniques are necessary, which limits the random weight sum method.
2. Not all points on PF* have corresponding weight vectors. Only the point on convex PF* can find a weight vector with which the weight sum has a minimum on this point. This phenomenon is illustrated in Fig. 6.20.[45] So we need other techniques to transform multiple objectives into one.
   For the convex PF* in Fig. 6.20, all the points on it and in its upper-right area are legal solutions. For dashed line $\alpha z_1 + (1-\alpha) z_2 = h$, its slope is $-\frac{\alpha}{1-\alpha} \in (-\infty, 0)$ if $0 < \alpha < 1$. The convex PF* is drawn using the equation $z_2 - 1 = \frac{1}{z_1 - 1}$. So every point on PF* has a slope $-\frac{1}{(z_1-1)^2} \in (-\infty, 0)$. With a decrease in h, the dashed line

[45] Readers are encouraged to draw the PF* with a concave or partial-convex-and-partial-concave shape to study this statement.

descends with the same slope. At the point at which PF* has the same slope as the dashed line. The dashed line is the tangent of PF*. The intersection point is the minimum of the weight sum function and also the corresponding nondominated MOP solution. According to the above discussion, any point on convex PF* will find a corresponding minimum for the weight sum function.

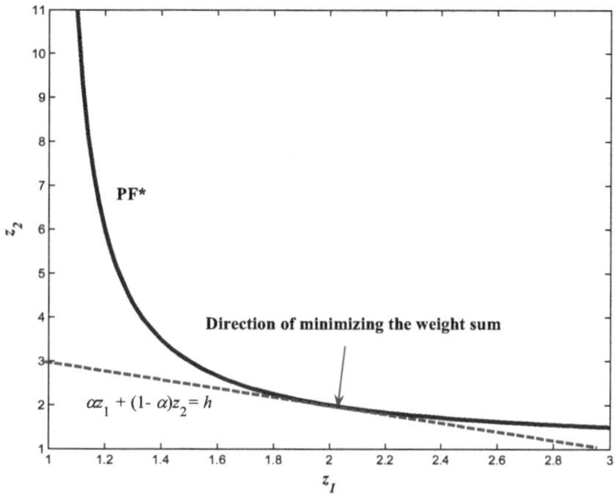

Fig. 6.20 Convex PF* and weight sum method

3. Even for convex PF*, evenly distributed samples of the weight vector might not correspond to evenly distributed solutions on PF*. Let us discuss the above example and let $\frac{\alpha}{1-\alpha} = \frac{1}{(z_1-1)^2}$. We use even samples of $0.1, 0.2, \cdots, 0.9$ on α and get the nondominated solutions of $(4.00, 1.33), (3.00, 1.50), \cdots, (1.33, 4.00)$. The distance between these nine points are $1.013, 0.497, 0.343, 0.290, 0.290, 0.343, 0.497, 1.014$, respectively.

Many researchers have attempted to overcome the above obstacles [26, 27]. In 2007, Zhang and Li suggested a MOEA based on decomposition (MOEA/D) to give the weight sum method a new life [28].

For the problem of scaling, Zhang and Li suggested using adaptive normalization for each objective. We can find the positive ideal point and negative ideal point in the current population. Then the normalized objective value is $\frac{z_k^i - z_k^+}{z_k^- - z_k^+}$, where z_k^i is the kth objective value of individual i, z_k^+ and z_k^- are k's values of positive and negative ideal points, respectively. Thus, every objective of the current population is mapped into the range $[0, 1]$.

The Tchebycheff approach and boundary intersection approach are used to handle the second drawback.[46] These approaches ensure that any point on PF* has a

[46] These approaches, together with the weight sum method, are decomposition methods by which a single-objective aggregation function is formed with multiple objectives. Then we can use single-

corresponding weight vector regardless of the shape of PF*. In these approaches, we also need weight vectors to aggregate multiple objectives into one, but not simply add them up. We will denote the aggregation function as $AF(\boldsymbol{\omega},\mathbf{z})$, where $\mathbf{z}$ is the normalized vector containing m objective values and $\boldsymbol{\omega}$ is the weight vector.

The initial idea of MOEA/D is that for two close weight vectors $\boldsymbol{\omega}^i$ and $\boldsymbol{\omega}^j$, their corresponding optimal solutions of the aggregation functions are close because the aggregation functions are continuous with $\boldsymbol{\omega}$. So if we can generate many evenly distributed weight vectors before optimization and find the closest neighbors for each vector according to Euclidean distance, we can do a single-objective search with different directions and use the information of the neighbors to help the search. The solution process of MOEA/D is as follows.

Solution Process of MOEA/D

Phase 1: Initialization.

Step 1.1: Assign the parameters for MOEA/D, such as $popsize$, closest neighbor number T, stop criteria, etc.

Step 1.2: Generate $popsize$ evenly distributed weight vectors $\boldsymbol{\omega}^1,\cdots,\boldsymbol{\omega}^{popsize}$, each having m components $\boldsymbol{\omega}^i = \omega_1^i,\cdots,\omega_m^i$.

Step 1.3: Calculate the Euclidean distances between any pair of weight vectors and thus find T closest weight vectors for each one. Set $B(i) = i_1,\cdots,i_T$, where $\boldsymbol{\omega}^{i_1},\cdots,\boldsymbol{\omega}^{i_T}$ are T closest neighbors of $\boldsymbol{\omega}^i$. We need to point out that $i \in B(i)$. The main driving force in MOEA/D are weight vectors, which is determined and whose neighbors are found in the initialization.

Step 1.4: Generate $popsize$ initial individuals $\mathbf{x}^1,\cdots,\mathbf{x}^{popsize}$ randomly and calculate their objective values. Calculate the positive and negative ideal points according to the initial individuals and normalize their objective values using techniques discussed above as $\mathbf{z}^1,\cdots,\mathbf{z}^{popsize}$. $gen = 0$.

Phase 2: Main loop. Repeat the following steps until the stop criteria are satisfied. In each loop, $popsize$ individuals are evolved in a sequential way using weight vectors. $i = 1$.

Step 2.1: Randomly select two indices, denoted as k and l, from $B(i)$. They correspond to individuals $\mathbf{x}^k$ and $\mathbf{x}^l$. Use variation operators to generate a new solution $\mathbf{y}$ from $\mathbf{x}^k$ and $\mathbf{x}^l$. Calculate its normalized objective values vector as $\mathbf{z}'$.

Step 2.2: For each index $j \in B(i)$, if $AF(\boldsymbol{\omega}^j,\mathbf{z}') \leq AF(\boldsymbol{\omega}^j,\mathbf{z}^j)$, $\mathbf{x}^j = \mathbf{y}$. In this way, we can maintain the best solution found thus far for the aggregation function generated by weight vector $\boldsymbol{\omega}^j$ and later use it to promote the local search for $\boldsymbol{\omega}^j$'s neighbors.

Step 2.3: $i = i+1$. Go back to step 2.1 until $i > popsize$.

---

objective optimization algorithms to solve the aggregation function. We neglect the implementation details of these decomposition methods and refer interested readers to [26, 27].

Step 2.4: Update the positive and negative ideal points with the current population. $g = g + 1$.

Phase 3: Submitting the final *popsize* individuals as the results of MOEA/D.

Zhang and Li also suggested using local search to guarantee that the offspring $\mathbf{y}$ will be a legal and feasible solution and they utilized the archive to contain the nondominated solutions found thus far. But these techniques are optional.

Apart from inheriting the fast convergence property of optimizing a single-objective aggregation function, MOEA/D established a good balance between exploration and exploitation. The convergence mechanisms of MOEA/D are the decomposition method, which ensures that the optimal solutions of aggregation functions are nondominated solutions, and the individual updating method in step 2.2, which ensures that $\mathbf{x}^i$ is the best solution found thus far for ω^i. The diversity mechanism of MOEA/D are (1) evenly distributed weight vectors, which direct the search; (2) the mating restriction in step 2.1, which limits the search to a small area; (3) a decomposition method that can handle the concave and convex shapes and provide evenly distributed solutions on PF*.

For Pareto-dominance-based MOEAs, their ability to find edge points is subject to variation operators. But MOEA/D's ability to find edge points could be adjusted by changing weight vectors. In addition, the property that the continuous aggregation function of ω will have close optimal solutions with close weight vectors might force the nondominated solutions to be distributed between edge points in an approximately even way.[47] These two factors make MOEA/D good at PF* exploration, especially when the shape of PF* is complex.

MOEA/D utilizes *popsize* evenly distributed weight vectors to search for *popsize* evenly distributed nondominated solutions on PF* and use neighbor information to accelerate the search. The closest neighbor number T of one weight vector is a critical control parameter. Too small T might limit the exploration ability but too large T might contribute worse parents with a high probability and thus limit the exploitation ability. Numerical experimental results done by Zhang and Li illustrate that T is not sensitive to benchmark problems.

In 2009, Li and Zhang adopted the directional mutation of DE in MOEA/D, together with other techniques, making MOEA/D-DE a more efficient MOEA [29].[48]

[47] Even though evenly distributed weight vectors would not guarantee evenly distributed nondominated solutions.

[48] The source code of MOEA/D-DE can be downloaded at: `http://cswww.essex.ac.uk/staff/zhang/`.

6.7 Performance Evaluation of Multiobjective Evolutionary Algorithms

The reason for a special section on performance evaluation besides the one in Chap. 3 is that situations in MOPs are quite different from those in single-objective problems. Every algorithm can maintain a group of nondominated individuals at the end of the run. Sometimes the result from one algorithm fully dominates the other, which is the simplest condition. But generally, some results from one algorithm dominate some from another algorithm, and *vice versa*.

Another reason for the special consideration on the performance evaluation is that we are interested in not only the convergence to PF* but also the distribution of the individuals along PF*. Adequately evaluating convergence and distribution is still an open problem in the field of MOEAs.

Benchmark problem design is also an interesting field because we want to conveniently generate problems with different shapes of PF*, different convergent difficulties, different dimensions, etc.

6.7.1 Benchmark Problems

In 2006, Husband *et al.* gave an excellent survey for generating multiobjective test problems [30]. They suggested that good benchmark MOPs should contain following characteristics.

1. The Pareto solution set should not reside at the edge of the feasible domain. We introduced in Chap. 4 the idea that truncating an infeasible individual to the edge of the feasible domain is one way of handling constraint problems. So this technique may gain advantages in optimizing benchmark problems that do not satisfy characteristic 1.
2. The Pareto solution set should not reside in the center of the domain. The intermediate crossover introduced in Chap. 2 may have advantages in optimizing benchmark problems that do not satisfy characteristic 2.
3. Benchmark MOPs should have a scalable number of variables so that the designer and the analyzer can generate arbitrary dimensional MOPs.
4. Benchmark MOPs should have a scalable number of objectives.
5. The variables of benchmark MOPs should have definition domains of different magnitudes. This characteristic tests the ability to change mutation strengths with different variables or the normalization ability of the algorithm.
6. The magnitudes of different objectives in PF* should be different.
7. The PF* of the problem can be expressed in explicit expression.

Besides all these requirements, Husband *et al.* also suggested that the following features need to be considered in designing benchmark MOPs.

1. **PF* geometry.** The geometry of the PF* in benchmark MOPs might show different shapes such as convex, concave, linear, disconnected, and combination of these shapes in one PF*.
2. **Parameter dependency.** If we can find the PF* by optimizing one variable after another, the variables of these kinds of problem are *separable* and the problem is *decomposable*.[49] By contrast, nonseparable variables or parameter dependency makes a more difficult problem.
3. **Bias.** If the evenly distributed points in the Pareto optimal solutions in P* map into the unevenly distributed points in PF*, we say this problem has *bias*. Real-world problems might always have a bias.
4. **Many-to-one mapping.** If different points in Pareto optimal set P* map into the same point in PF*, this problem is a *many-to-one mapping*. The extreme condition is that PF* has a flat region, which means small perturbations of the decision variables do not change the objective values. Problems with many-to-one mapping, especially with a flat region, are difficult for MOEAs.
5. **Modality.** We introduced modality in Chap. 5. An objective function with only one optimum is unimodal, which is easy to solve. *Multimodal* in a multiobjective environment means that a point in a small definition region definition is the efficient solution but is dominated by real Pareto optimal solutions. This point is called a local optimal solution in MOPs. Multimodal MOPs are difficult for MOEAs, which has been proved by Li and Zhang using numerical experiments [29].

We need to mention that for generating a test suite of benchmark MOPs, it is not good to select all the "difficult" problems. The purpose of designing MOEAs is to solve real-world MOPs. Not all of them are extremely difficult. So it is necessary to make a tradeoff between different benchmark problems and real-world problems. Also, according to the No Free Lunch theorem discussed in sect. 3.6.1, the average performance of any pair of algorithms across all possible problems is identical. So if an algorithm is quite good at solving "difficult" problems, it might exhibit overtraining-like characteristics in artificial neural networks and might not solve "easy" problems properly.

Different benchmark MOP suites have been suggested to emphasize different aspects. We will list some of these problems as the benchmark MOPs in Appendix. These problems are selected from the suites listed below.[50]

1. Van Veldhuizen summarized the multiobjective test problems before 1999 and selected seven of them as the benchmark [31].
2. **ZDT.** In 1999, Deb suggested a way to construct multiobjective test problems systematically [32]. In Deb's method, there is a function h to control the shape of PF*, a function g to test the MOEAs' ability to converge to PF*, and a function f_1 to test the MOEAs' ability to distribute the individuals along PF*. In 2000, Zitzler *et al.* used Deb's method to generate six benchmark MOPs [33].

[49] Those two terms were discussed in Sect. 3.2.2.1.

[50] Readers with an interest in constructing the benchmark problem are encouraged to read these papers in the listed sequence.

3. **DTLZ.** In 2001, Deb *et al.* developed ZDT to nine scalable benchmark problems [34].
4. **OKA.** In 2004, Okabe *et al.* suggested another way to generate benchmark MOPs with an arbitrary Pareto optimal set shape and PF* shape [35]. Apart from two examples to illustrate the effectiveness of the method, Okabe *et al.* also introduced a way to measure the convergence difficulty in OKA.
5. **WFG.** In 2005 and 2006, Huband *et al.* suggested a new scalable benchmark MOP suite with nine problems that contain and consider the characteristics and features discussed above [30, 36].
6. In 2006, Iorio and Li pointed out that rotation might introduce difficulties for MOEAs and suggested four rotated benchmark MOP examples [37].
7. In 2006, Deb *et al.* addressed the importance of parameter dependencies for designing MOP benchmark problems and developed their ZDT and DTLZ through variable linkage [38].
8. In 2007 and 2009, the IEEE Congress on Evolutionary Computation held special sessions on multiobjective optimization and multiobjective optimization with constraints, respectively. The technical reports illustrate the corresponding benchmark problems [39, 40].
9. In 2009, Li and Zhang provided a new way of generating MOP benchmark problems with arbitrary prescribed PF* shapes and gave nine examples [29].

6.7.2 Performance Indices

After determining which benchmark MOPs to optimize, we need to make a careful decision on how to evaluate the performance of different MOEAs. Those criteria are *performance indices* (PI).

As was discussed above, we want MOEAs converge to PF*, so the convergence is often the first thing taken into consideration. Convergence performance can be further divided into two groups: how many individuals belong to PF* and what is the overall evaluation of convergence to PF*.

One of the reasons for using MOEAs in MOPs is to distribute the final individuals evenly along PF*. So users are also interested in the distribution. Apart from that, spread performance evaluates MOEAs' ability to capture the edge points of PF*, which is sometimes taken into consideration.

So we can give the taxonomy of PIs in Fig. 6.21.[51]

There might be several situations in which to evaluate an algorithm. First, we know PF* of the benchmark MOPs and want to evaluate how good the algorithm is while optimizing the benchmark MOPs. Second, PF* is not known *a priori*, but we want to compare the solutions of many algorithms. So the nondominated solu-

---

[51] In 2003, Okabe *et al.* suggested an excellent survey on PI [41] and Zitzler proposed a very theoretical analysis on PI also in 2003 [42]. Readers interested in designing and analyzing PIs are urged to read them.

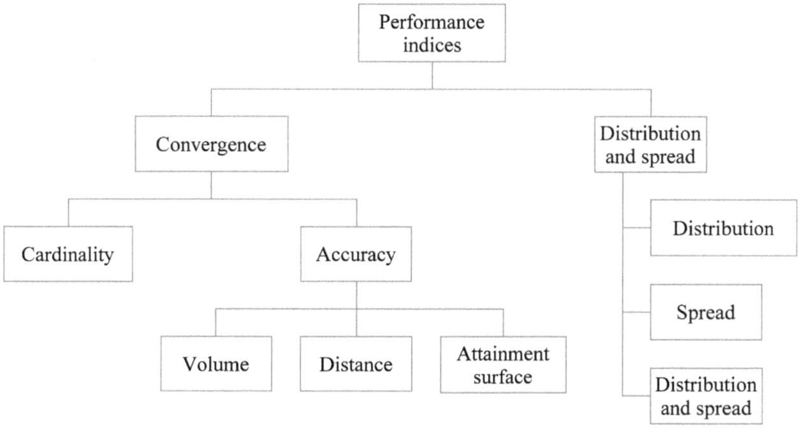

Fig. 6.21 PI taxonomy

tions in the union of these final solutions constitute a *reference set* (RS). [52] Results from different algorithms are evaluated with a RS. The PI for these two situations is called a *unary index* (input of the PI is one solution set). For an algorithm, all the nondominated solutions in the final archive are defined as set S.

The third situation is that we would just like to compare two results (S_1 and S_2), or we want to compare many different results pair-by-pair, then we need the PI to point out which one is better, sometimes how much better. The PI for the third situation is called a *binary index*, whose inputs are two solution sets.

In the comparison of two algorithms, the number of objective evaluations for each algorithm should be the same to ensure the same search endeavor over the objective space and their archives should have the same capacity to ensure the same ability to report the nondominated solutions. For algorithm 1, all the nondominated solutions in the archive are called S_1. S_2 contains all the nondominated solutions found by algorithm 2.

We also need to point out again that the selection of the PI to be introduced is based on the importance and the authors' interests.

6.7.2.1 Cardinality-based Performance Indices

If we already know PF*, or RS, we can evaluate the quality by counting how many individuals in S are in PF*.

In 1999, Van Veldhuizen suggested a unary PI called *error ratio* (*ER*) [31].

[52] The acronym RS was also used by scatter search in Sects. 2.4.2.1 and 6.6.1.2. In this section, RS always implies the nondominated solution union.

$$ER(S) = \frac{\sum_{i=1}^{|S|} e_i}{|S|} \tag{6.33}$$

where $e_i = \begin{cases} 0 & z_i \in \text{PF*} \\ 1 & z_i \notin \text{PF*} \end{cases}$, z_i is solution i in S, and $|S|$ is the cardinality of S.[53] ER can be calculated based on RS too. ER is to evaluate how bad a result is. So the smaller ER is, the better.

In 1999, Zitzler suggested a binary PI called *coverage* (C) [43].

$$C(S_1, S_2) = \frac{\left|\left[s_2 \in S_2 \middle| \exists s_1 \in S_1 : s_1 \preceq s_2 \right]\right|}{|S_2|} \tag{6.34}$$

$C(S_1, S_2)$ is the percent of the individuals in S_2 who are weakly dominated by S_1. The larger $C(S_1, S_2)$ is, the better S_1 outperforms S_2 in C. It is necessary to point out that generally $C(S_1, S_2) + C(S_2, S_1) \neq 1$.

There is one drawback to $C(S_1, S_2)$. Let us take the solutions in Fig 6.22 as an example,[54] where S_1 is represented by squares and S_2 is represented by cycles. It is clear that $C(S_1, S_2) = C(S_2, S_1) = 0.5$. But we may prefer S_1 because it is lower than S_2 in most areas. How to demonstrate such difference?

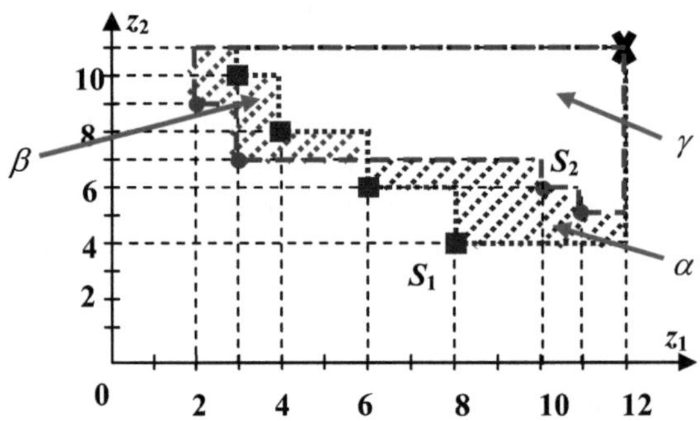

Fig. 6.22 Two solution sets to be compared

6.7.2.2 Volume-based Performance Indices

In 1999, Zitzler suggested a unary PI called *hypervolume* (HV) [43].[55] The hypervolume is the size of the space dominated by S. In calculating HV, we need to point

[53] We will omit the explanation for such operators in the following content of this section.

[54] We will discuss the objective space of 2-D minimal problem as examples in this subsection. Most of these examples could be expanded to higher dimensions.

[55] This PI was originally called S metric by Zitzler, where S is for "size of the dominated space." It is now often referred to as *hypervolume* or *hyperarea*.

out the reference point to compute, illustrated by $\times$ in Fig. 6.22. Then the size of the space dominated by S_1 is the size enclosed by the dotted line, and that by S_2 is the size enclosed by the dashed line. The larger HV of a solution set is, the better it is. For the example in Fig. 6.22, $HV(S_1) = 1\times 9 + 2\times 8 + 2\times 6 + 2\times 4 = 45$ and $HV(S_2) = 2\times 10 + 2\times 9 + 1\times 2 + 1\times 1 = 41$. So S_1 is better than S_2 in HV. Knowles and Corne reported in 2003 that the selection of the reference point might influence the relative relationship between two sets of nondominated solutions [20].

To make the difference more obvious, in 1999 Zitzler suggested another binary PI called *coverage difference* (D) [43].

$$\begin{cases} D(S_1, S_2) = HV(S_1 + S_2) - HV(S_2) \\ D(S_2, S_1) = HV(S_1 + S_2) - HV(S_1) \end{cases} \tag{6.35}$$

In Fig. 6.22, $HV(S_1 + S_2) = 2\times 10 + 2\times 9 + 1\times 6 + 2\times 4 = 52$. $D(S_1, S_2) = 11$ is area α and $D(S_2, S_1) = 7$ is area β. The larger $D(S_1, S_2)$ is, the better S_1 outperforms S_2 in D.

In 1999, Van Veldhuizen suggested a similar PI called *hypervolume ratio* (HR) [31].

$$HR(S_1, S_2) = \frac{HV(S_1)}{HV(S_2)} \tag{6.36}$$

The larger $HR(S_1, S_2)$ is, the better S_1 outperforms S_2 in HR.

Because hypervolume-based PI is widely used and the calculation of HV is complicated in high-dimensional situations, many efficient ways of calculating HV have been suggested recently [20, 44, 45].

6.7.2.3 Distance-based Performance Indices

Distance-based PI evaluate the performance of the solutions according to the distance to PF*, or RS.

In 1999, Van Veldhuizen suggested a unary PI called *generational distance* (GD) [31]. First we need to define the minimum distance from S to PF* as

$$d_i = \min_{p \in \mathrm{PF}*} \left\{ \sqrt{\sum_{k=1}^{m} \left(z_k^i - z_k^p\right)^2} \right\} \tag{6.37}$$

where z_k^i is the kth objective value of the ith individual and d_i is its minimum Euclidean distance to PF*. Other norms, i.e., distance measure method, could also be used. GD is defined as follows:

$$GD(S) = \frac{\left(\sum_{i=1}^{|S|} d_i^q\right)^{1/q}}{|S|} \tag{6.38}$$

where q is a parameter. If $q = 1$, GD is equal to Deb's γ PI [8] and Zitzler's M_1^* PI [43]. The smaller GD for one solution set is, the better it is in approaching PF*.

Laumanns introduced the concept of ε-dominance, discussed in Sect. 6.6.2.2, [19] and in 2003 Zitzler *et al.* developed a definition of ε-dominance for performance evaluation [42].

If $z_l^i \leq \varepsilon z_l^j, \forall l \in \{1, 2, \cdots, m\}$, we say *i* ε*-dominates* *j*, expressed as $i \prec_\varepsilon j$.[56] Figure 6.23 illustrates the meaning of Zitzler's ε-dominance in two-objective situations.[57] We neglect the difference between dominance and strong dominance in the following discussion, so as to neglect the vertical and horizontal line extended from point $\mathbf{z}^i$. If $\varepsilon = 1$, ε-dominance is the same as Pareto dominance. For $\varepsilon > 1$, the area dominated by $\mathbf{z}^i$ is enlarged because $\mathbf{z}^i$ is only required to dominate the enlarged version of $\mathbf{z}^i$, i.e., $\varepsilon\mathbf{z}^i$. For $\varepsilon < 1$, the area dominated by $\mathbf{z}^i$ is shrunk. So generally speaking, ε-dominance relaxes the area of Pareto dominance ε times.

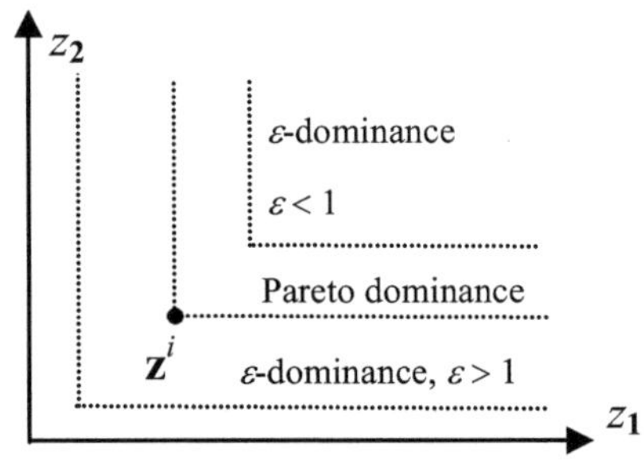

Fig. 6.23 ε-dominance suggested by Zitzler *et al.*

Zitzler *et al.* suggested a binary ε*-dominance-based PI* as follows:

$$I_\varepsilon(S_1, S_2) = \inf_{\varepsilon \in R} \left\{ \forall z^2 \in S_2,\ \exists z^1 \in S_1 : z^1 \prec_\varepsilon z^2 \right\} \tag{6.39}$$

Equation 6.39 means we need to to find the minimum ε so that for every solution z^2 in S_2, there will always be at least one solution z^1 in S_1 that ε-dominates z^2. According to Fig. 6.23 and Eq. 6.39, if $I_\varepsilon(S_1, S_2) > 1$, at least one solution in S_2 is Pareto dominated by at least one solution in S_1. For S_1, the smaller the $I_\varepsilon(S_1, S_2)$ value is, the better; for S_2, the larger the $I_\varepsilon(S_1, S_2)$ value is, the better.

Zitzler *et al.* also suggested a pragmatic way to get $I_\varepsilon(S_1, S_2)$. For any two solutions z^1 and z^2 from S_1 and S_2, respectively, z_i^1 / z_i^2 calculates ε in the ith objective. If $\max\limits_{1 \leqslant i \leqslant m} \frac{z_i^1}{z_i^2} < 1$, then z^1 Pareto dominates z^2. Otherwise, z^1 is dominated by z^2 or z^1 and z^2 are nondominated by each other. So we can compare every individual pair from S_1 and S_2 by

[56] This definition is not in the original form, i.e., Eq. 6.26 or Eq. 6.27.

[57] Compare it with Fig. 6.14.

$$\varepsilon_{z^1,z^2} = \max_{1 \le i \le m} \frac{z_i^1}{z_i^2}, \forall z^1 \in S_1, \forall z^2 \in S_2 \tag{6.40}$$

Equation 6.40 calculates the extent of superiority of z^1 over z^2 in the worst objective. We can use Eq. 6.40 to illustrate the relationship between z^1 and z^2. We need to mention that $\varepsilon_{z^1,z^2} = 1$ means z^1 covers z^2.[58]

Provided we know the relationship between every two individuals in S_1 and S_2, Eq. 6.41 finds the best individual in S_1 for dominating z^2. We can use Eq. 6.41 to illustrate the quality of z^2. If $\varepsilon_{z^2} < 1$, then there is at least one individual in S_1 that dominates z^2.

$$\varepsilon_{z^2} = \min_{z^1 \in S_1} \varepsilon_{z^1,z^2}, \forall z^2 \in S_2 \tag{6.41}$$

Then we can get the quality of every individual in S_2 using Eq. 6.41. If S_1 totally dominates S_2, the ε_{z^2} of every individual in S_2 should be less than 1. So we can use the best individual (with the largest ε_{z^2}) to illustrate the quality of S_2 compared to S_1 using ε-dominance as follows:

$$I_\varepsilon(S_1, S_2) = \max_{z^2 \in S_2} \varepsilon_{z^2} \tag{6.42}$$

For the example given by Fig. 6.24, $I_\varepsilon(S_1, S_2) = 0.875$ and $I_\varepsilon(S_2, S_1) = 1.5$, which means S_1 is better than S_2 in ε-dominance-based PI.[59]

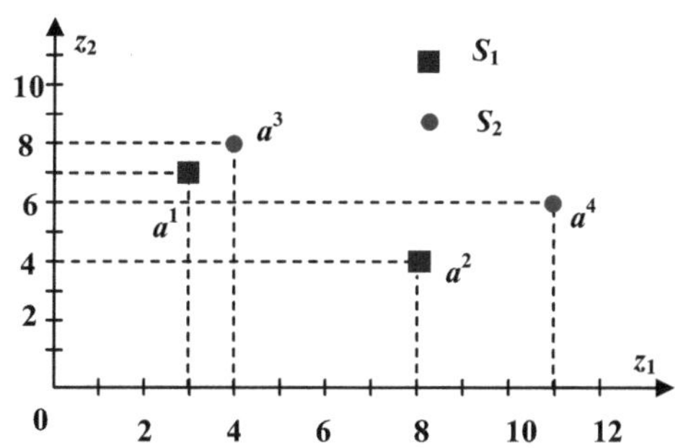

Fig. 6.24 Example of ε-dominance-based PI

To draw conclusions using an ε-dominance-based PI, we need to calculate both $I_\varepsilon(S_1, S_2)$ and $I_\varepsilon(S_2, S_1)$ [46].

- If $I_\varepsilon(S_1, S_2) \le 1$ and $I_\varepsilon(S_2, S_1) > 1$, S_1 is better than S_2 in ε-dominance-based PI, and *vice versa*.
- If $I_\varepsilon(S_1, S_2) > 1$ and $I_\varepsilon(S_2, S_1) > 1$, then at least one solution in S_2 dominates at least one solution in S_1 and *vice versa* or all the solutions in S_1 and S_2 are nondominated by each other; therefore no conclusion can be drawn regarding which set is better.

---

58 Either z^1 equals z^2 or z^1 dominates z^2. Why might $\varepsilon_{z^1,z^2} = 1$ mean that z^1 dominates z^2?

59 We encourage readers to calculate these two PI values using Eqs. 6.40–6.42.

- If $I_{\varepsilon}(S_1, S_2) = 1$ and $I_{\varepsilon}(S_2, S_1) = 1$, then $S_1 = S_2$.
- For any solution set S, $I_{\varepsilon}(\mathrm{PF}^*, S) \leq 1$ and $I_{\varepsilon}(S, \mathrm{PF}^*) \geq 1$.

6.7.2.4 Attainment Surface-based Performance Indices

In 1996, Fonseca and Fleming suggested a way to display nondominated solutions in an archive after the termination of the algorithm [47], and in 2000 Knowles and Corne made a clear description for comparing two algorithms using attainment surface [6]. In a 2-D situation, instead of connecting the solutions with lines, Fonseca and Fleming used a dashed/dotted line to enclose the dominated area of nondominated solutions, which resembles hypervolume illustrated in Fig. 6.22. They called such a dashed/dotted line the *attainment surface*.

Generally we need to run a MOEA on a benchmark MOP several times (at least 20 times, 30 or more are recommended). How do we evaluate these runs? Averaging is not a good idea in MOPs. Union is also not so good.

Fonseca and Fleming suggested a 50% attainment surface to illustrate the average performance of a MOEA in many runs. First, we draw the attainment surfaces for every run. Figure 6.25 illustrates three runs. The solid square, the square with the small hole, and the square with the large hole are four nondominated solutions for each run. The dashed line, dotted line, and the dashed-and-doted line illustrate different attainment surfaces.

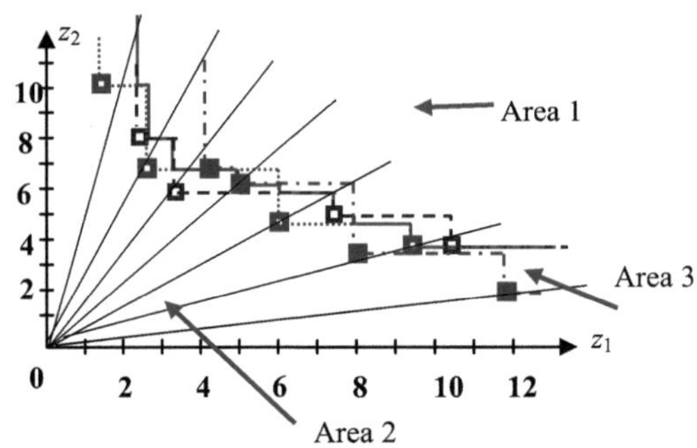

Fig. 6.25 Determining the 50% attainment surface

Three attainment surfaces divide the objective space into three parts. Area 1 is dominated by all the nondominated solutions in all runs. Area 2 dominates all the nondominated solutions in all runs. Area 3 includes areas that are dominated by some runs and dominates other runs.

We can draw several auxiliary straight lines from the origin going in different directions in the first quadrant. Every auxiliary line intersects with all the attainment surfaces. Starting from the origin toward the direction of the auxiliary line, the first met point is the best one and the last met point is the worst one. Then in these intersections, we can determine which point is the median point, i.e., in the middle of the performance rank.

If we draw enough auxiliary lines, we will get all the median points and their corresponding partial attainment surface lines. Connecting these partial attainment surface lines together will generate the 50% attainment surface, which is the average performance of a MOEA. The solid line in Fig. 6.25 is the 50% attainment surface for the three runs of the algorithm.

On comparing two algorithms, each of them has many runs on a benchmark MOEA. We draw all the attainment surfaces as in Fig. 6.26. Squares (solid or with holes) represent the nondominated solutions generated by algorithm 1 in two runs. Circles (solid or with holes) represent the nondominated solutions generated by algorithm 2 in two runs. Similar to the 50% attainment surface, we draw many auxiliary straight lines from the origin in different directions in the first quadrant. The intersections of every auxiliary line can be used to make a statistical test to determine whether or not the intersections for one algorithm are better than those of the others with statistical significance.[60] Besides a rigorous statistical hypothesis testing procedure, we can do the comparison in an empirical way. Suppose both algorithms 1 and 2 run the problem s times. For the lth auxiliary line, we get $2s$ intersections and we are only concerned with the first s intersections starting from the origin, i.e., the best s intersections. In these s points, if 70% of them belong to algorithm 1, we conclude that algorithm 1 wins in line l; if 70% of them belong to algorithm 2, we conclude that algorithm 2 wins in line l.[61] If neither of the algorithms wins, we say that there is no significant conclusion for this auxiliary line. For all L lines, algorithm 1 wins a times and algorithm 2 wins b times. If $a > b$, we say that generally algorithm 1 outperforms algorithm 2, and *vice versa*. It is necessary to point out that $a + b \neq L$. Using attainment-surface-based PI, we cannot only determine which algorithm is better but also, perhaps more importantly, know which algorithm wins in which region.

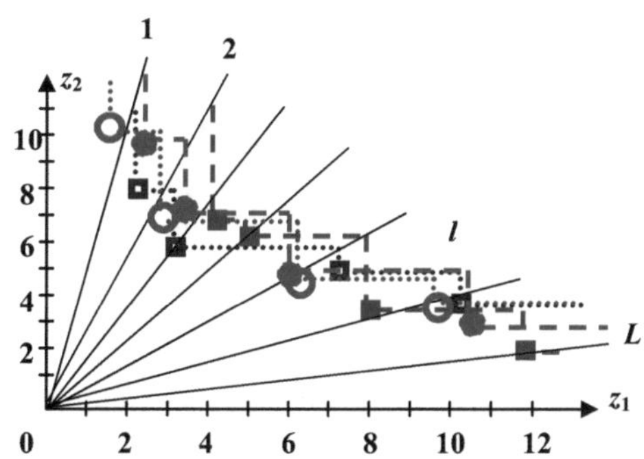

Fig. 6.26 Comparing results from different algorithms using attainment surface

---

60 We introduced the statistical comparison in Chap. 3. Readers interested in drawing statistical conclusions are referred to [48].

61 The larger the percentage,the clearer the statistical significance.

6.7.2.5 Distribution Performance Indices

We will introduce three distribution PI examples.

The first one was suggested by Schott in 1995 [49]. Schott called it *spacing* (*SP*), which is the standard deviation of the closest distances. Schott used the 1-norm, instead of 2-norm (Euclidean distance), for evaluating the distance.

$$d_i = \min_{s_j \in S \cap s_j \neq s_i} \sum_{k=1}^{m} \left| z_k^i - z_k^j \right| \tag{6.43}$$

where d_i is the smallest distance from the ith individual in S. Then the average smallest distance of all individuals is calculated as follows:

$$\bar{d} = \frac{\sum_{i=1}^{|S|} d_i}{|S|} \tag{6.44}$$

SP is the standard deviation of d_i.

$$SP(S) = \sqrt{\frac{1}{|S|-1} \sum_{i=1}^{|S|} \left(d_i - \bar{d}\right)^2} \tag{6.45}$$

A smaller *SP* might mean better distribution. But Okabe *et al.* pointed out that the smallest distance can be used twice, which might lead to wrong conclusion [41].[62]

The second distribution PI, suggested by Zitzler as M_2^* in 1999, is based on the niche concept [43].

$$M_2^*(S) = \frac{1}{|S|-1} \sum_{i=1}^{|S|} \left| s_j \in S \left| \left\| s_i - s_j \right\| > \sigma \right. \right| \tag{6.46}$$

where $\|\|$ is a means of distance, such as Euclidean distance, and σ is the radius of the niche. For the ith solution in S, we first find out how many individuals are far from it and use it as a representative of sparsity. Then we calculate the average isolated individual number to demonstrate the overall distribution. Larger M_2^* means better distribution.[63]

The third PI, suggested by Li and Zhang, considers convergence and distribution by inverting the meaning of generational distance (*IGD*) as follows [29]. We first need to get PF* or RS and select N solutions evenly on PF* or RS as representations. For each nondominated solution in PF* or RS, the closest individual in S and the corresponding distance can be found as follows:

$$d_i = \min_{p \in S} \left\{ \sqrt{\sum_{k=1}^{m} \left(z_k^{*i} - z_k^p\right)^2} \right\} \tag{6.47}$$

[62] Readers are encouraged to calculate the *SP* for $(0,8)$, $(1,7)$, $(7,1)$, $(8,0)$.

[63] What is the largest value of M_2^*?

where z_k^{*i} is the kth objective value of the ith solution in PF* or RS. *IGD* is defined as follows:

$$IGD(S) = \frac{\sum_{i=1}^{N} d_i}{N} \tag{6.48}$$

As can be seen from Eq. 6.48, in the early stage of evolving, *IGD* mainly evaluates the convergence ability of MOEAs. Later, while the archive or population is approaching PF*, smaller *IGD* requires the individuals in *S* being distributed similarly to the *N* selected solutions, which promotes the even distribution of the results. The minimum of *IGD* is zero, which means that the MOEA found every representation of PF* or RS. So larger number of *N* is preferred.

6.7.2.6 Spread Performance Indices

Spread is used to evaluate the ability of the algorithm in extreme conditions (one weight is 1 and the others are 0). In 1999, Zitzler suggested M_3^* to evaluate the spread [43].[64]

$$M_3^*(S) = \sqrt{\sum_{i=1}^{m} \max\{\|u_i - v_i\| \, |u, \, v \in S\}} \tag{6.49}$$

where $\|\|$ is a way of measuring distance. We first find the largest distance in different objectives and calculate their sum square. The larger M_3^*, the better the spread.

6.7.2.7 Distribution and Spread Performance Indices

In 2000, Deb *et al.* proposed a way to evaluate distribution and spread simultaneously for two-objective MOPs [8]. Let us take Fig. 6.27 as an example, where squares are solutions of *S* and circles are edge points generated from PF* or RS.

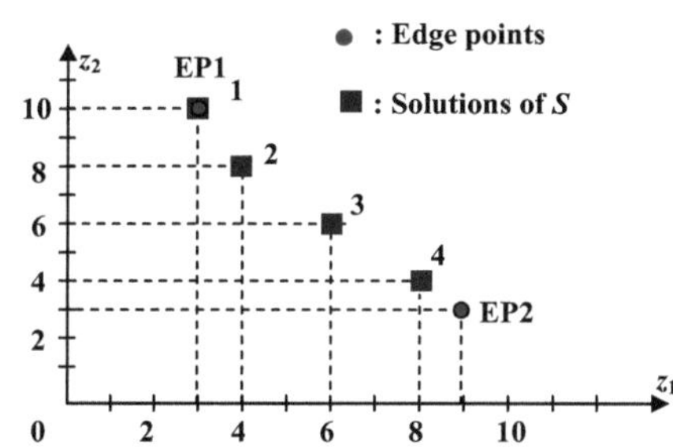

Fig. 6.27 Example of distribution and spread PI

[64] In Sects. 6.4 and 6.5, we ask readers to consider the preferences for different rank schemes. Do those that prefer the side (not the center) have advantages in finding the edge points?

We first calculate the distance between consecutive solutions, i.e., $d(1,2), d(2,3), d(3,4)$ in Fig. 6.27, and denote it d_i.[65] Then we calculate the average of the consecutive distances as $\bar{d} = \frac{\sum_{i=1}^{|S|-1} d_i}{|S|-1}$. The distribution and spread PI (Δ) are defined as follows:

$$\Delta(S) = \frac{d_f + d_l + \sum_{i=1}^{|S|-1} |d_i - \bar{d}|}{d_f + d_l + (|S|-1)\bar{d}} \tag{6.50}$$

where d_f and d_l represent the smallest distance between the edge points and the boundary solutions of S, i.e., the distance between point 1 and EP1 is d_f $(=0)$ and the distance between point 4 and EP2 is d_l $(= d(4, \mathrm{EP2}))$ in Fig. 6.27. If $d_f = d_l = 0$, S has a good spread property. Based on that, if solutions in S distribute evenly, $\Delta(S) = 0$. By contrast, if solutions in S are crowded into one small group, even if they are distributed evenly, $\Delta(S)$ is getting larger.[66] A smaller Δ means a better distribution and spread.

In 2003, Leung and Wang proposed another way, *U-measure*, to evaluate distribution and spread simultaneously [50]. They suggested an efficient way to calculate the smallest distance to others for individual i, i.e., d_1^i. The U-measure also requires the smallest distance d_2^i to the edge points, like d_f and d_l in Δ.

A good distribution and spread of S means that $\{d_1^i\}$ are similar and $\{d_2^j\}$ approaches 0. Leung and Wang uses an innovative way to combine these two objectives by generating a new set $\{d'^i\}$ with $D = |S| + m$ individuals. For the smallest distances between individuals $\{d_1^i\}$, $d'^i = d_1^i$; for the minimal distance to the edge points, $d'^j = d_2^j + \sum_{i=1}^{|S|} d_1^i \Big/ |S|$. By this transformation, the requirements for distribution and spread become the requirement that $\{d'^i\}$ be similar.

Then Leung and Wang define the ideal distance, which is the average of $\{d'^i\}$:

$$d_{\text{ideal}} = \frac{\sum_{i=1}^{D} d'^i}{D} \tag{6.51}$$

If the nondominated individuals are evenly distributed and occupy the edge points, $d'^i \big/ d_{\text{ideal}} \to 1$ for all $i = 1, \cdots, D$, then the U-measure calculates the discrepancy among $\{d'^i\}$:

$$U(S) = \frac{1}{D} \sum_{i=1}^{D} \left| \frac{d'^i}{d_{\text{ideal}}} - 1 \right| \tag{6.52}$$

[65] Unlike Eq. 6.43.

[66] $\sum_{i=1}^{|S|-1} |d_i - \bar{d}| = 0$, but d_f and d_l are not zero.

The smaller U is, the better the distribution and spread.

By the end of this subsection, we need to mention that [42]:

- No single PI is able to account for all aspects of the quality of MOEAs.
- With unary indices, we can only say that algorithm 1 is no worse than algorithm 2 if the PI of algorithm 1 is better than that of algorithm 2.

Sect. 6.6.2 introduced several techniques to maintain an archive with limited capacity. In 2003, Knowles and Corne suggested an archive maintenance method using performance index [20]. For any newly generated individual i, while the archive is full, it can enter the archive if either of the two following conditions satisfied. (1) It dominates any member of the archive. (2) PI(s) are improved by adding the new indidivual and removing one member from the archive.

6.8 Objectives vs. Constraints

There are two problems in the relationship of objectives and constraints. How are constraints handled in a MOP environment? Can constraints be regarded as objectives and MOEAs used to solve single-objective constrained problems? We will discuss these problems separately.

6.8.1 Handling Constraints in Multiobjective Optimization Problems

Generally speaking, the methods discussed in Chap. 4 can be used to handle the constraints in MOPs. But if we could implant the feasibility requirement into the binary tournament selection, which is used in almost all MOEAs, then the constraint handling problem might be solved in a more flexible way.

In NSGA-II [8], Deb *et al.* suggested an intuitive idea to modify the Pareto rank procedure in Fig. 6.10 and many MOEAs have adopted their idea in implementing the algorithm. We just need to make a small change in Pareto dominance by turning it into *constrained dominance*.[67] In objective space, if any of the following conditions is true, point i constrained-dominates point j.

1. Point i is feasible and point j is infeasible.
2. Both of them are infeasible, but the overall constraint violation of point i is less than that of point j.
3. Both of them are feasible; point i Pareto dominates point j.

After determining the constrained-dominance pairwise, the rank can be allocated by Fig. 6.6. Then all the feasible solutions will have selective advantages both in

[67] This idea comes from Deb's constraint handling technique published in 2000 [51].

the Pareto ranking procedure and in the tournament selection, which accelerates the convergence toward the feasible region.

6.8.2 Multiobjective Evolutionary Algorithms for Constraint Handling

In real-world problem modeling, designers need to satisfy many requirements. Some of them are *boundary requirements*, which are often modeled as constraints, and some of them are *tendency requirements*, which are often modeled as objectives. But there exist some requirements that could be modeled as both objectives and constraints. Let us take the cost of a product as an example. Of course, lower cost means more competitive advantages in the market. But the designer knows that costs cannot be reduced without any side effects, such as lower quality. Yes, we could model the quality as another objective and make it a MOP. But sometimes quality (and other factors) is hard to express with formulas of design variables. Why not model the costs as a constraint to ensure that they are below the upper bound?

Another viewpoint is from MOEAs. We have discussed many effective algorithms that can handle multiple objectives. Why not fully use the power of MOEAs to make them handle the constrained satisfaction problems?

Considerations from these two sides promoted the field of applying the MOEAs to solve the *constrained optimization problem* (COP).[68] Fonseca and Fleming formulated this idea formally in 1998 [52]. In 2002, Mezura-Montes and Coello Coello provided a numerical comparison of multiobjective-based techniques to handle constraints [53] and in 2006 gave an excellent survey in this field [54].

If we want to use MOEAs to solve COPs, constraints need to be transformed into objective(s). In Mezura-Montes and Coello Coello's classification, there are two categories of transformation.

The first category is to transform the COP into a two-objective problem. One objective is the original objective and the other is the overall violation of the constraints. We want to minimize both of them simultaneously.

The second category is to transform the COP into a $(k+1)$-objective problem, where k is the number of constraints and every objective related to one constraint is the violation of that constraint. We want to minimize all of them simultaneously.

Here we need to present the *constrained optimization problem* (COP) again for convenience.

$$\begin{aligned} &\min\ f(\mathbf{x}),\ \mathbf{x}\in\mathbb{R}^n \\ &\text{s.t.}\quad g_i(\mathbf{x})\leqslant 0, \quad i=1,\cdots,q \\ &\qquad\ \ h_j(\mathbf{x})=0, \quad j=q+1,\cdots,k \end{aligned} \tag{6.53}$$

[68] We are only interested in single-objective COP here.

If a point $\mathbf{x}$ satisfies $g_i(\mathbf{x}) = 0$ for inequality constraint i, we say that constraint i is *active* at point $\mathbf{x}$. All equality constraints are considered active at feasible region F.

Runarsson and Yao presented an intensive discussion in 2005 [55] on the above ideas and concluded that a search bias toward the feasible region must be introduced in optimizing the MOPs formulated by the procedure discussed above. Their opinion is illustrated by Fig. 6.28.[69] The Pareto front of the two-objective problem has been illustrated. The global optimum must be feasible. So it must reside in the line $constraint - violation = 0$, which is the thick horizontal line in Fig. 6.28. And the objective of the problem is to find as small an objective value as possible. So the circle at the intersection of the feasible line and the Pareto front is the point we want, which means we do not need the evenly distributed solutions at the end of the algorithms, but one edge point on the PF*. That is the meaning of Runarsson and Yao's bias toward the feasible region.

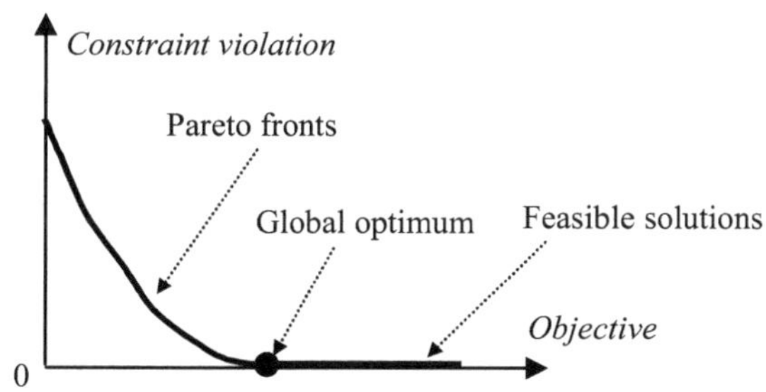

Fig. 6.28 Using MOEAs to solve constraint problems

6.8.2.1 Transforming Constrained Problems into $(1+k)$-objective Problems

For q inequality constraints, we can do the transformation in the following way:

$$f_i(\mathbf{x}) = \max\left(0, g_i(\mathbf{x}) \leqslant 0\right), \quad i = 1, \cdots, q \tag{6.54}$$

For $k - q$ equality constraints, we can define

$$f_i(\mathbf{x}) = |h_i(\mathbf{x})|, \quad i = q+1, \cdots, k \tag{6.55}$$

Or we can first transform equality constraints into inequality constraints:

$$|h_i(\mathbf{x})| - \varepsilon < 0, \quad i = q+1, \cdots, k \tag{6.56}$$

where ε is a predefined small tolerance. Then Eq. 6.54 can be used to transform the above inequality constraint into an objective.

[69] It seems that Fig. 6.28 only addresses the first category, but the second one has the same conclusion.

After the procedure, we get an unconstrained MOP with $k+1$ minimum objectives.

As mentioned above, bias should be taken in account in treating these $k+1$ objectives, where the first objective is the original objective.

In 2003, Angantyr *et al.* proposed a way to use Pareto rank to solve $(k+1)$-objective problems [56]. They first rank the population according to the first objective. A value $rank_1(\mathbf{x}_j)$ is given for individual j. The smaller, the better.

Then Goldberg's Pareto rank method (Fig. 6.6) is carried out on the remaining k objectives. A value $rank_2(\mathbf{x}_j)$ is given for individual j. The smaller, the better.

Angantyr *et al.* used the following equation to express their consideration for exploration and exploitation in the constrained solution space:

$$\varphi(\mathbf{x}_j) = \frac{feasize}{popsize} rank_1(\mathbf{x}_j) + \frac{popsize - feasize}{popsize} rank_2(\mathbf{x}_j) \tag{6.57}$$

where $\varphi(\mathbf{x}_j)$ is the fitness value for individual $\mathbf{x}_j$, $popsize$ is the number of individuals in the population, and $feasize$ is the number of feasible individuals in the population. Equation 6.57 can be discussed with the following considerations.

1. If there is no feasible individual in the population, $\varphi(\mathbf{x}_j) = rank_2(\mathbf{x}_j)$, the search is directed toward the feasible region.
2. If there are feasible and infeasible individuals in the population, the feasible ones might have less selective advantage than the infeasible ones according to the percentage of feasible individuals.
3. If the majority of the population is feasible individuals, the search is directed toward the unconstrained optimum.

Considerations 2 and 3 might cause the oscillation in the evolving process if the unconstrained optimum is infeasible. But Angantyr *et al.* claimed that the oscillation was helpful for exploring the global feasible optimum.

In 2003, Aguirre *et al.* suggested another way to handle $k+1$ objectives based on PAES [57]. The algorithm is a $(1+1)$-ES, maintaining an archive A, and only requires Pareto dominance. The solution process is as follows:

One Generation of the Algorithm Suggested by Aguirre et al.

Phase 1: Generating new individual c by mutating parent h on every variable dimension i using σ_i. Unfamiliar readers are referred to Sects. 2.3.1 and 3.2.2.3.

Phase 2: Elitism. Maintain the best feasible solution found thus far. No need to insert it in the evolving process.

Phase 3: Determining whether c could replace h and whether c could enter A. The determining procedure was elaborated in Sect. 6.5.3.

Phase 4: Selecting randomly from A to reinitialize h for every g generations. Angantyr *et al.* suggest that $g = 10$.

Phase 5: Shrinking the search domain for every r generations. Angantyr *et al.* suggest that $r = 1$ or 2.

Step 5.1: Select the top 15% from A. For every constraint, the infeasible individuals are eliminated until the remaining individuals of A is $15\%A$. If there are more than $15\%A$ feasible individuals in A, select the best $15\%A$ and discard others.

Step 5.2: Determine the upper bound $\bar{x}_i(t)$ and lower bound $\underline{x}_i(t)$ in generation t for every variable i $(i = 1, \cdots, n)$ using the $15\%A$ individuals. Suppose the interval for variable i is $width_i(t) = \bar{x}_i(t) - \underline{x}_i(t)$.

Step 5.3: Shrink the variable interval using a factor $0 < \beta < 1$ by $width_i(t+1) = width_i(t)\beta$ and generating $\bar{x}_i(t+1)$ and $\underline{x}_i(t+1)$ accordingly.

Step 5.4: Calculate the standard deviation for generating new individuals by $\sigma_i(t+1) = width_i(t+1)/\sqrt{n}$, $i = 1, \cdots, n$.

In this way, the algorithm shrinks the mutation interval for every r generations to exploit the global optimal solution and maintains a balance between the feasible and infeasible individuals.

6.8.2.2 Transforming Constrained Problems into Two-objective Problems

In 2006, Cai and Wang suggested a way to transform COPs into two-objective problems [58], one of them being the original objective and the other one the overall constraint violation. Their main idea could be expressed as follows:

- A good NLP problem solver needs to direct the search toward the feasible region effectively. So information contained in the infeasible solutions is quite helpful, especially when the global optimal solution is at the edge of a feasible region or the proportion of feasible region to whole region is small. So they use an archive A to store the "best" infeasible individuals (with small constraint violation) when all the offspring are infeasible.
- To find the global optimal solution, the algorithm should employ effective ways to generate, compare, and replace individuals. So Cai and Wang implemented their algorithm under the general ES framework.

The algorithm contains two groups of individuals: P, which is equal to the population in GAs, and A, which stores the "best" infeasible individuals close to the feasible region.

The solution process is as follows.

One Generation of the Algorithms Suggested by Cai and Wang

Phase 1: Improving P.

Step 1.1: Select μ individuals from P randomly to form the parents in ES. Then delete the μ individuals from P.

Step 1.2: Generate λ individuals from μ individuals using simplex cross-over (SPX) to form the offspring in ES. SPX has been discussed in Sect 3.2.2.1.

Step 1.3: Determine the nondominated individuals in λ offspring using Pareto dominance on the two-objective problem. Suppose there are m' nondominated offspring.

Step 1.4: For every nondominated offspring in m', try to find one individual in μ parents to be replaced. If an offspring does not dominate any parent, there is no replacement. If an offspring only dominates one parent, it replaces the parent. If an offspring dominates more than one feasible parent, it replaces the worst one. Otherwise (the offspring dominates some infeasible parents), it replaces one randomly.

Step 1.5: Combine the parent set with P.

Phase 2: Updating A. If there is no feasible solution in the λ offspring, find the infeasible individual with the lowest degree of constraint violation and add it to A.

Phase 3: Inserting individuals of A into P for every m generations. For every m generations, randomly select n individuals from A and replace n randomly selected individuals in P. Then empty A.

Phase 1 is a $(\mu+(m',\lambda))$-ES. There is an alternative way, which is to randomly select only one nondominated offspring to replace one parent. The second way is a $(\mu+(1,\lambda))$-ES.

Besides the above two ideas, Cai and Wang also analyzed the difficulty of equality constraints. If we do not transform them into inequality constraints (Eq. 6.56), the ratio of feasible region to solution region is nearly zero, which means feasible points are far fewer in number than infeasible points in the neighborhood of the global optimal solution.

In the procedure above, if μ parents do not contain any feasible points and the nondominated offspring include feasible solutions but they do not dominate the infeasible parents, the feasible solutions cannot enter P.[70] To make things worse, phase 3 in the above procedure will insert infeasible individuals for every m generations. So in the final stage of the evolution, the population is approaching the global optimal solution, which resides in the line of equality constraints, and it will gradually lose all the feasible solutions step by step.

Cai and Wang made a pragmatic yet effective change in the above procedure to avoid such a tragedy. They define a condition: individuals in P are all infeasible and

[70] Consider this statement with the help of Fig. 6.28.

their objective values are similar, which means P is approaching the global optimal solution but it is hard to contain the feasible solution due to equality constraints. If the condition holds, replacement rules for m' nondominated offspring replacing parents, i.e., step 1.4, are changed to one similar to Deb's idea discussed in Sect. 6.8.1. For every nondominated offspring, randomly select one individual among the parents and begin the competition. Feasible offspring will replace infeasible parents. Infeasible offspring with smaller constraint violations will replace the worse infeasible parents. Apart from that, if the condition holds, phases 2 and 3 will not be carried out. In this way, a strong preference for feasible individuals is set up, no more infeasible individuals will be inserted into P, and the tragedy discussed above will be avoided.

The archive scheme directs the evolution toward the feasible region. The Pareto dominance in the replacement procedure directs the evolution toward feasible solutions with good objective values. SPX explores the solution space effectively. The tragedy-avoiding mechanism handles the equality constraint in a flexible way. All these factors make the algorithm suggested by Cai and Wang "remarkably outperforms" state-of-the-art COEAs including the stochastic ranking.[71]

6.9 Application Example

There are so many real-world applications of MOEAs on MOPs. Here we just introduce the one on the design of superconducting magnetic energy storage (SMES) solenoids suggested by Zhao and Yu in 2008 [59].

The SMES system has an attractive potential in power systems and other areas. A solenoid coil, one important configuration of SMES, can be simply built and provide high energy density per unit of the conductor.

Many factors need to be considered in the design of solenoid SMES.

1. **Coil volume.** The volume determines the occupation of area, which in turn determines the installation and using cost.
2. **Uniformity of magnetic field.** In some applications (such as magnetic resonance imaging, MRI), solenoids are required to provide a uniformly distributed magnetic field.
3. **Energy requirement.** The energy determines the requirement of the user.
4. **Stray field.** The SMES should be safe for people far enough away from it.
5. **Superconductivity requirement.** In order to maintain superconductivity, the current density J and the magnetic field B must ensure that the working point in the $B-J$ plane lies below the critical quenching curve.

The design variables are the inner radius of the solenoid coil (R_i), the outer radius of the solenoid coil (R_o), the height of the solenoid coil (h), and the current density fed into the solenoid coil (J) (Fig. 6.29).

---

[71] Stochastic ranking was discussed in Chap. 4.

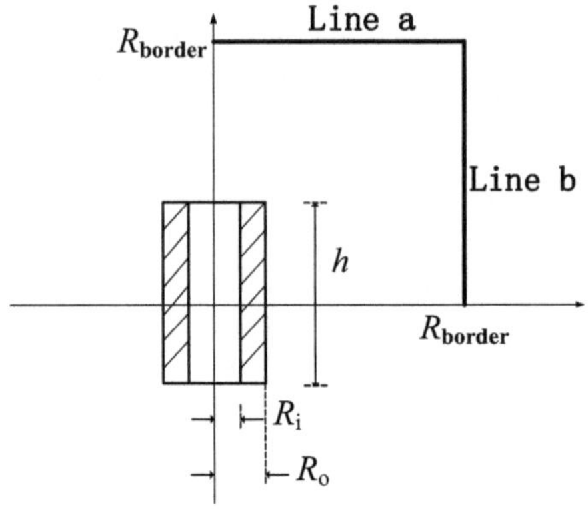

Fig. 6.29 Solenoid coil

The volume can be calculated easily with R_{i}, R_{o}, and h.

The uniformity of the magnetic field can be calculated by

$$U = \frac{|B_{\text{m}} - B_0|}{B_0} \tag{6.58}$$

where B_{m} represents the maximum of the magnetic flux density in the solenoid and B_0 is the magnetic flux density at the center of the solenoid. An evenly distributed magnetic field means small U.

Figure 6.29 illustrates the calculation of the stray field. Lines a and b are borderline for evaluating the stray field. m points are picked with the same space along lines a and b. The average stray field is calculated using

$$B_s^2 = \frac{\sum_{i=1}^{m} |B_{si}|^2}{m} \tag{6.59}$$

where B_{si} is the leak magnetic intensity at point i.

The linear approximation of the real critical quenching curve can be considered a quench condition and can be expressed as follows. It must not be violated at any point in the coils.

$$|J_C| = (-6.4\,|B_{\text{m}}| + 54.0) \quad \text{MA/m}^2 \tag{6.60}$$

We take factors 1 and 2 as objectives and 3 to 5 as constraints and get a two-objective design model as

$$\begin{aligned} &\min\ V_s \\ &\min\ U = \frac{|B_m - B_0|}{B_0} \\ &s.t\ \frac{|E - E_0|}{E_0} < \varepsilon \\ &B_s < B_{s0} \\ &J < \sigma J_c \end{aligned} \tag{6.61}$$

where V_s is the solenoid volume, E is the energy stored in the solenoid, E_0 is the design target of energy, ε represents the acceptable tolerance of energy error, B_{s0} is the safe magnet stray field, and $0 < \sigma < 1$ is the safe factor of the current density.

According to Zhao and Yu's calculation, the design model is highly constrained, which means that the ratio of the feasible region to the search space is very small. So we adopt $(\mu + \lambda)$-ES to solve the problem.

In the replacement stage of ES, we divide the $(\mu + \lambda)$ individuals into four groups [feasible Pareto (FP) solutions, feasible dominated (FD) solutions, infeasible Pareto (IP) solutions, and infeasible dominated (ID) solutions] and use binary tournament selection with replacement to determine the μ parents in the next generation.

A comparison of the group is summarized in Table 6.1.

Table 6.1 The regulation of the binary tournament selection

	FP	FD	IP	ID
FP	Select the FP with a larger crowding distance	Select FP	Select FP	Select FP
FD	Select FP	Select the FD nearer to FP	Select FD or IP by a certain rate	Select FD
IP	Select FP	Select IP or FD by a certain rate	Select the IP with less constraint violation level	Select IP
ID	Select FP	Select FD	Select IP	Neither

With $(\mu + \lambda)$-ES and comparison rules in Table 6.1, the feasible Pareto front is found after 200 generations with $\mu = 50$. The real solenoid SMES design parameters are determined by picking one from the nondominated solutions and improving it manually.

6.10 Summary

Basically there are two ways to handle MOPs. If users can provide some kind of preference information, such as weights on objectives, goals of objectives, etc., the preference-based approach discussed in Sect. 6.2 is quite efficient.

On the other hand, if the designer and the decision maker need to pick one solution from many nondominated solutions, MOEAs should try to converge toward PF* and distribute the individuals evenly along PF*. In Sect. 6.4, we introduce how to evaluate the quality and distribution of individuals during the evolution, which is extremely important for designing and analyzing MOEAs.

Classical MOEAs such as NSGA-II and SPEA2 were discussed in depth in Sect. 6.5 and some interesting state-of-the-art considerations were introduced in Sect. 6.6. Readers should learn and borrow the innovative ideas from those sections.

In a multiple-objective environment, comparisons between the results of two algorithms are hard to make because we have multiple objectives to compare: convergence, distribution, and spread. Many performance indices were introduced in Sect. 6.7.

Transforming constraints into objectives and using MOEAs to solve the transformed MOP is a wonderful idea for treating single-objective constrained problems. In Sect. 6.8, we give the classification first, and then discuss three inspiring algorithms.

After reading this chapter, you should understand the terms used in MOEAs, have a full understanding of at least one classical MOEA, know some improvement techniques, know how to make unbiased comparison of algorithms, and improve your constraint-handling skills.

In all, designing a MOEA is the art of tradeoff between converging toward PF* and distributing evenly along PF*.

Suggestions for Further Reading

Deb published the first MOEA monograph in 2001 [60]. Coello Coello *et al.* published their monograph on MOEAs in 2002 and revised it for a second edition in 2007 [61]. The third monograph on MOEAs was by Knowles *et al.* in 2008 [62]. In these three books, topics are discussed in more detail and more areas covered. Books on the application have been written by Coello Coello *et al.* [63] and Tan *et al.* [64].

There are some comparison papers [33, 65–67] suitable for reading to learn both which algorithm is better in numerical experiments and how to design, implement, and summarize comparisons of MOEAs.

We introduce the treatment of two-objective problems in this chapter for pedagogical reasons. But real-world MOPs might contain multiple objectives. Numerical experiments indicate that the performances of MOEAs do not scale well with respect to the number of objectives. Interested readers are referred to [34, 68].

We have not introduced the fast nondominated sorting approach to NSGA-II because we wished to focus on the quality and diversity of solutions at first for students reading this textbook. We encourage readers to check [8] for this approach before carrying out their implementation. Jensen suggested a faster approach to nondominated sorting in 2003 [69].

If readers have a greater interest in evaluating quality (rank) during evolution, they are suggested to read the paper published by Pierro *et al.* in 2007 [70]. Another paper by Chan *et al.*, published in 2008, discusses how to design new operators that explore and exploit more efficiently for PF* [46].

We only discussed archive maintenance techniques with limited capacity in Sect. 6.6.2. In 2003, Fieldsend *et al.* suggested an efficient data structure, i.e., dominated trees, to search, add, and delete elements in archive quickly [71]. With the help of dominated trees, their algorithm does not need to limit the size of the archive, which promotes search speed and quality prominently.

The concept of the orthogonal GA has been introduced in the Suggested Readings of Chap. 3. In 2004, Zeng *et al.* expanded the orthogonal GA into the field of MOPs [72], which can be regarded as another example of expanding single-objective algorithms into MOPs.

In 2005 Knowles and Corne gave a good introduction to combine memetic algorithms with MOEAs, i.e., utilizing local search to promote PF* search and even distribution [73], in 2009 Ishibuchi *et al.* implemented a biased neighborhood to assign larger probabilities to more promising neighbors [74], and in 2009 Goh *et al.* published a book on this topic [75].

Those readers interested in the estimation of distribution algorithms, introduced in Sect. 3.5.2.6, will find its application to MOPs in a paper published in 2008 [76].

We have not discussed the issue of MOEAs in an uncertain environment, which is extremely important in real-world applications because noise is everywhere. For handling noise in MOEAs, readers are referred to [77] and [78], published in 2002 and 2007 by Buche *et al.* and Goh *et al.*, respectively. For designing robust MOEAs, readers might be interested in reading [79] by Deb and Gupta and [80] by Paenke *et al.*, published in 2006 respectively. In 2006 Knowles published a paper to estimate the solution landscape for MOPs [81] and this paper won the IEEE Transactions on Evolutionary Computation Outstanding Paper Award. For dynamic problems, objective values change with time, and two papers published by Farina *et al.* and Goh and Tan in 2004 and 2009, respectively, are suitable for reading [18, 82].

The research on archive maintenance, PI, constraint handling, uncertain environments, expanding other algorithms into MOPs, and applications has contributed to the rapid increase in the number of papers indexed by SCI from 2003 until now.

For those readers interested in other discussions on PI, we suggest three papers published in 2001, 2003, and 2008, respectively [83–85].

Using MOEAs to deal with COPs has attracted more attention recently. Apart from Cai and Wang's elaborate suggestion, which is introduced in Sect. 6.8.2.2, Venkatraman and Yen also proposed a two-objective optimization approach for COPs in 2005 [86]. They first force the population to evolve toward feasible region using a linear-rank-based approach by only considering the constraint violation. After at least one feasible solution has been found, the algorithm proceeds to the second phase, simultaneously optimizing the objective function and constraint violation. In 2007 and 2008, Wang *et al.* utilized deterministic crowding[72] to maintain diversity and iteratively select half of the nondominated solutions according to the constraint violation to push individuals toward a feasible domain respectively in two-objective optimization approach for COPs [87, 88].

The must-read papers for MOEAs are [8] for NSGA-II, [10] for SPEA2, [23] for elaborately transforming MOPs into two-objective problems and utilizing the drawback of VEGA elaborately, [30] for test problem design and analysis, [41] for PIs, [54] for constraint handling, and [33] or [29] for comparison.

Exercises and Potential Research Projects

6.1. Implement at least one classical MOEA introduced in Sect. 6.5 and use at least three MOPs in Appendix to test its performance. Techniques introduced in Sect. 3.6.2 are required for drawing conclusions.

6.2. Adopt at least one improving technique introduced in Sect. 6.6 to your algorithm and do a comparison between the original one and the improved one using several PIs introduced in Sect. 6.7. Techniques introduced in Sect. 3.6.2 are required for drawing conclusions.

6.3. Consider a way to improve Deb's crowding distance for diversity evaluation without any parameter and use the benchmark problems and PIs to verify your idea. Techniques introduced in Sect. 3.6.2 are required for drawing conclusions.

6.4. What is the convergence, distribution, and elitism mechanism of PESA and PAES?

6.5. Summarize the essence of [46] on a single sheet of paper to illustrate its new operator.

6.6. Summarize the essences of [70] on a single sheet of paper to illustrate its new ranking scheme.

6.7. Summarize the essences of [24] on a single sheet of paper to illustrate its new dynamic *popsize* control method.

6.8. Is it possible that $I_\varepsilon(S_1, S_2) < 1$ and $I_\varepsilon(S_2, S_1) < 1$ in ε-dominance discussed in Sect. 6.7.2.3? Give examples to support your statement.

6.9. Summarize and review the edge-points-keeping methods introduced in this chapter.

---

[72] DC is introduced in Sect. 5.4.1.

6.10. Compare the archive maintenance technique discussed in Sects. 6.6.2.1 and 6.6.2.2. If possible, implement them in your programming environment and use the benchmark problems and PIs to do numerical comparison. Techniques introduced in Sect. 3.6.2 are required for drawing conclusions.

6.11. What's the difference between MOEA/D introduced in Sect. 6.6.3.2 and run *popsize* times of single-objective EAs, each optimizing one of the *popsize* single-objective aggregation functions generated by *popsize* weight vectors?

6.12. Implement at least one method of applying MOEAs to COPs introduced in Sect. 6.8.2 and use the benchmark problems in Appendix to test its performance. Techniques introduced in Sect. 3.6.2 are required for drawing conclusions.

References

1. Schaffer JD (1985) Multiple objective optimization with vector evaluated genetic algorithms. In: Proceedings of the 1st international conference on genetic algorithms, pp 93–100
2. Goldberg DE (1989) Genetic algorithms in search, optimization, and machine learning. Addison-Wesley, Boston, MA
3. Fonseca C, Fleming P (1993) Genetic algorithms for multiobjective optimization: Formulation, discussion and generalization. In: Proceedings of the fifth international conference on genetic algorithms, pp 416–423
4. Horn J, Nafpliotis N, Goldberg D (1994) A niched pareto genetic algorithm for multiobjective optimization. In: Proceedings of the genetic and evolutionary computation conference, pp 82–87
5. Corne DW, Knowles JD, Oates MJ (2000) The pareto envelope-based selection algorithm for multiobjective optimization. In: Proceedings of the international conference on parallel problem solving from nature, pp 839–848
6. Knowles JD, Corne DW (2000) Approximating the nondominated front using the pareto archived evolution strategy. Evol Comput 8:149–172
7. Osyczka A, Kundu S (1996) A modified distance method for multicriteria optimization, using genetic algorithms. Comput Ind Eng 30(4):871–882
8. Deb K, Pratap A, Agarwal S *et al* (2002) A fast and elitist multiobjective genetic algorithm: NSGA-II. IEEE Trans Evol Comput 6(2):182–197
9. Srinivas N, Deb K (1994) Multiobjective optimization using nondominated sorting in genetic algorithms. Evol Comput 2:221–248
10. Zitzler E, Laumanns M, Thiele L (2001) SPEA2: improving the strength pareto evolutionary algorithm. Tech. rep. 103, ETH Zurich, Switzerland
11. Corne DW, Jerram NR, Knowles JD *et al* (2001) PESA-II: region-based selection in evolutionary multiobjective optimization. In: Proceedings of the genetic and evolutionary computation conference, pp 283–290
12. Goldberg DE (1989) Sizing populations for serial and parallel genetic algorithms. In: Proceedings of the third international conference on Genetic algorithms, pp 70–79
13. Coello Coello CA, Pulido GT (2001) A micro-genetic algorithm for multiobjective optimization. In: Proceedings of the genetic and evolutionary computation conference, pp 126–140
14. Iorio AW, Li X (2006) Incorporating directional information within a differential evolution algorithm for multi-objective optimization. In: Proceedings of the ACM annual conference on Genetic and evolutionary computation, pp 691–698
15. Nebro AJ, Luna F, Alba E *et al* (2008) AbYSS: adapting scatter search to multiobjective optimization. IEEE Trans Evol Comput 12(4):439–457

16. Tan K, Yang Y, Goh C (2006) A distributed cooperative coevolutionary algorithm for multiobjective optimization. IEEE Trans Evol Comput 10(5):527–549
17. Tan KC, Khor EF, Lee TH *et al* (2003) An evolutionary algorithm with advanced goal and priority specification for multi-objective optimization. J Artif Intell Res 18:183–215
18. Goh C, Tan KC (2009) A competitive-cooperative coevolutionary paradigm for dynamic multiobjective optimization. IEEE Trans Evol Comput 13(1):103–127
19. Laumanns M, Thiele L, Deb K *et al* (2002) Combining convergence and diversity in evolutionary multiobjective optimization. Evol Comput 10(3):263–282
20. Knowles J, Corne D (2003) Properties of an adaptive archiving algorithm for storing nondominated vectors. IEEE Trans Evol Comput 7(2):100–116
21. Deb K, Mohan M, Mishra S (2005) Evaluating the ε-domination based multi-objective evolutionary algorithm for a quick computation of Pareto-optimal solutions. Evol Comput 13(4):501–525
22. Toffolo A, Benini E (2003) Genetic diversity as an objective in multi-objective evolutionary algorithms. Evol Comput 11(2):151–167
23. Lu H, Yen G (2003) Rank-density-based multiobjective genetic algorithm and benchmark test function study. IEEE Trans Evol Comput 7(4):325–343
24. Yen G, Lu H (2003) Dynamic multiobjective evolutionary algorithm: adaptive cell-based rank and density estimation. IEEE Trans Evol Comput 7(3):253–274
25. Miettinen K (1999) Nonlinear multiobjective optimization. Kluwer, Holland
26. Ishibuchi H, Murata T (1998) A multi-objective genetic local search algorithm and its application to flowshop scheduling. IEEE Trans Syst Man Cybern C Appl Rev 28(3):392–403
27. Jaszkiewicz A (2002) Genetic local search for multi-objective combinatorial optimization. Eur J Oper Res 137(1):50–71
28. Zhang Q, Li H (2007) MOEA/D: a multiobjective evolutionary algorithm based on decomposition. IEEE Trans Evol Comput 11(6):712–731
29. Li H, Zhang Q (2009) Multiobjective optimization problems with complicated pareto sets, MOEA/D and NSGA-II. IEEE Trans Evol Comput 13(2):284–302
30. Huband S, Hingston P, Barone L *et al* (2006) A review of multiobjective test problems and a scalable test problem toolkit. IEEE Trans Evol Comput 10(5):477–506
31. Van Veldhuizen DA (1999) Multiobjective evolutionary algorithms: classifications, analyses, and new Innovations. Ph.D. thesis, Air Force Institute of Technology, OH
32. Deb K (1999) Multi-objective genetic algorithms: Problem difficulties and construction of test problems. Evol Comput 7:205–230
33. Zitzler E, Deb K, Thiele L (2000) Comparison of multiobjective evolutionary algorithms: empirical results. Evol Comput 8(2):173–195
34. Deb K, Thiele L, Laumanns M *et al* (2001) Scalable test problems for evolutionary multiobjective optimization. Tech. rep. 2001001, Kanpur Genetic Algorithms Laboratory. Indian Institute of Technology
35. Okabe T, Jin Y, Olhofer M *et al* (2004) On test functions for evolutionary multi-objective optimization. In: Proceedings of the international conference on parallel problem solving from nature, pp 792–802
36. Huband S, Barone L, While L *et al* (2005) A scalable multi-objective test problem toolkit. In: Coello Coello CA, Aguirre AH, Zitzler E (eds) Evolutionary multi-criterion optimization Springer, Berlin Heidelberg New York, pp 280–295
37. Iorio AW, Li X (2006) Rotated test problems for assessing the performance of multi-objective optimization algorithms. In: Proceedings of the ACM annual conference on genetic and evolutionary computation, pp 683–690
38. Deb K, Sinha A, Kukkonen S (2006) Multi-objective test problems, linkages, and evolutionary methodologies. In: Proceedings of the 8th ACM annual conference on genetic and evolutionary computation, pp 1141–1148
39. Huang VL, Qin AK, Deb K *et al* (2007) Problem definitions for performance assessment on multi-objective optimization algorithms. Tech. rep., Nanyang Technological University, Indian Institute of Technology, Swiss Federal Institute of Technology, University Dortmund, The University of Western Australia, Singapore

40. Zhang Q, Zhou A, Zhaoy S *et al* (2008) Multiobjective optimization test instances for the cec 2009 special session and competition. Tech. rep. CES-487, University of Essex, Nanyang Technological University, Clemson University, Singapore
41. Okabe T (2003) A critical survey of performance indices for multi-objective optimization. In: Proceedings of the IEEE congress on evolutionary computation, pp 878–885
42. Zitzler E, Thiele L, Laumanns M *et al* (2003) Performance assessment of multiobjective optimizers: an analysis and review. IEEE Trans Evol Comput 7(2):117–132
43. Zitzler E (1999) Evolutionary algorithms for multiobjective optimization: methods and applications. Ph.D. thesis, Swiss Federal Institute of Technology (ETH) Zurich, Switzerland
44. While L, Hingston P, Barone L *et al* (2006) A faster algorithm for calculating hypervolume. IEEE Trans Evol Comput 10(1):29–38
45. Bradstreet L, While L, Barone L (2008) A fast incremental hypervolume algorithm. IEEE Trans Evol Comput 12(6):714–723
46. Chan T, Man K, Tang K *et al* (2008) A jumping gene paradigm for evolutionary multiobjective optimization. IEEE Trans Evol Comput 12(2):143–159
47. Fonseca CM, Fleming PJ (1996) On the performance assessment and comparison of stochastic multiobjective optimizers. In: Proceedings of the international conference on parallel problem solving from nature, pp 584–593
48. Conover WJ (1999) Practical nonparametric statistics, 3rd edn. Wiley, New York
49. Schott J (1995) Fault tolerant design using single and multicriteria genetic algorithm optimization. Master thesis, MIT, MA
50. Leung Y, Wang Y (2003) U-measure: a quality measure for multiobjective programming. IEEE Trans Syst Man Cybern A 33(3):337–343
51. Deb K (2000) An efficient constraint handling method for genetic algorithms. Comput Methods Appl Mech Eng 186:311–338
52. Fonseca C, Fleming P (1998) Multiobjective optimization and multiple constraint handling with evolutionary algorithms. I: a unified formulation. IEEE Trans Syst Man Cybern A 28(1):26–37
53. Mezura-Montes E, Coello Coello CA (2002) A numerical comparison of some multiobjective-based techniques to handle constraints in genetic algorithms. Tech. rep. EVOCINV-03-2002, Evolutionary Computation Group at CINVESTAV-IPN, Mexico
54. Mezura-montes E, Coello Coello CA (2006) A survey of constraint-handling techniques based on evolutionary multiobjective optimization. In: Proceedings of the PPSN workshop on multiobjective problem solving from nature
55. Runarsson T, Yao X (2005) Search biases in constrained evolutionary optimization. IEEE Trans Syst Man Cybern C 35(2):233–243
56. Angantyr A, Andersson J, Aidanpaa J (2003) Constrained optimization based on a multiobjective evolutionary algorithm. In: Proceedings of the IEEE congress on evolutionary computation, 1560–1567
57. Aguirre AH, Rionda SB, Coello Coello CA *et al* (2004) Handling constraints using multiobjective optimization concepts. Int J Numer Methods Eng 59(15):1989–2017
58. Cai Z, Wang Y (2006) A multiobjective optimization-based evolutionary algorithm for constrained optimization. IEEE Trans Evol Comput 10(6):658–675
59. Yuan Z, Xinjie Y (2008) Pareto competition based evolution strategy for two-objective optimization design of SMES solenoids. IEEE Trans Appl Superconduct 18(2):1513–1516
60. Deb K, Kalyanmoy D (2001) Multi-objective optimization using evolutionary algorithms. Wiley, New York
61. Coello Coello CA, Lamont GB, Van Veldhuizen DA (2007) Evolutionary algorithms for solving multi-objective problems, 2nd edn. Springer, Berlin Heidelberg New York
62. Knowles J, Corne D, Deb K (2008) Multiobjective problem solving from nature: from concepts to applications. Springer, Berlin Heidelberg New York
63. Coello Coello CA, Lamont GB, Coello CA (2004) Applications of multi-objective evolutionary algorithms. World Scientific, Singapore
64. Tan KC, Khor EF, Lee TH (2005) Multiobjective evolutionary algorithms and applications. Springer, Berlin Heidelberg New York

65. Zitzler E, Thiele L (1998) Multiobjective optimization using evolutionary algorithms - a comparative case study. In: Proceedings of the international conference on parallel problem solving from nature, pp 292–304
66. Zitzler E, Thiele L (1999) Multiobjective evolutionary algorithms: a comparative case study and the strength pareto approach. IEEE Trans Evol Comput 3(4):257–271
67. Jaszkiewicz A (2002) On the performance of multiple-objective genetic local search on the 0/1 knapsack problem - a comparative experiment. IEEE Trans Evol Comput 6(4):402–412
68. Khare V, Yao X, Deb K (2002) Performance scaling of multi-objective evolutionary algorithms. Tech. rep. 2002009, KanGAL, Indian Institute of Technology Kanpur, India
69. Jensen M (2003) Reducing the run-time complexity of multiobjective EAs: the NSGA-II and other algorithms. IEEE Trans Evol Comput 7(5):503–515
70. Pierro F, Khu S, Savić D (2007) An investigation on preference order ranking scheme for multiobjective evolutionary optimization. IEEE Trans Evol Comput 11(1):17–45
71. Fieldsend J, Everson R, Singh S (2003) Using unconstrained elite archives for multiobjective optimization. IEEE Trans Evol Comput 7(3):305–323
72. Zeng SY, Kang LS, Ding LX (2004) An orthogonal multi-objective evolutionary algorithm for multi-objective optimization problems with constraints. Evol Comput 12(1):77–98
73. Knowles J, Corne D (2005) Memetic algorithms for multiobjective optimization: issues, methods and prospects. In: Hart WE, Krasnogor N, Smith JE (eds) Recent advances in memetic algorithms. Springer, Berlin Heidelberg New York, pp 313–352
74. Ishibuchi H, Hitotsuyanagi Y, Tsukamoto N *et al* (2009) Use of biased neighborhood structures in multiobjective memetic algorithms. Soft Comput 13(8):795–810
75. Goh C, Ong Y, Tan KC (2009) Multi-objective memetic algorithms. Springer, Berlin Heidelberg New York
76. Zhang Q, Zhou A, Jin Y (2008) RM-MEDA: a regularity model-based multiobjective estimation of distribution algorithm. IEEE Trans Evol Comput 12(1):41–63
77. Buche D, Stoll P, Dornberger R *et al* (2002) Multiobjective evolutionary algorithm for the optimization of noisy combustion processes. IEEE Trans Syst Man Cybern C 32(4):460–473
78. Goh C, Tan K (2007) An investigation on noisy environments in evolutionary multiobjective optimization. IEEE Trans Evol Comput 11(3):354–381
79. Deb K, Gupta H (2006) Introducing robustness in multi-objective optimization. Evol Comput 14(4):463–494
80. Paenke I, Branke J, Jin Y (2006) Efficient search for robust solutions by means of evolutionary algorithms and fitness approximation. IEEE Trans Evol Comput 10(4):405–420
81. Knowles J (2006) ParEGO: a hybrid algorithm with on-line landscape approximation for expensive multiobjective optimization problems. IEEE Trans Evol Comput 10(1):50–66
82. Farina M, Deb K, Amato P (2004) Dynamic multiobjective optimization problems: test cases, approximations, and applications. IEEE Trans Evol Comput 8(5):425–442
83. Wu J, Azarm S (2001) Metrics for quality assessment of a multiobjective design optimization solution set. J Mech Design 123(1):18–25
84. Farhang-Mehr A, Azarm S (2003) An information-theoretic entropy metric for assessing multi-objective optimization solution set quality. J Mech Design 125(4):655–663
85. Lizarraga-Lizarraga G, Hernandez-Aguirre A, Botello-Rionda S (2008) G-Metric: an m-ary quality indicator for the evaluation of non-dominated sets. In: Proceedings of the 10th ACM annual conference on genetic and evolutionary computation, pp 665–672
86. Venkatraman S, Yen G (2005) A generic framework for constrained optimization using genetic algorithms. IEEE Trans Evol Comput 9(4):424–435
87. Wang Y, Cai Z, Guo G *et al* (2007) Multiobjective optimization and hybrid evolutionary algorithm to solve constrained optimization problems. IEEE Trans Syst Man Cyberne B: Cybern 37(3):560–575
88. Wang Y, Cai Z, Zhou Y *et al* (2008) An adaptive tradeoff model for constrained evolutionary optimization. IEEE Trans Evol Comput 12(1):80–92

Chapter 7
Combinatorial Optimization

Abstract Previous chapters discuss *parameter optimization*, i.e., we need to find the optimal values of variables so that the objective function has the maximum/minimum value. Many real-world problems are not like this. We often need to select some elements from a set or arrange the sequence of some events with constraints so that the objective function has the maximum/minimum value. These problems belong to *combinatorial optimization*. We will introduce three examples, explain their respective properties, illustrate how EAs solve them, and summarize design-effective algorithms for them.

7.1 Introduction

7.1.1 Combinatorial Optimization

Combinatorial optimization is optimization derived from discrete mathematics,[1] especially combinatorics. So we often face the concepts of enumeration, combination, and permutation. Suppose we work on a set S whose size is $|S| = n$.

An *enumeration* of set S is an exact listing of all its element.

A *combination* is an *unordered* collection of distinct elements, usually of a prescribed size k and taken from S. The number of possible combinations in such a situation is given by Eq. 7.1:

$$C_n^k = \binom{n}{k} = \frac{n!}{k!(n-k)!} \tag{7.1}$$

where $n!$ denotes the factorial n. For example, the number of combinations of two elements from set $S = \{w,x,y,z\}$ is $C_4^2 = \binom{4}{2} = 6$, which is exactly $\{w,x\}$, $\{w,y\}$, $\{w,z\}$, $\{x,y\}$, $\{x,z\}$, $\{y,z\}$.

---

[1] Discrete mathematics is the branch of mathematics dealing with objects that only have distinct, separate values, often characterized by integers.

A *permutation* is a *sequence* containing k distinct elements from S. The number of possible permutations in such a situation is given by Eq: 7.2.

$$P_n^k = \frac{n!}{(n-k)!} \tag{7.2}$$

The number of permutations of two elements from set $S = \{w,x,y,z\}$ is $P_4^2 = 12$, which is exactly $\{w \to x\}$, $\{x \to w\}$, $\{w \to y\}$, $\{y \to w\}$, $\{w \to z\}$, $\{z \to w\}$, $\{x \to y\}$, $\{y \to x\}$, $\{x \to z\}$, $\{z \to x\}$, $\{y \to z\}$, $\{z \to y\}$.

An optimization problem related to the optimal combination of the elements is called a *grouping problem* [1]. *Bin packing* is an example of this. Suppose we have n items. Item i has capacity c_i. We also have a lot of bins with the same size B. The problem is to determine the minimum number of bins so that every item can be packed into one bin and the sum of capacities of the items in every bin is no more than B. In the perspective of a grouping problem, what we need to do is a proper grouping of n items that is subject to a size constraint and with a minimum number of groups.

Optimization related to the optimal permutation of the elements is called a *scheduling problem* [2]. *Flow-shop scheduling* is an example. Suppose we have n jobs. Each job has to be processed on each of the m machines, i.e., from machine 1 to machine m in the sequence order. The operation time of job i on machine k is p_{ik}. The problem is to determine the minimum total processing time by adjusting the sequence of jobs. In the perspective of a scheduling problem, what we need to do is to schedule n jobs so as to make the minimum total processing time.[2]

For each combinatorial optimization problem, there might be several ways to represent its solutions. The straightforward expression of the solution of bin packing is n integers, and the definition domain of each integer is $[1,k]$, where k is the maximum allowed number of bins. It can also be expressed as a permutation of n integers. The permutation $3 \to 5 \to 2 \to 1 \to 4$ means that we put items in this order into one bin until its size has been exceeded, then put them into another bin, etc.

If the variables of the optimization are restricted to integers, we often call this problem *integer programming*. Further, if the definition domain of the variables is only $\{0,1\}$, it is a *0/1 programming*.

Perhaps the most distinctive part of combinatorial optimization compared with continuous optimization, introduced in Chaps. 4–6, lies in the absence of the search direction due to the lack of a native neighborhood definition, which makes the solution landscape hard to search. In continuous optimization, we can define a threshold $\varepsilon > 0$ and assign all the points within the range $(x-\varepsilon, x+\varepsilon)$ to the neighbors of x. Then many types of random sampling techniques can be used to search the neighbor. But in combinatorial optimization, the neighborhood definition is representation dependent, which will be illustrated more specifically with the problems in the following part of this chapter.

---

[2] For the sake of simplicity, here we use the First In First Out rule implicitly when there is more than one job waiting to be processed on one machine. So we just discuss permutation flow-shop scheduling. Interested readers are referred to [2].

Equation 7.3 is an example of integer programming, which is illustrated by Fig. 7.1.

$$\begin{aligned} &\max\ z = 20x_1 + 10x_2 \\ &\text{s.t. } 5x_1 + 4x_2 \le 24 \\ &\quad 2x_1 + 5x_2 \le 13 \\ &\quad x_1,\ x_2 \ge 0,\ x_1,\ x_2 \in N \end{aligned} \tag{7.3}$$

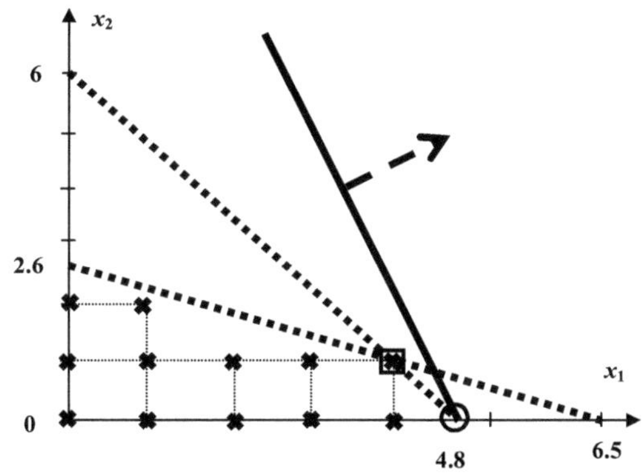

Fig. 7.1 An example of integer programming

In Fig. 7.1, two dotted lines and the axes limit the 12 feasible solutions, represented by the cross in the figure, in which point $(4,1)$ is the optimal solution, represented by the square in the figure.

Equation 7.3 is an integer linear programming. If we could "relax" the constraint of integer, the problem would be so easy that we could solve it on the 2-D plane introduced as follows. The solid line and the arrow in Fig. 7.1 represent the line $h = 20x_1 + 10x_2$, where h is a constant and the increasing direction of h. The intersection of the line and the feasible region is the optimal solution of the relaxed 2-D linear programming, i.e., $(4.8,0)$, which is represented by the circle in the figure. But we cannot get the optimal solution by simply rounding up or rounding down, which generates infeasible solution $(5,0)$ and nonoptimal solution $(4,0)$, respectively. So the idea of "relax" and "round up/down" does not work.[3]

Another intuitive idea is to enumerate all the feasible solutions and find the optimum because there will always be a limited number of feasible solutions. Yes, it is simple in Fig. 7.1, but let us consider flow-shop scheduling. The number of solutions for n jobs is $n!$. If $n = 50$, which is not a large problem, $n! \approx 10^{64}$. Currently the fastest supercomputer, IBM's Sequoia, can calculate approx. 10^{16} times per second. It will take Sequoia about 10^{48} s to finish the enumeration, which is about 10^{40} years!!

[3] But the result of the relaxed problem can help the search. See [3] for details.

So "crude" enumeration does not work even for medium-sized combinatorial optimization problems. We need extra insight to limit the search space. Before discussing the solving algorithms, we need to discuss something about the difficulty extent of problems.

7.1.2 NP-complete and NP-hard Problems

In combinatorial optimization, we often need to solve the problem instance with dimension n using some algorithm. It is quite natural to estimate the time of the algorithm to solve the instance. This time is related to both n and the specific algorithm. For flow-shop scheduling with n jobs, if we use permutation to represent a solution and enumerate every permutation to find the optimal solution, this "crude" algorithm will take $n!$ steps to solve it.[4] In order to compare the time complexities of different algorithms on the same problem instance, we often use $O(n!)$, where O stands for "order," meaning that the *time complexity* of the algorithm is on the order of $n!$.

Another question related to the efficiency and efficacy of the algorithm is how hard the problem itself is, which leads to the concept of NP-complete and NP-hard.[5]

We need to first transform the optimization problem into a *decision problem*. If it can be solved by a *deterministic Turing machine* using an algorithm with polynomial time complexity, i.e., $O(n^k)$, this problem belongs to the class of P, where P stands for "polynomial time complexity".[6] Problems in class P have algorithms with polynomial time complexity.

If the decision problem of a problem can be solved by a *nondeterministic Turing machine* using an algorithm with polynomial time complexity, this problem belongs to the class of NP, where NP stands for "nondeterministic polynomial time complexity." It has been proved that $P \subseteq NP$, but most computer scientists believe that $P \neq NP$, even though there is no prove of that. Then we need to classify further for those problems belong to $NP - P$.

If we can design an algorithm with polynomial time complexity to transform any solution of the decision problem D_1 of one problem to one solution of the decision problem D_2 of another problem, we denote $D_1 \propto D_2$ with the meaning that D_2 is at least as hard as D_1. This *polynomial transformation* is sometime called *reduction*.

If decision problem C belongs to NP and there exist polynomial algorithms to transform every problem in NP into C, we say C is *NP-complete* or C is a *NP-complete problem.*[7] If a known NP-complete problem can be reduced to an NP problem C, we could also say that C is NP-complete because of the transitive property of polynomial transformation. That is to say, if one of the NP-complete problems

---

[4] We often use the key step number to represent the time requirement of an algorithm.

[5] It is the key part of computational complexity theory.

[6] We do not have content to explain the detail of the above statement. Interested readers are referred to [4].

[7] It means that C is no easier than any other NP problem.

can be solved by a polynomial algorithm, so can all of them. But up to now, any polynomial algorithm for any NP-complete problem has yet to be suggested. That is why many researchers think $P \neq NP$.

If a problem H does not belong to NP but there exists an NP-complete problem C that can be polynomially transformed into H, we say H belongs to the class of *NP-hard* or H is a *NP-hard problem* with the meaning that H is at least as hard as C. The relationship between these classes is illustrated by Fig. 7.2.

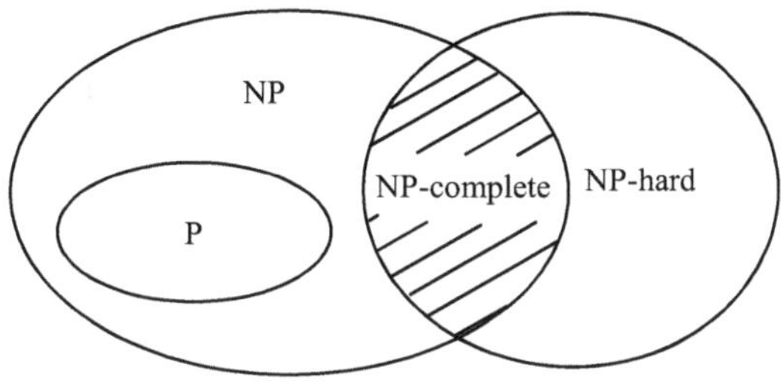

Fig. 7.2 The relationship between P, NP, NP-complete, and NP-hard

Problems discussed in this chapter and most of the practical combinatorial optimization problems are all NP-complete or NP-hard [4],[8] which means that they might be the hardest problems. Right now, no algorithm with polynomial time complexity can guarantee that an optimal solution will be found.

7.1.3 Evolutionary Algorithms for Combinatorial Optimization

Algorithms for combinatorial optimization can be classified into the following three categories. The chart of Fig. 7.3 illustrates these statements in which n is the dimension of the problem.

1. **Exact algorithms.** These algorithms guarantee the optimality of the results. But they do not have polynomial time complexity for NP-complete problems. So they suffer from the "curse of dimension" or the "dimension explosion." This category contains dynamic programming, branch and bound, cutting plane method, etc.
2. **Heuristic algorithms.** These algorithms can either generate a not-so-bad solution fast, i.e., *construction algorithm*,[9] or improve the current solution fast, i.e., *local search algorithm*.[10] The quality of solutions provided by heuristic algorithms is relatively low in the worst case. Their time complexity is low. Heuristic algorithms are problem-dependent. So we will introduce several of them in the following sections.

[8] Crescenzi and Kann maintain a Web site containing a compendium of the NP-complete optimization problems `http://www.csc.kth.se/~viggo/problemlist/`.

[9] Sometimes called the *construction method* or *generation method*.

[10] Sometimes called the *local search method*.

3. **EAs.** These algorithms will achieve a near-optimal solution, especially after being combined with heuristic algorithms, in a relatively short time because their time complexity is not directly linked with the dimension of the problem.

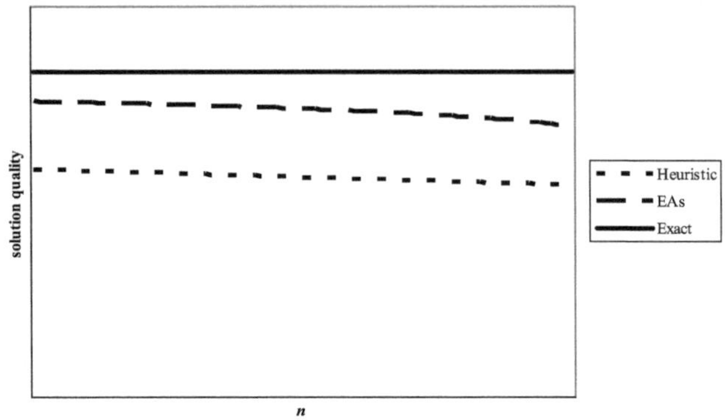

(a)

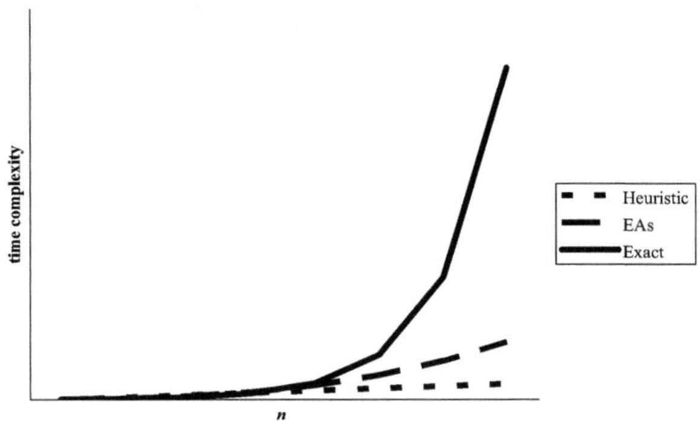

(b)

Fig. 7.3 Solution quality and time complexity of different algorithms: (a) Solution quality, and (b) Time complexity

As can be seen from Fig. 7.3a, exact algorithms have a constant and the highest solution quality, followed by EAs and then heuristic algorithms. The solution quality of EAs will decrease if we keep the control parameter, i.e., *popsize*, *maxgen*, etc., unchanged. Heuristic algorithms almost keep the same solution quality. In Fig. 7.3b, the increased speed of exact algorithms is apparently faster than that of the others, which makes the exact algorithms unusable for large (or even intermediate-size) combinatorial optimization.

The following points need to be focused on when designing and analyzing an EA for combinatorial optimization.

- The decoding and repair mechanism of the EA. Because combinatorial optimization problems are often highly constrained, special decoding techniques and repair methods are the most common strategies. We will introduce permutation code, integer code, random key code, and their corresponding decoding mechanisms in the follow sections.
- Special variation operators to generate decodable chromosomes. For permutation code, ordinary crossover and mutation operators might generate illegal chromosomes. We will discuss this in detail in Sect. 7.3.
- The way of selecting and utilizing local search methods. The search space of combinatorial optimization problems are rather huge, so we need memetic algorithms to get high-quality solutions within a limited time. Different local search methods will be introduced in the following sections and their combination with EAs will be discussed.

The above properties separate the main topic of this chapter from that of Chaps. 4–6, in which the methods to maintain the proper population diversity in a selection or replacement process are the main concern.

Figure 7.4 illustrates the number of papers indexed by the SCI on EAs-based combinatorial optimization,[11] which means this field is rather hot.

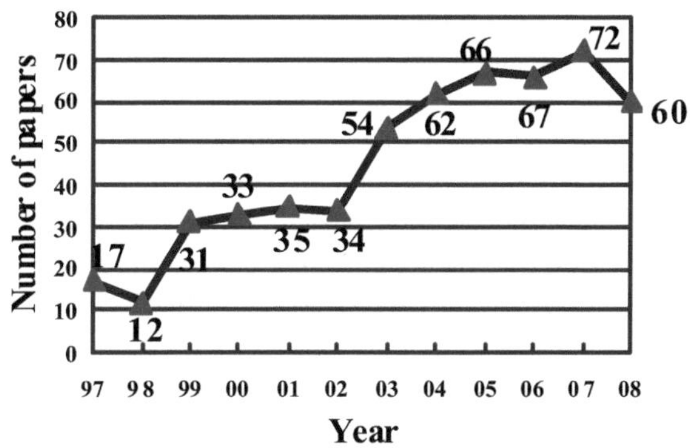

Fig. 7.4 Number of papers indexed by SCI on EAs-based combinatorial optimization

[11] TS = ((“combinatorial optimization”) AND (“genetic algorithm” OR “genetic algorithms” OR “evolutionary computation” OR “evolutionary computing” OR “evolutionary algorithms” OR “evolutionary intelligence”)). The SCI index “TS” is for the search topic in the title, the keywords, and the abstract.

7.2 Knapsack Problem

7.2.1 Problem Description

The *knapsack problem* considers a robber containing a knapsack with limited capacity in a department store with n items. The robber wants to maximize the profit while subject to the constraint of the knapsack's capacity as follows:

$$f(\mathbf{x}) = \max \sum_{i=1}^{n} p_i x_i \tag{7.4}$$

$$\begin{aligned} \text{s.t.} \quad & \sum_{i=1}^{n} w_i x_i \le W \\ & x_i \in \{0,1\},\ i = 1,2,\cdots,n \end{aligned} \tag{7.5}$$

where p_i and w_i are the profit and weight of item i, respectively, and W is the capacity of the knapsack. $x_i = 1$ means item i is put into the knapsack and $x_i = 0$ means item i is not put into the knapsack.

The knapsack problem has diverse practical applications such as cargo loading, project selection, assembly line balancing, etc.

The knapsack problem can be regarded as grouping items into two classes, those being put into the knapsack and those being discarded. So it is a grouping problem.

There are other variations of the standard knapsack problem.

If there is more than one item i in a department store, i.e., $x_i \in \{0,1,\cdots,b_i\}$, where b_i is the maximum number of item i, it is called a *bounded knapsack problem.*

If we consider the constraint of the knapsack capacity in more detail, i.e., k constraints such as the length, width, height, etc., and give the corresponding properties for each item, i.e., $(w_{i1},\cdots,w_{ik})$ for item i with k-dimensional properties, the constraint, Eq. 7.5, can be rewritten as follows:

$$\begin{aligned} \text{s.t.} \quad & \sum_{i=1}^{n} w_{ij} x_i \le W_j \quad j = 1,\cdots,k \\ & x_i \in \{0,1\},\ i = 1,2,\cdots,n \end{aligned} \tag{7.6}$$

The explanation of Eq. 7.6 is that the robber cannot take items too long, too wide, or too high. We can also interpret Eq. 7.6 as a k robbers problem. Each robber has one knapsack. These k knapsacks have different capacities W_j and there is more than one item i in the department store. Whenever the robbers decide to take item i, it is put into k knapsacks. They want to maximum their profits with the constraints of each knapsack. Combining Eqs. 7.4 and 7.6 forms a *multidimensional knapsack problem* [3, 5].

If we consider the profit of an item in more detail, i.e., m profits such as price, satisfaction level, commemorative meaning, etc., and give corresponding properties for each item, i.e., $(p_{i1},\cdots,p_{im})$ for item i with m profit properties, the objective, Eq. 7.4 can be rewritten as follows:

$$f_j(\mathbf{x}) = \max \sum_{i=1}^{n} p_{ij} x_i \quad j = 1, \cdots, m \tag{7.7}$$

The explanation of Eq. 7.7 is that the robbers not only want the money but also expect some spiritual harvests. Combining Eqs. 7.7 and 7.5 forms a *multiobjective knapsack problem.*

In 1999 Zitzler and Thiele use Eqs. 7.6 and 7.7 as a benchmark problem for testing MOEAs and many researchers have worked on it since then [6–8].

In the following section, we will only discuss a model of Eqs. 7.4 and 7.5.

7.2.2 Evolutionary Algorithms for Knapsack Problem

IN 2005, Raidl and Gottlieb studied six representations of the multidimensional knapsack problem with different variation operators [3]. They suggested that a good EA for combinatorial optimization should have the following characteristics.

- **Locality.** Small steps in the operation domain corresponds to small steps in the definition domain, which promotes a meaningful local search around a solution.
- **Heritability.** The offspring generated by a crossover operator should contain meaningful features of their parents, which means that the merits of the previous search could be maintained.
- **Heuristic bias.** The map from the operation domain to the definition domain is better when it contains some heuristic which promote the intensive search of high fitness value region.[12]

Before we discuss the genetic code for the knapsack problem, an intuitive heuristic method for constructing solutions and repairing infeasible solutions needs to be introduced. It can be regarded as both a construction method and a local search method.

Let us begin the discussion with an example of Eqs. 7.4 and 7.5, in which $W = 9$ and $n = 7$. Other parameters are listed in Table 7.1.

Table 7.1 Example data for knapsack problem

Item	1	2	3	4	5	6	7
p_i	6	5	8	9	6	7	3
w_i	2	3	6	7	5	9	4
p_i/w_i	3	1.67	1.33	1.29	1.2	0.78	0.75

[12] Reconsider the effect of different $\mathbf{r}_0$ in Sect. 4.2.2.

It is natural for a "reasonable" robber to pick an item with a large profit and small weight. So we can use the *profit/weight ratio* p_i/w_i for item *i* to represent this preference, as represented in Table 7.1 [9].

While constructing a new solution, we just need to pick items in a "greedy" way, i.e., pick the one with the highest value of p_i/w_i, then the second, ..., until the knapsack capacity is not enough. In this way, we can generate the solution $\{1,2,7\}$ with a profit of 14.[13]

While repairing the infeasible solutions, we can take the one with the smallest value of p_i/w_i in the knapsack first, then the second, ..., until the solution is a feasible one.

We will introduce three code schemes in this section as follows:

- Binary code
- Variable-length code
- Permutation code

Binary Code

The *binary code* for the knapsack problem is very straightforward; just use a locus to represent the item and an allele to represent whether the item is selected (1) or not (0). An example chromosome of the seven-item knapsack problem is illustrated by Fig. 7.5.

1	2	3	4	5	6	7
0	1	0	1	0	1	0

Fig. 7.5 Binary code for knapsack problem

The upper numbers in Fig. 7.5 represent the order number of the items and the lower numbers represent the selection of the items. But this code scheme might generate infeasible solutions. We will introduce two methods for it.

It is not difficult to understand that the optimal solution of the knapsack problem must be at or near the boundary of the feasible region, i.e., the sum of weights of the items in the knapsack should be the same as or close to the knapsack capacity. Olsen suggested a way to penalize the solution over *and* under the capacity of the knapsack [10].

Olsen first defines the scale of the penalization as follows:

$$\delta = \min\left\{W, \left|\sum_{i=1}^{n} w_i - W\right|\right\} \tag{7.8}$$

According to Eq. 7.8, the maximum value of δ is W. Then the penalty function is defined as follows:

[13] But the optimal solution is $\{1,4\}$ with a profit of 15.

$$p(\mathbf{x}) = 1 - \frac{\left| \sum_{i=1}^{n} w_i x_i - W \right|}{\delta} \tag{7.9}$$

The graphical illustration of the above penalty function is Fig. 7.6. This penalty function penalizes not only the infeasible solutions but also the feasible solutions that are not at the boundary of the feasible region.[14]

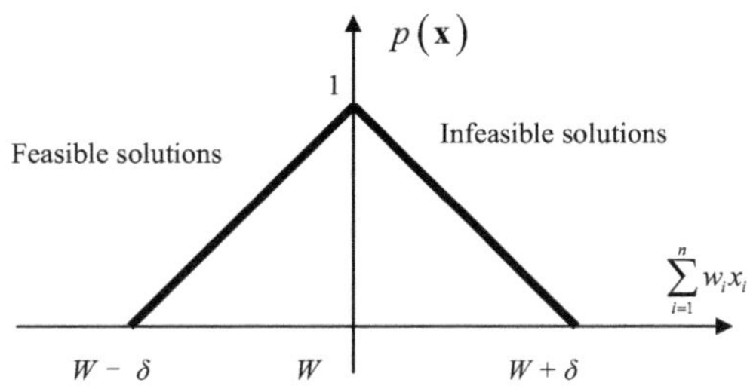

Fig. 7.6 Penalty function for knapsack problem

The fitness function is the time of the objective function and the penalty function as follows:

$$fitness(\mathbf{x}) = f(\mathbf{x})\, p(\mathbf{x}) \tag{7.10}$$

Repair is another way to handle infeasible individuals. It is basic idea, using the profit/weight ratio, was introduced previously. Here we just discuss an example with $W = 90$ and $n = 7$. Other parameters are listed in Table 7.2.

Table 7.2 Example of repair in knapsack problem

Item	1	2	3	4	5	6	7
p_i	40	60	10	10	3	20	60
w_i	40	50	30	10	10	40	30
p_i/w_i	1	1.2	0.33	1	0.33	0.5	2

Individuals represented by Fig. 7.5 are infeasible. The sum of weights of items 2, 4, and 6 is 100, which is larger than $W = 90$. According to the profit/weight ratios listed in Table 7.2, we can delete item 6, which has the smallest profit/weight ratio in the three selected items and thus makes (0101000) a feasible solution.

Repair is often used to change infeasible solutions to feasible solutions. The method to deal with solutions out of the definition domain, introduced at the end of Sect. 4.2, is also a repair method. Repair might limit the population in a small solution area, i.e., it has bias, which needs to be avoided by designers.

[14] For solutions rather far from W, we can just assign its fitness value as zero.

Variable-length Code

Hinterding suggested *variable-length code* for the knapsack problem [11]. The length of the chromosome could change during the evolving process. The gene locus has no meaning and the allele of the gene means the order number of the item. That is $(1,6,5)$ means items 1, 5, and 6 are in the knapsack.

Variable-length code has special methods in maintaining the feasibility of individuals. In initialization, we can generate random n number permutations. We check from the first item in the permutation whether the constraint will be violated if it is selected. If adding the current item does not violate the knapsack capacity, it is selected. If not, we discard it and check the next item until the knapsack is full or we reach the end of the permutation. One feasible individual can be generated by one random permutation. Repeat this method until $popsize$ feasible individuals are generated.

In crossover, the following steps are necessary to keep the feasibility of the offspring, which is illustrated as an example in Fig. 7.7.[15] The capacity of the knapsack is $W = 90$ and $n = 7$. Other parameters are listed in Table 7.2.

1. **Step 1:** Select a random point in parent 1.
2. **Step 2:** Select a random fragment in parent 2.
3. **Step 3:** Insert the fragment of parent 2 into the point of parent 1 and delete the duplicate items to get an offspring, which might be infeasible.
4. **Step 4:** Use the profit/weight ratio to repair the offspring if it is infeasible.

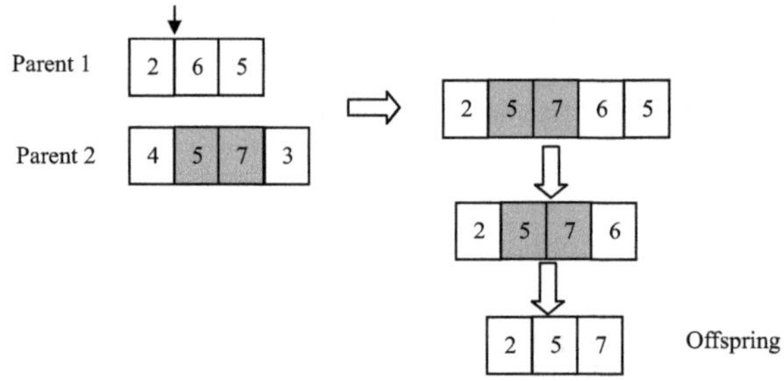

Fig. 7.7 Example of crossover for variable-length code

In the example illustrated by Fig. 7.7, the result after step 3 is $(2,5,7,6)$, which is infeasible. According to the profit/weight ratios in Table 7.2, we first try to delete item 5 but it does not work because $(2,7,6)$ is still infeasible. So we try item 6, thus making $(2,5,7)$ feasible.

The mutation operator is simple. Randomly delete an item from the chromosome and randomly add an item, that does not exist in the chromosome. Then use step 4 introduced above to repair the mutant, if necessary, to be a feasible individual.

[15] It is a two-parents-one-offspring crossover operator.

Permutation Code

Any permutation of n numbers could be interpreted as the sequence of picking items. Let us revisit the example illustrated by Table 7.2. Here we assume that $W = 100$ and $n = 7$.

For individual $(1,6,4,7,3,2,5)$, the decoding procedure is as follows.

We put item 1 into the knapsack first. Item 6 is OK because $40+40 = 80 < 100 = W$. Item 4 is picked because $40+40+10 = 90 < 100 = W$. But item 7 is not acceptable. We keep on searching until we meet item 5, which means that $40+40+10+10 = 100 = W$. So the decoding results for individual $(1,6,4,7,3,2,5)$ is item 1, 6, 4, and 5.

In this way, every n permutation can be decoded as a feasible solution.

Ordinary crossover and mutation operators will destroy the permutation, so special variation operators are the main consideration in permutation code, which will be discussed in detail in Sect. 7.3.

Another thing we need to mention is that individuals (1,6,4,7,3,2,5), (1,4,6,7,3,2, 5), (6,4,1,7,3,2,5), ..., (6,4,1,7,3,5,2) will all be decoded as the same feasible solution with items 1, 6, 4, and 5, which means that the map between the operation domain and the definition domain is many-to-one. The many-to-one map might waste the search ability of EAs and cause low efficiency.

Finally, let us compare the search space of the binary code and the permutation code with examples. In the binary code of n items, the search space contains 2^n solutions. In the permutation code of n items, the search space contains $n!$ solutions. The numerical comparison of n from 1 to 10, which are rather small numbers, is illustrated by Table 7.3

Table 7.3 Comparison of the search space of binary code and permutation code

n	1	2	3	4	5	6	7	8	9	10
2^n	2	4	8	16	32	64	128	256	512	1024
$n!$	1	2	6	24	120	720	5040	40320	362880	3628800

It is quite obvious that the search space of the permutation code is far more than that of the binary code.

But sometimes we need to make a tradeoff between feasibility preservation and limiting the search scale. Combinatorial optimization problems might have hard constraints so that the ordinary techniques introduced in Chap. 4 do not work; we will introduce such examples in the following sections. But if we can design the decoding process so that every permutation corresponds to a feasible solution, we might take this code scheme even if it is a many-to-one map.

7.3 Traveling Salesman Problem

7.3.1 Problem Description

The *traveling salesman problem* (TSP) is perhaps the most famous combinatorial optimization problem and is used as a benchmark for many optimization methods.

Suppose there is a salesman who would like to start from his home city and visit other cities for commodity promotion. He would like to visit all other cities once and return to his home city. The problem is how to arrange the visiting sequence of these cities so that the total distance/traveling expenses are minimized.

TSP has diverse practical applications such as the vehicle routing problem, very-large-scale integration (VLSI) design, printed circuit board design, etc.

Is such a clear and seemingly easy problem worth so much attention? Yes! We will look at why this is so.

By the language of *graph theory*, we can formulate TSP in a more formal way. A *graph* $G = (V,E)$ contains a set of *vertex* V and a set of *edge* E. An edge $e \in E$ links two vertices $v_1, v_2 \in V$ and can be represented as $e = (v_1, v_2)$. If the edge $e = (v_1, v_2)$ has no information of direction, G is an *undirected graph*. Otherwise, i.e., $e_1 = (v_1, v_2) \neq (v_2, v_1) = e_2$, G is a *directed graph*. The cardinality of V, i.e., $|V|$, is the *order* of graph G and the cardinality of E, i.e., $|E|$, is the *size* of graph G. The number of edges connected to a vertex is the *degree* of that vertex.

If two edges $e = (v_1, v_2)$ and $e^{'} = (v_1, v_2)$ share the same vertices, i.e., there are two edges linking vertices v_1 and v_2, we say e and $e^{'}$ are *parallel*. Graphs without parallel edges are called *simple graphs*. A *complete graph* is a simple graph in which every pair of distinct vertices is connected by an edge. Figure 7.8 is an example of an undirected simple graph whose order and size are both 4.

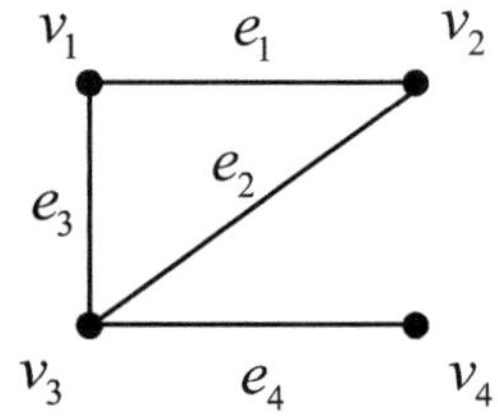

Fig. 7.8 An example of graph

For a graph $G^{'} = (V^{'}, E^{'})$, if $V^{'} \subseteq V$ and $E^{'} \subseteq E$, we say that $G^{'}$ is a *subgraph* of G. $(\{v_1, v_3, v_4\}, \{e_3, e_4\})$ is a subgraph of the graph illustrated by Fig. 7.8.

A *path* in graph G is a subgraph of G with a start vertex and an end vertex. We can "walk" along the edges linking the vertices[16] from the start vertex to the end vertex. If there exists at least a path for every two vertices in graph G, we call G a *connected graph*. Otherwise, it is a *disconnected graph*. The example in Fig. 7.8 is a connected graph.

---

[16] These edges and vertices are all elements of the subgraph path.

A *cycle* or *loop* is a path such that the start vertex and the end vertex are the same. A special case of a cycle is $e = (v, v)$, i.e., there exist an edge e that starts at vertex v and ends at vertex v. It is called a *self-cycle*. We generally do not consider graphs with self-cycles.

A *tree* is a connected subgraph of G containing all the vertices and without any cycle.

A path with no repeated vertices is called a *simple path* and a cycle with no repeated vertices besides the start/end vertex is a *simple cycle*.

A *Hamiltonian path* is a simple path in an undirected graph that visits each vertex exactly once. A *Hamiltonian cycle* is a simple cycle in an undirected graph that visits each vertex exactly once and also returns to the start vertex.

If we regard the vertices as cities and let the edges represent the distance information between cities, then TSP could be interpreted as needing to find a Hamiltonian cycle in an undirected simple complete graph with the minimum distance.[17] Such an example is given by Fig. 7.9, in which thick lines form a solution. In TSP, we often use *city* instead of *convex*, *road* instead of *edge*, and *tour* instead of *cycle*.

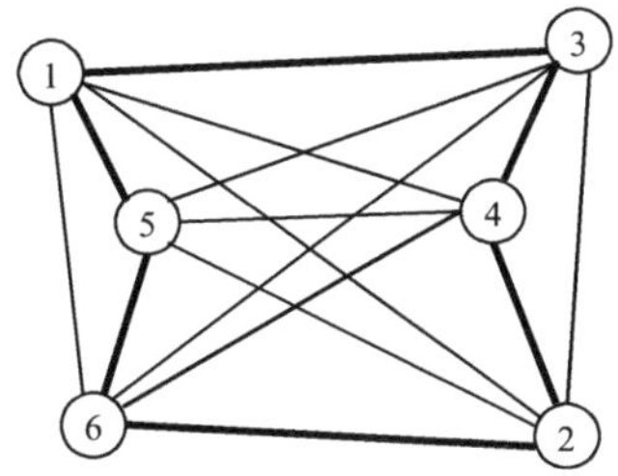

Fig. 7.9 An example of TSP

TSP can be regarded as grouping all edges into two classes, those in the solution and other remaining edges, with the constraint that those edges in the solution should form a Hamiltonian cycle. So it is a grouping problem.

The above problem is a *symmetric TSP* because we suppose the distance from v_i to v_j is the same as that of v_j to v_i, i.e., the graph is undirected. If the graph is directed, we could represent some other considerations, such as the the road from city i to city j is uphill so the traveling expenses are different. In this situation, the problem is an *asymmetric TSP*. We could use *distance matrix* $\mathbf{D}$ with $n \times n$ nonnegative elements d_{ij} to represent the distance information of TSP with n cities. It easy to see that the distance matrices of symmetric TSP and asymmetric TSP are symmetric and asymmetric, respectively.

If cities of TSP are in a metric space, i.e., the elements of the distance matrix satisfy Eq. 7.11, we call this problem *metric TSP*. Otherwise, it is *nonmetric TSP*. Further, if the elements of distance matrix d_{ij} is the Euclidean distance between city i and city j, it is a *Euclidean TSP*. Furthermore, if the cities are in a 2-D Euclidean space, it is called a *2-D Euclidean TSP*.

[17] If the graph for a real TSP is not complete, we can add edges with a very large distance to make it complete.

$$d_{ij} \leq d_{ik} + d_{kj} \tag{7.11}$$

There is a TSPLIB that contains a library of sample instances for the TSPs and their current best solutions maintained by Reinelt.[18] The largest city number in TSPLIB is 85,900. This problem came from a VLSI application that arose in Bell Laboratories in the late 1980s and was solved by Applegate *et al.*[19] Right now, they are working on a TSP with 100,000 cities.

For reasons of simplicity, we will consider a symmetric Euclidean TSP in what follows unless otherwise specified, but most of the techniques could be used in asymmetric TSP.[20]

7.3.2 Heuristic Methods for Traveling Salesman Problem

As has been introduced above, construction methods are used for generating solutions from scratch that can be used to generate initial individuals with relatively higher quality, and local search methods are for improving current solutions that can be used in MAs.

7.3.2.1 Construction Methods

Closest Neighbor Method

We can construct a solution for TSP by considering the closest city of the current city sequentially. The specific steps are illustrated as follows.

Closest Neighbor Construction Method for Traveling Salesman Problem

Step 1: Start from any city and take it as the current city.

Step 2: Select the closest city to the current city that has not been visited. Go there and make that city the current city.

Step 3: Repeat step 2 until all cities have been visited. Then return to the first city.

Suppose we start from city 1 of Fig. 7.9. The solution we get is illustrated by Fig. 7.10, in which the arrows represent the visiting sequence.

---

18 `http://comopt.ifi.uni-heidelberg.de/software/TSPLIB95/`.

19 `http://www.tsp.gatech.edu/pla85900/index.html`.

20 There are other representation methods for TSP, such as 0/1 programming. Interested readers are referred to sects. 8–10 of [12].

The closest neighbor method is a deterministic construction. But we could generate different solutions by picking different initial cities.

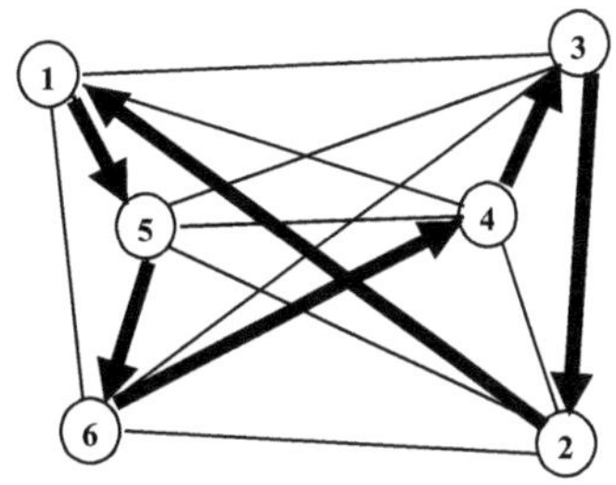

Fig. 7.10 The result of the closest neighbor construction method for TSP

Closest Insertion Method

Another way to construct a solution for TSP is to absorb a new city into the current cycle each time with a minimum distance increase until we get the Hamiltonian cycle for the problem. The specific steps are illustrated as follows.

Closest Insertion Construction Method for Traveling Salesman Problem

Step 1: Use the closest neighbor construction method, or other methods, to construct a cycle of three cities. Cycle $(6,5,1,6)$ is constructed in this way in Fig. 7.11.

Step 2: Select the closest city, which has not been visited, to the current cycle. Suppose there are m cities in the current cycle. For each of the other cities, we calculate its distance to these m cities and pick the smallest value to represent its distance to the current cycle. Then the city closest to the current cycle is selected. In Fig. 7.11, city 4 is selected because it has the smallest distance to the current cycle.

Step 3: Find the best way to absorb the selected city into the cycle with the smallest distance increase. Suppose there are m edges in the current cycle; try to break each edge and absorb the selected city into the cycle. There are m ways to absorb it. We compare the distance of these m new cycles and find the one with the smallest distance to be the current cycle. The distances of cycles $(6,5,4,1,6)$, $(6,4,5,1,6)$, and $(6,5,1,4,6)$ are evaluated to determine the current cycle in Fig. 7.11.

Step 4: Repeat steps 2 and 3 until all the cities have been included in the cycle.

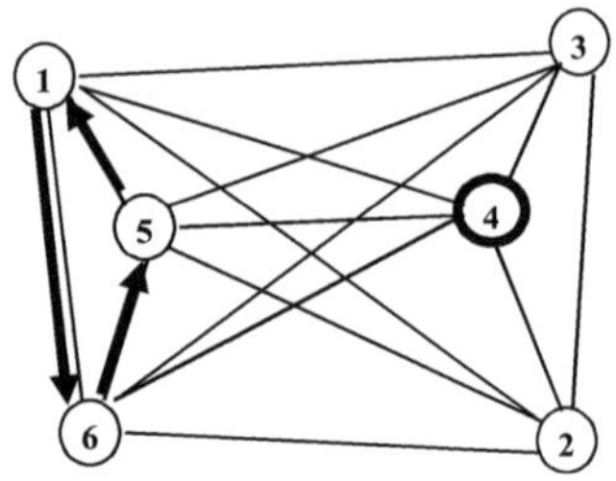

Fig. 7.11 The procedure of the closest insertion construction method for TSP

7.3.2.2 Local Search Methods

2-opt

2-opt was suggested by Croes in 1958 [13]. It eliminates two edges that do not share the cities from the current solution and add edges to form a new solution. We may try every two such edges and change the current solution to the best improved one if it exists. So 2-opt is the abbreviation of "local optimal by 2-edge perturbation."

For the example given by Fig. 7.12a, edge e_i between vertices v_w and v_x and edge e_j between vertices v_y and v_z are in the solution. The dashed lines are other edges of the solution. After removing e_i and e_j, there is only one way to construct a new solution, i.e., add edge e_k between vertices v_w and v_z and edge e_l between vertices v_x and v_y.[21] That forms the solution represented by Fig. 7.12b.

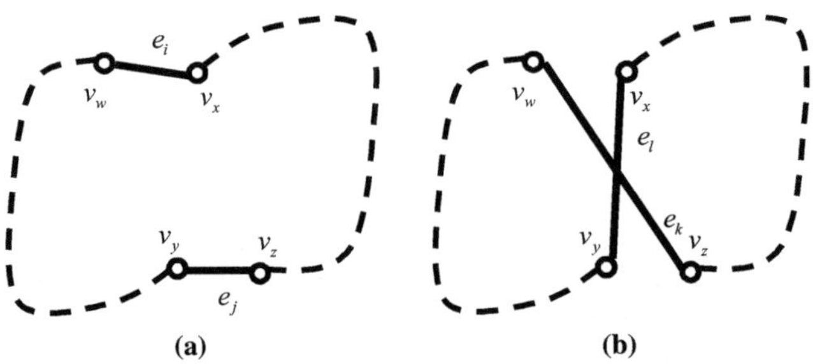

Fig. 7.12 An example of one step in 2-opt: (a) solution before edge perturbation, and (b) solution after edge perturbation

[21] Adding the edge between vertiecs v_w and v_y and the edge between vertices v_x and v_z will form local cycles and thus generate illegal solutions.

3-opt and Lin–Kernighan Algorithm

3-opt was suggested by Lin for symmetrical TSPs in 1965 [14]. Similar to 2-opt, 3-opt tries to remove three edges that do not share cities and reconnect them as a new solution. With one set of three edges, seven new solutions could be generated in which three solutions are actually 2-opt results.[22] We try every three such edges and change the current solution to the best improved one if it exists. So 3-opt is the abbreviation of "local optimal by 3-edge perturbation."

Generally, the local search results of 3-opt are better than those of 2-opt.

Starting from 3-opt, we can generalize it as k-opt. But what is the best k value for our TSP? Lin and Kernighan developed 3-opt and suggested the so-called Lin–Kernighan algorithm in 1973 [15]. The Lin–Kernighan algorithm has special considerations in limiting the search space and adaptively determining k. In 1997 Johnson and McGeoch used modern data structures to develop the Lin–Kernighan algorithm [16]. So sometimes it is called Lin–Kernighan–Johnson algorithm.

Among these local search methods, the Lin–Kernighan algorithm is the clear winner.[23]

7.3.3 Evolutionary Algorithm Code Schemes for Traveling Salesman Problem

TSP is actually a touchstone for combinatorial optimization EAs. So there are several code methods for TSP. Before discussing the specific code methods, we need to determine the real size of the search space for an n-city TSP.

Theorem 7.1. *For an n-city symmetric TSP, there are* $\frac{(n-1)!}{2}$ *different solutions.*

Proof. We can prove the theorem by mathematical induction. For the smallest TSP, $n = 3$, it is easy to know that there are $1 = \frac{2!}{2}$ solutions. Suppose there are different $\frac{(n-2)!}{2}$ solutions for $(n-1)$ cities. We can get the solution for an n-city TSP in the following way. For every edge of one solution for $(n-1)$ cities, we break it and then absorb city n into the cycle. This means that for each solution of an $(n-1)$-city TSP, we have $(n-1)$ different ways to generate the tour for an n-city TSP. Because the $\frac{(n-2)!}{2}$ solutions for $(n-1)$ cities are different, there are $\frac{(n-1)!}{2}$ different solutions for n cities.

[22] The proof of such a statement is left as an exercise.

[23] Neto provided C source code for the Lin–Kernighan algorithm at the Web site `http://www.cs.toronto.edu/~neto/research/lk/`.

7.3.3.1 Edge Code

For an n-vertex complete graph, it has $\frac{n(n-1)}{2}$ edges.[24] The initial idea is that we can number them, and select n edges from them to form a Hamiltonian cycle, which is *edge code*. An example edge code of an eight-city TSP is $(14,28,18,23,27,7,6,16)$. These eight edges form a Hamiltonian cycle.

But the main drawback of edge code is ensuring that these n edges form a Hamiltonian cycle and preventing the variation operators from breaking these Hamiltonian cycles. These questions are very hard to answer, so edge code is rarely used.

The operation domain of edge code is the number of combinations of n edges in $\frac{n(n-1)}{2}$, i.e., $C^n_{\frac{n(n-1)}{2}} = \binom{\frac{n(n-1)}{2}}{n}$.

7.3.3.2 Binary Code

Since representing solutions of TSPs with edges directly is hard, we turn to solutions with vertices, i.e., cities. One type is to use *binary code* to represent city numbers. For an n-city TSP, we need $\lceil \log_2 n \rceil$ binary numbers to represent the city, where $\lceil\ \rceil$ is a rounding-up function. An example binary code of a six-city TSP is $(000\,010\,001\,100\,011\,101)$, which means tour $1-3-2-5-4-6-1$.

The standard variation operators will destroy the feasibility of offspring or mutants. So extra repair techniques are necessary.

Generally, every gene, e.g., city 3 (010) in the above example, has n possibilities. So for an n-city tour, the operation domain of the binary code is n^n.

7.3.3.3 Random Key Code

Random key code is quite useful in a hard constrained situation, which will be illustrated more clearly in Sect. 7.4. For an n-city TSP, we could generate n uniformly distributed random numbers in the range $(0,1)$ and then rank them from small to large. The locus of the smallest random number is the first city being visited, and then the locus of the second smallest random number is the second city being visited, etc. An example random key code of a six-city TSP is $(0.9501,0.2311,0.4568,0.4860,0.8913,0.7621)$, which corresponds to tour $2-3-4-6-5-1-2$.

The standard variation operators will always generate legal offspring. That is the reason for random key code being widely used in hard constrained combinatorial optimization. But the problem is that a lot of information contained in the parents will sometimes be lost by standard single-point crossover. For example, another chromosome is $(0.4565,0.0185,0.8214,0.6154,0.4447,0.7919)$, which corresponds to tour $2-5-1-4-6-3-2$. A single-point crossover of these two chromosomes at the

---

[24] The proof of this statement is left as an exercise.

intermediate location will generate $(0.9501, 0.2311, 0.4568, 0.6154, 0.4447, 0.7919)$ and $(0.4565, 0.0185, 0.8214, 0.4860, 0.8913, 0.7621)$, which corresponds to tour $2-5-3-4-6-1-2$ and $2-1-4-6-3-5-2$. The roads in the parents are almost lost in the offspring, i.e., the good information contained in the parent is lost. The reason for the low efficiency of random key code is that the sequence of the city in the tour is not determined solely by its own value in the chromosome but also by the relative magnitude to others.

The operation domain for random key code is a real value in the range $(0,1)$, which contains infinite numbers.

7.3.3.4 Path Code

Perhaps the most straightforward representation for TSP is *path code*, i.e., the sequence of cities being visited as a Hamiltonian path is used for representing a tour. For an n-city TSP, we could use an n permutation to represent a tour. An example path code of a six-city TSP is $(2-3-6-1-4-5)$, which is corresponding to tour $2-3-6-1-4-5-2$.

Sure the standard variation operators will break the permutation. Researchers have suggested a lot special techniques for generating legal offsprings and mutants with path code, which will be elaborated in Sect. 7.3.4.

The operation domain of the edge code is n number permutation, i.e., $P_n^n = \frac{n!}{(n-n)!} = n!$.

It's necessary to mention that there are $2n$ redundancy in path representation [17]. Chromosomes $(2-3-6-1-4-5)$, $(3-6-1-4-5-2)$, $(6-1-4-5-2-3)$, $(1-4-5-2-3-6)$, $(4-5-2-3-6-1)$, $(5-2-3-6-1-4)$, $(5-4-1-6-3-2)$, $(4-1-6-3-2-5)$, $(1-6-3-2-5-4)$, $(6-3-2-5-4-1)$, $(3-2-5-4-1-6)$, and $(2-5-4-1-6-3)$ all represent the same tour. Representation redundancy might cause unnecessary search effort between individuals corresponding to the same phenotype and thus waste the search resource.

But we would like to remind the reader again that random key code and path code are all tradeoffs between feasibility preserving and search scaling limiting.

7.3.3.5 Adjacent Code

Adjacent code and the following ordinal code were all suggested by Grefenstette *et al.* [18]. *Adjacent code* is similar to path code, i.e., also an n permutation, in which allele j in locus i means city j just follows city i in the tour, i.e., they are "adjacent." An example adjacent code of a six-city TSP is $(2-3-6-1-4-5)$. We could decode it in this way. Locus 1 has a value of 2, which means cities $1-2$ are in the tour. Then we go to locus 2. It has a value of 3, which means cities $1-2-3$ are in the tour. Then locus 3 with a value of 6 means cities $1-2-3-6$ are in the tour. Finally we can get the tour $1-2-3-6-5-4-1$.

The variation operators for path code can be used in adjacent code and the operation domain of the edge code is also $n!$.

The problem with adjacent code is that not all permutations can generate a legal tour. For example, $(2-3-1-6-4-5)$ means two cycles, $1-2-3-1$ and $4-6-5-4$. So special considerations, e.g., repair, are necessary when adjacent variation operators generate new individuals for adjacent code.

7.3.3.6 Ordinal Code

We can use a standard city order list for n-city TSP, such as $1-2-3-4-5-6$ for six cities. Ordinal code also uses an n permutation to represent a solution. We let allele j in locus i represent that city i in the tour is the city has the jth number, i.e., its "ordinal," in the current city order list. After we pick a city, it is deleted from the order list. So for *ordinal code* chromosome $(3,2,4,2,2,1)$ in a six-city TSP, its decoding process could be illustrated by Table 7.4.

Table 7.4 The decoding process of chromosome $(3,2,4,2,2,1)$ in a six-city TSP

Sequence	Current order list	Current working gene	Current tour
1	1-2-3-4-5-6	(**3**,2,4,2,2,1)	3
2	1-2-4-5-6	(3,**2**,4,2,2,1)	3-2
3	1-4-5-6	(3,2,**4**,2,2,1)	3-2-6
4	1-4-5	(3,2,4,**2**,2,1)	3-2-6-4
5	1-5	(3,2,4,2,**2**,1)	3-2-6-4-5
6	1	(3,2,4,2,2,**1**)	3-2-6-4-5-1-3

According to the above decoding process, it is not difficult to determine that the constraint for allele j of locus i is that $1 \leq j \leq n-i+1$.

The standard variation operators will generate legal offspring. But the problem is that the heritability of ordinal code is not good, similar to random key code, because the sequence of the city in the tour is not only determined by its own value in the chromosome.[25]

The first gene has n possible values. The second gene has $(n-1)$ possible values, etc. So the operation domain of the ordinal code is also the number of n permutation, i.e., $n!$.

---

[25] We leave this statement as an exercise.

7.3.3.7 Matrix Code

Apart from one-dimensional representation for TSP, several 2-D representations, i.e., *matrix code*, have been suggested by researchers [19, 20]. Special decoding mechanisms are necessary. These code schemes suffer from bad heritability or require a repair process to maintain the legality of the offspring. So we neglect the introduction to them.

Before ending this subsection, we need to compare the search space of the operation domain for different code schemes. The numerical comparison is listed in Table 7.5.[26]

Table 7.5 The numerical comparison of the search space of the operation domain for different code schemes for TSP

Code	Space \ n	3	4	5	6	7	8	9	10
Real	$\frac{(n-1)!}{2}$	1	3	12	60	360	2520	20160	181440
Edge	$C^{n}_{\frac{n(n-1)}{2}} = \begin{pmatrix} \frac{n(n-1)}{2} \\ n \end{pmatrix}$	1	15	252	5005	10^5	10^6	10^7	10^9
Binary	n^n	27	256	3125	10^4	10^5	10^7	10^8	10^{10}
Path Ordinal Adjacent	$n!$	6	24	120	720	5040	10^4	10^5	10^6

As can be seen from Table 7.5, even though path code has $2n$ redundancy, its search space is relatively small compared to other coding methods, which means with the same search ability of EAs, path code has more chances to find good solutions. That is why path code has become the most often used code scheme.[27]

7.3.4 *Variation Operators for Permutation Code*

In this subsection, we will focus on several variation operators for permutation code, which might be the most often used code scheme apart from real code. Mutation operators will be introduced first for their simplicity.

[26] Some of the large numbers are represented in the approximate form of scientific notation to save space without loss of generality.

[27] The shortcomings of adjacent code and ordinal code have been explained above.

7.3.4.1 Mutation Operators

Picking one gene at random and changing its allele into that of another city will definitely break a permutation. So at least we need to handle them pairwise.

Exchange Mutation

The *exchange mutation* (EM) was suggested by Banzhaf [21]. It is also called *swap mutation*.

In EM, we can just pick two loci randomly and exchange their alleles. Thus permutation is kept and perturbation is achieved. For example, we pick loci 3 and 6 in path code $(1-2-3-4-5-6-7-8-9)$ for a nine-city TSP and get $(1-2-6-4-5-3-7-8-9)$ as in Fig. 7.13.

Fig. 7.13 An example of exchange mutation for permutation code

Insertion Mutation

The *insertion mutation* (ISM) was suggested by Michalewicz [22].

In ISM, we can just pick one locus randomly and insert its allele into a random place. For example, we pick locus 6 in path code $(1-2-3-4-5-6-7-8-9)$ and insert its allele into the place between loci 2 and 3 for a nine-city TSP and get $(1-2-6-3-4-5-7-8-9)$ as in Fig. 7.14.

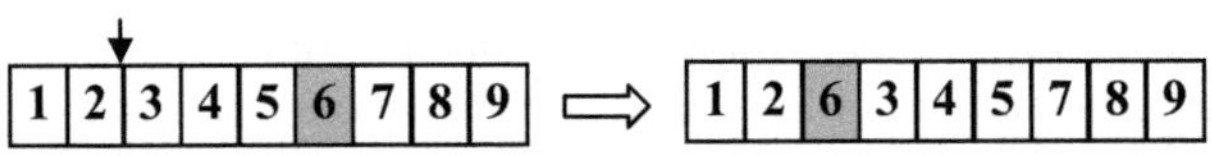

Fig. 7.14 An example of insertion mutation for permutation code

Displacement Mutation

The *displacement mutation* (DM) was also suggested by Michalewicz [22].

DM is just a segment form of ISM; we just pick one subtour randomly and insert it into a random place. For example, we pick subtour $4-5-6$ in path code $(1-2-3-4-5-6-7-8-9)$ and insert it into the place between loci 8 and 9 for a nine-city TSP and get $(1-2-3-7-8-4-5-6-9)$ as in Fig. 7.15.

Fig. 7.15 An example of displacement mutation for permutation code

Simple Inversion Mutation

The *simple inversion mutation* (SIM) was suggested by Grefenstette *et al.* [18].

In SIM, we can just pick one subtour randomly and inverse it. For example, we pick subtour $4-5-6$ in path code $(1-2-3-4-5-6-7-8-9)$ and inverse it and get $(1-2-3-6-5-4-7-8-9)$ as in Fig. 7.16.

Fig. 7.16 An example of simple inversion mutation for permutation code

The SIM in Fig. 7.16 is identical to the tour change in Fig. 7.17. SIM is the same as a step of local search in 2-opt.[28]

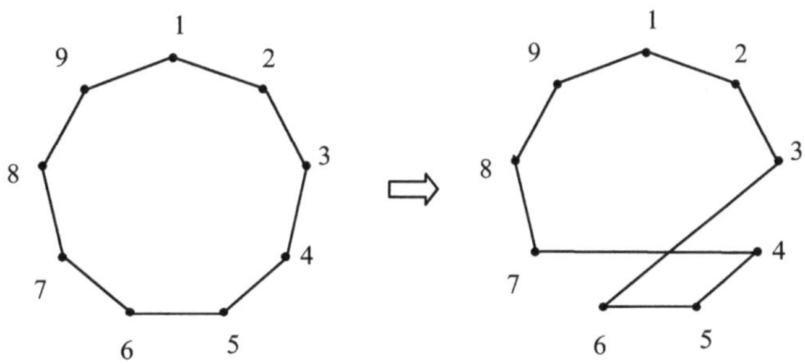

Fig. 7.17 The relationship between simple inversion mutation and a step of local search in 2-opt

Inversion Mutation

Fogel combined the characteristics of SIM and DM and suggested the *inversion mutation* (IM) [23].

We just pick one subtour randomly and insert it inversely into a random place. For example, we pick subtour $4-5-6$ in path code $(1-2-3-4-5-6-7-8-9)$

[28] It is quite clear by comparing Figs. 7.17 and 7.12.

and insert its allele into the place between loci 8 and 9 for a nine-city TSP and get $(1-2-3-7-8-6-5-4-9)$ as in Fig. 7.18.

Fig. 7.18 An example of inversion mutation for permutation code

7.3.4.2 Crossover Operators

Crossover operators, i.e., single-point crossover, multiple-point crossover, and uniform crossover, will break a permutation. So we need to consider the real link situation of cities in crossover for permutation code.

Partially Mapped Crossover

The *partially mapped crossover* (PMX) was suggested by Goldberg and Lingle [24]. It can be implemented with the following steps.

Partially Mapped Crossover for Permutation Code

Step 1: Select two random places in a chromosome, e.g., between loci 2 and 3 and between loci 6 and 7 in Fig. 7.19.

Step 2: Exchange the subtours formed by these two places and get the temporary offspring 1' and 2'. They might break the permutation.

Step 3: Determine the partial maps in these two subtours. The partial map is the allele pair at same locus of the subtours. Thus $1 \leftrightarrow 6 \leftrightarrow 3$, $4 \leftrightarrow 9$, and $5 \leftrightarrow 2$ are partial maps.

Step 4: Use the partial maps to map the conflict allele to the legal allele. For example, in the first locus of offspring 1', allele 1 is in conflict with the allele in the exchanged subtour, which needs to be kept, so we try to map it to 6 according to the partial maps, but 6 is still in conflict. Then we should map it again using partial maps to get the legal allele 3.

PMX keeps one subtour and changes others according to partial maps for maintaining permutation. So only the city connection relationships in the subtour, e.g., $3-4-5-6$ and $6-9-2-1$ in Fig. 7.19, and those do not in conflict with the subtour, e.g., $7-8$ in Fig. 7.19, will be kept.

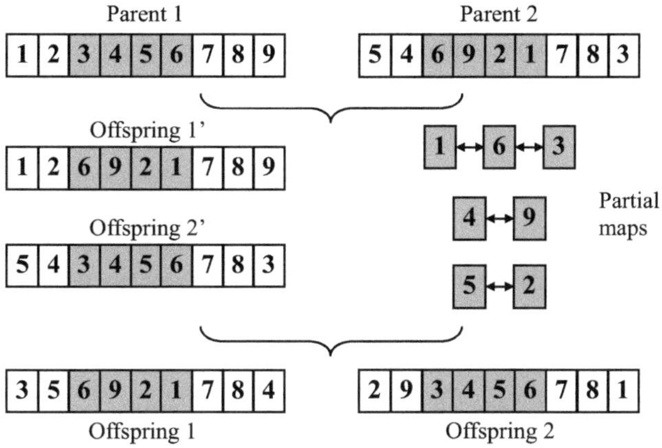

Fig. 7.19 An example of partially mapped crossover for permutation code

Order Crossover

The *order crossover* (OX) was suggested by Davis [25]. It can be implemented with the following steps. For simplicity, we only describe the process of generating offspring 1. Offspring 2 is left as an exercise.

Order Crossover for Permutation Code

Step 1: Select two random places in a chromosome, e.g., between loci 2 and 3 and between loci 6 and 7 in Fig 7.20. Mark the second place on parent 2.

Step 2: Generate partial offspring 1 by copying the subtour from parent 1.

Step 3: Mark the allele in parent 2 that already exists in the partial offspring 1.

Step 4: From the marked place in parent 2, sequentially pick the allele that is not in partial offspring 1 and put it sequentially into the locus start from the marked place.

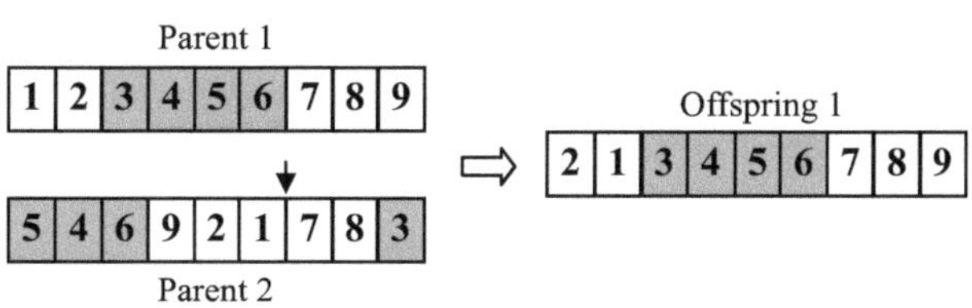

Fig. 7.20 An example of order crossover for permutation code

OX not only keeps the exchanged subtour but also tries to keep the city connection after the marked point. Apart from that, OX keeps the precedence relationship between cities, which might contribute to the search.

Position-base Crossover

The *position-based crossover* (PBX) was suggested by Syswerda [26]. It can be implemented with the following steps. For simplicity, we only describe the process of generating offspring 1. Offspring 2 is left as an exercise.

Position-based Crossover for Permutation Code

Step 1: Select several loci randomly from parent 1, e.g., loci 2, 5, 6, and 9 in parent 1 in Fig. 7.21.

Step 2: Generate partial offspring 1 by putting the selected alleles from parent 1 into their corresponding loci.

Step 3: Mark the alleles in parent 2 that have been selected in parent 1.

Step 4: From the beginning of parent 2, sequentially select the allele that has not been marked and put it into the vacancy of partial offspring 1 from the beginning.

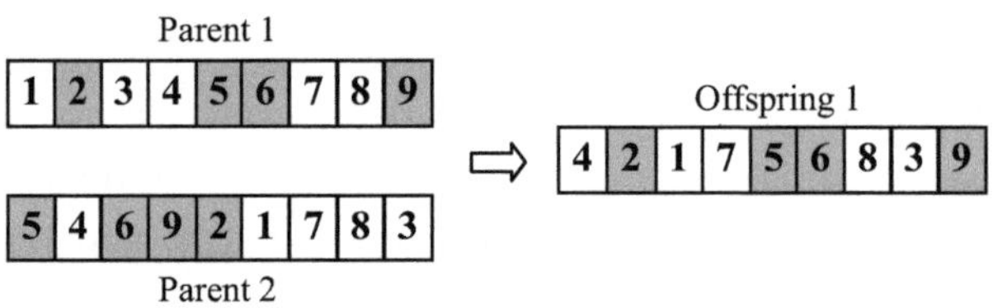

Fig. 7.21 An example of position-based crossover for permutation code

PBX keeps the position of several alleles from one parent and the precedence relationship from another parent. The position in the chromosome for TSP has little meaning because of the redundancy of permutation code. So generally PBX only keeps the precedence relationship of parents unless the selected loci are connected.

Syswerda also suggested another crossover operator, *order-based crossover* (OBX), which is almost the same as PBX.

Cycle Crossover

The *cycle crossover* (CX) was suggested by Oliver *et al.* [27]. They use the concept of virtual cycle to maintain a permutation and thus keep the position of parents in a deterministic way. It can be implemented with the following steps. For simplicity, we only describe the process of generating offspring 1. Offspring 2 is left as an exercise.

Cycle Crossover for Permutation Code

Step 1: Generate the virtual cycle from the first unused locus in parent 1. Suppose we are at locus i with allele j in parent 1; there is an instant transfer machine that can transfer us from locus i of parent 1 to that of parent 2, whose allele is k. The instant transfer machine will then send us instantly back to locus k in parent 1, whose allele is the next step of our virtual cycle. Let us start the virtual cycle from city 1 in parent 1 in Fig 7.22. The instant transfer machine will transfer us to locus 1 of parent 2 and send us to locus 5 of parent 1 instantly. With this instant transfer machine, we can find the virtual cycle formed by cities $1-5-2-4-9-1$.

Step 2: Copy the alleles in the virtual cycle in parent 1 to their corresponding loci in offspring 1.

Step 3: Generate the virtual cycle from the first unused locus in parent 2 with the procedure introduced above. $6-3$ is the virtual cycle generated in Fig. 7.22.

Step 4: Copy the alleles in the virtual cycle in parent 2 to their corresponding loci in offspring 1.

Step 5: Repeat the above steps until all the loci have been marked.

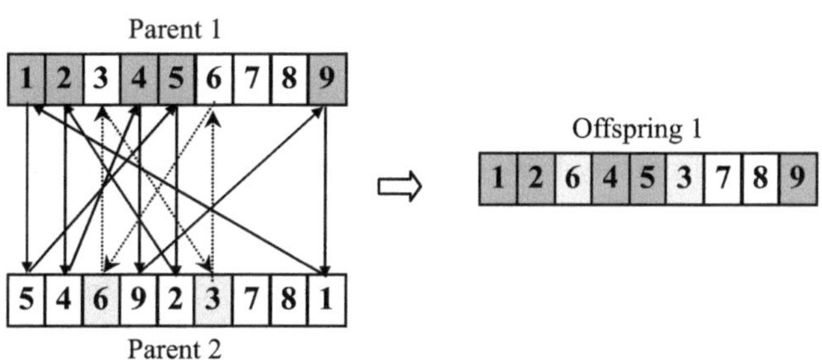

Fig. 7.22 An example of cycle crossover for permutation code

CX uses virtual cycle to maintain the permutation and keep the position of the related alleles. For TSP, it means to keep the precedence relationship between the related cities.

Edge Recombination Crossover

The *edge recombination crossover* (ERX) was suggested by Whitley *et al.* [28] and we adopt the way of edge-3 in Eiben and Smith's book to introduce the method [29].

The real information contained in the permutation code of TSP is the edge between two cities. So it is better to maintain the same edge in both parents. That is the basic idea of ERX.

For a six-city TSP, suppose parent 1 is $(1-2-3-4-5-6)$ and parent 2 is $(4-1-2-6-3-5)$. Then we can generate the edge table in Table 7.6. The edge column in the table is the cities with edges on the corresponding city. A $^+$ means that this edge is in both parents and thus should be kept in the offspring.

Table 7.6 Edge table for $(1-2-3-4-5-6)$ and $(4-1-2-6-3-5)$

City	Edges	City	Edges
1	$2^+, 4, 6$	4	$1, 3, 5^+$
2	$1^+, 3, 6$	5	$3, 4^+, 6$
3	$2, 4, 5, 6$	6	$1, 2, 3, 5$

ERX starts by randomly selecting one city and then checks the edge table to determine the next city according to the following rules.

- If the current city has a common edge, go to the city with the common edge.
- Otherwise go to the city with the shortest edge list.[29]
- If there is more than one city with the shortest edge list, ties are split by randomly going to one of them.

After determining the next city, update the edge table by deleting all the cities already visited. Repeat the above procedure until all the cities are visited.

Suppose we start from city 1; we can use Table 7.7 to help in the construction of offspring.

We can draw the parents and offspring of ERX in Fig. 7.23. Only the connections between cities 1 and 3 are new because city 3 is the last choice in the above procedure. So the edge information in parents is maintained by ERX, which has made it the best crossover operator for a rather long time.

[29] The reason for selecting the city with the shortest edge list is to decrease the possibility of incompletion in the construction process. Readers are encouraged to verify this statement by deleting this rule and finishing Table 7.7 again.

Table 7.7 ERX table for $(1-2-3-4-5-6)$ and $(4-1-2-6-3-5)$

Edge table	Current city	Choices	Reason	Next city	Partial tour
$(1: 2^+, 4, 6)$, $(2: 1^+, 3, 6)$ $(3: 2, 4, 5, 6)$, $(4: 1, 3, 5^+)$ $(5: 3, 4^+, 6)$, $(6: 1, 2, 3, 5)$	1	2, 4, 6	Common edge	2	(1-2)
(2: 3, 6) (3: 2, 4, 5, 6), $(4: 3, 5^+)$ $(5: 3, 4^+, 6)$, (6: 2, 3, 5)	2	3, 6	Shortest edge list	6	(1-2-6)
(3: 4, 5, 6), $(4: 3, 5^+)$ $(5: 3, 4^+, 6)$, (6: 3, 5)	6	3, 5	Random	5	(1-2-6-5)
(3: 4, 5), $(4: 3, 5^+)$ $(5: 3, 4^+)$	5	3, 4	Common edge	4	(1-2-6-5-4)
(3: 4), (4: 3)	4	3	Last city	3	(1-2-6-5-4-3)

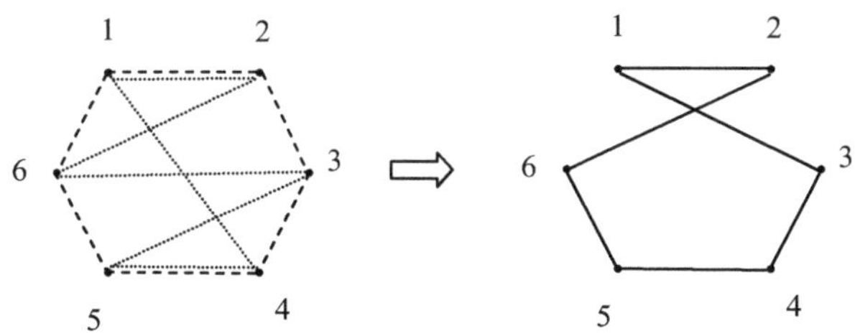

Fig. 7.23 Parents and offspring of ERX

Edge Assembly Crossover

Johnson[30] published a paper in *Nature* magazine in 1987 calling for the combination of problem-specific techniques in coding and variation schemes [30], which caused the invention of the following crossover operators.

The *edge assembly crossover* (EAX), which was suggested by Nagata and Kobayashi in 1997 [31], is such a crossover operator. EAX contains three steps.

1. Generate the "AB-cycles" by parents A and B.
2. Generate the "E-set" from "AB-cycles."
3. Generate an offspring using parent A, "E-set," and the greedy method.

For two parents A and B, illustrated by the solid line and the dashed line in Fig. 7.24, we can start at any city and go along the road belonging to A and B in turn until we come back to the original city, which forms a cycle. If in a cycle the edge number belonging to A is the same as that belonging to B, we call it an AB-cycle, i.e., a cycle formed by roads in A and B in turn with the same road number.

[30] The coauthor of the Lin–Kernighan–Johnson algorithm.

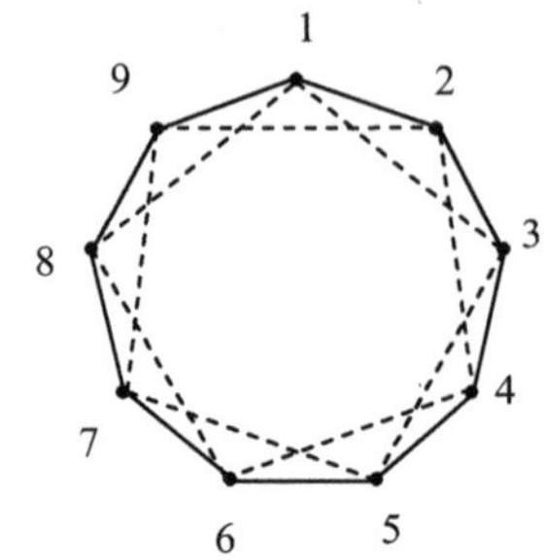

Fig. 7.24 Parents A and B of EAX

After one AB-cycle has been determined, its roads are removed from Fig. 7.24. Iteratively operating the above process will generate several AB-cycles, which is illustrated in Fig. 7.25.

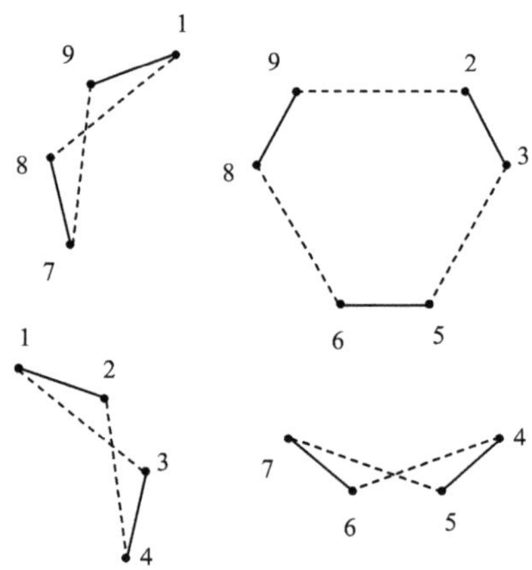

Fig. 7.25 AB-cycles in Fig. 7.24

Then a deterministic and a random approach were suggested by Nagata and Kobayashi to select elements from an AB-cycle set to form an E-set. We can just simplify the process as selecting each AB-cycle into an E-set with probability 0.5. Suppose in the four AB-cycles of Fig. 7.25 only the one in the upper right is selected for the E-set.

After that, a temporary offspring is generated by copying parent A. Then every road in the E-set is used to modify the temporary offspring. If it belongs to parent A and appears in the temporary offspring, it is deleted from the temporary offspring. If it belongs to parent B and does not appear in the temporary offspring, it is added to the temporary offspring. This procedure will form several subtours, which is illustrated by Fig. 7.26.

Then a greedy method is adopted to generate offspring that are short in length. We start from the subtour with the smallest number of roads, such as the lower left subtour in Fig. 7.26, and check which subtour to combine it with and how to combine them in a greedy way. Suppose we are considering the edges (v_q, v_{q+1})

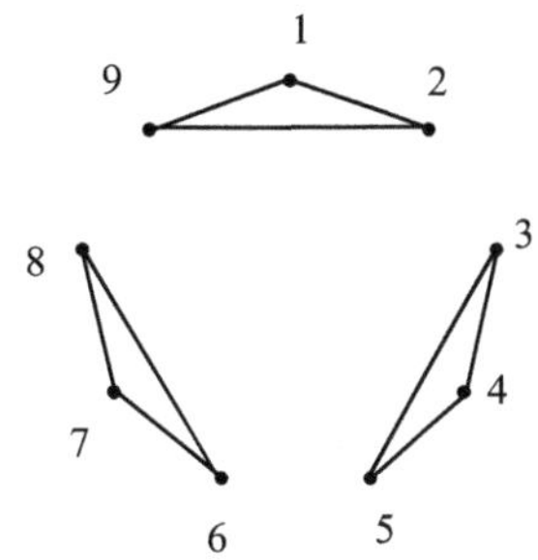

Fig. 7.26 Subtours generated by parent A and E-set

and (v_r, v_{r+1}) in Fig. 7.27, where (v_q, v_{q+1}) is a road in the considered subtour and (v_r, v_{r+1}) is a road in an other subtour.[31]

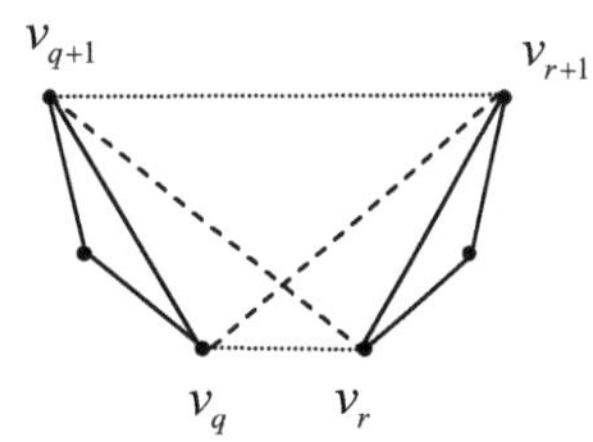

Fig. 7.27 Greedy method to generate offspring

The decrease of distance by combining these two subtours is as follows:

$$cut\,(q,r) = d\,(v_q, v_{q+1}) + d\,(v_r, v_{r+1}) \tag{7.12}$$

where $d()$ is the length of the road. The increase of distance by combining these two subtours is as follows:

$$link\,(q,r) = \min\left\{d\,(v_q, v_r) + d\,(v_{q+1}, v_{r+1}), d\,(v_q, v_{r+1}) + d\,(v_{q+1}, v_r)\right\} \tag{7.13}$$

The meaning of Eq. 7.13 is to find the way with smallest distance increase. The first part, i.e., $d\,(v_q, v_r) + d\,(v_{q+1}, v_{r+1})$, is illustrated by dotted lines in Fig. 7.27, and the second part, i.e., $d\,(v_q, v_{r+1}) + d\,(v_{q+1}, v_r)$, is illustrated by dashed lines in Fig. 7.27.

We start from the subtour with the smallest number of roads and find the edge (v_q, v_{q+1}) in it and the edge (v_r, v_{r+1}) in the other subtours to minimize the total length increase expressed by Eq. 7.14. Then these two subtours are integrated:

$$\min_{q,r}\left\{link\,(q,r) - cut\,(q,r)\right\} \tag{7.14}$$

Repeat the above integration procedure until we get a full tour, i.e., the offspring.

[31] Here the definition of node v_q, v_{q+1}, v_r, and v_{r+1} is for the simplicity of the following statement.

EAX requires a lot of computational resources for determining the AB-cycle, the E-set, and offspring. But it uses edges from parents as much as possible and has greedy methods to do local searches in the crossover process, which makes it a highly efficient search operator.

Nagata and Kobayashi used a steady state EA with a parent-offspring competition in the replacement process.

Distance Preserving Crossover

In 1995 Boese finished a technical report for studying the solution landscape of TSP [32]. The very important conclusion for TSPs is that the local optimal solutions are close to each other and also close to the global optimal solution, which means that the solution landscape roughly has a unimodal shape with small ripples. So generating offspring that are close to parents is good for the search in a TSP solution landscape.

The *distance preserving crossover* (DPX) was suggested by Freisleben and Merz in 1996 [33].

The distance between two tours defined by Freisleben and Merz is the number of roads in which two tours differ, corresponding to Hamming distance in binary code. For example, Fig. 7.28 lists two parents. Both of them connect 5 with 3 and 3 with 9. But in parent 1, 9 is connected to 1, which is different from parent 2. In this way, we can find out that their distance is 4, corresponding to $9-1$, $7-8$, $4-6$, and $6-2$ in parent 1 and $1-2$, $9-4$, $8-6$, and $6-7$ in parent 2.

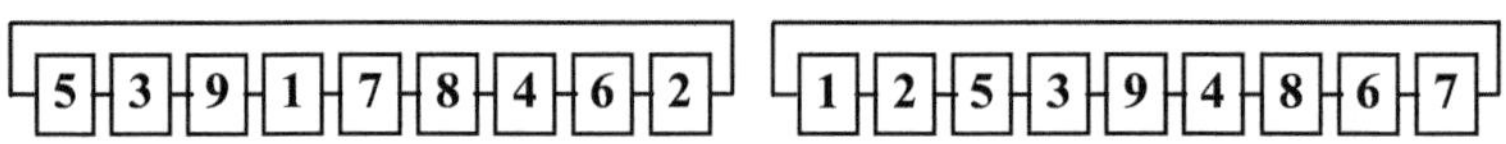

Fig. 7.28 Parents of DPX

Parent 1 is then subdivided by the different roads into several subtours, i.e., $|5-3-9|1-7|8-4|6|2|$.[32]

Then we can start from any city in parent 1, e.g., city 5, and get the offspring with following rules.

- If the current city is connected with another city in the subtour, go to that city. After finishing a subtour, delete it from the list of current subtours, such as $(5-3-9)$ in Table 7.8.
- Otherwise find out the cities that might be the edge cities of the other subtours, according to the current subtour. Delete the cities that are connected to either parent. City 1 is deleted because of $9-1$ in parent 1 and city 4 is deleted because of $9-4$ in parent 2. Then use a greedy method to find the minimum distance in

[32] The subtour number has strong ties with the distance. What is their relationship?

$d(9,7)$, $d(9,8)$, $d(9,6)$, and $d(9,2)$. Suppose city 6 is the closet city to city 9 in the choices. Then delete subtour 6 from the list of current subtours.[33]

Table 7.8 DPX table for $(5-3-9-1-7-8-4-6-2)$ and $(1-2-5-3-9-4-8-6-7)$

Edge table	Current city	Choice	Reason	Next city	Partial tour
(1: 2⁺, 4, 6), (2: 1⁺, 3, 6) (3: 2, 4, 5, 6), (4: 1, 3, 5⁺) (5: 3, 4⁺, 6), (6: 1, 2, 3, 5)	1	2, 4, 6	Common edge	2	(1-2)
(2: 3, 6) (3: 2, 4, 5, 6), (4: 3, 5⁺) (5: 3, 4⁺, 6), (6: 2, 3, 5)	2	3, 6	Shortest edge list	6	(1-2-6)
(3: 4, 5, 6), (4: 3, 5⁺) (5: 3, 4⁺, 6), (6: 3, 5)	6	3, 5	Random	5	(1-2-6-5)
(3: 4, 5), (4: 3, 5⁺) (5: 3, 4⁺)	5	3, 4	Common edge	4	(1-2-6-5-4)
(3: 4), (4: 3)	4	3	Last city	3	(1-2-6-5-4-3)

Table 7.8 illustrates the crossover process for generating offspring $(5-3-9-6-1-7-4-8-2)$. It is easy to verify that the distance from the offspring to either of the parents is 4, which means that DPX could generate an offspring that is equidistant to both of its parents.[34]

DPX uses the duplicated roads in the parents, preserves the distance to the parents, and utilizes a greedy method to connect the subtours into a solution. According to Johnson's paper [30] and Boese's report [32], these properties promote searchs in the TSP solution landscape.

Freisleben and Merz also used the Lin–Kernighan algorithm for the offspring of DPX. This further improves the algorithm's search ability and makes it the benchmark of EA-based TSP algorithms.

Natural Crossover

Jung and Moon divided EAs for TSPs into two categories: pure EAs, which do not utilize local search methods and thus have the requirement that variation operators be able to generate good offspring, and hybrid EAs, which are actually MAs and thus have the requirement that variation operators have strong search abilities to provide the local search method with a promising initial solution.

[33] In this way, the offspring can keep the same distance to both parents.

[34] It is a little bit like UNDX in real code.

Based on this idea, Jung and Moon suggested a *natural expression* and a *natural crossover* (NX) for 2-D Euclidean TSPs in 2002 [34].

There is no code for a solution of 2-D Euclidean TSPs. Jung and Moon just used a real path to represent solutions of 2-D Euclidean TSPs. In Fig. 7.29, solid lines represent one solution and dashed lines represent another solution. They are parents 1 and 2 of NX, respectively.

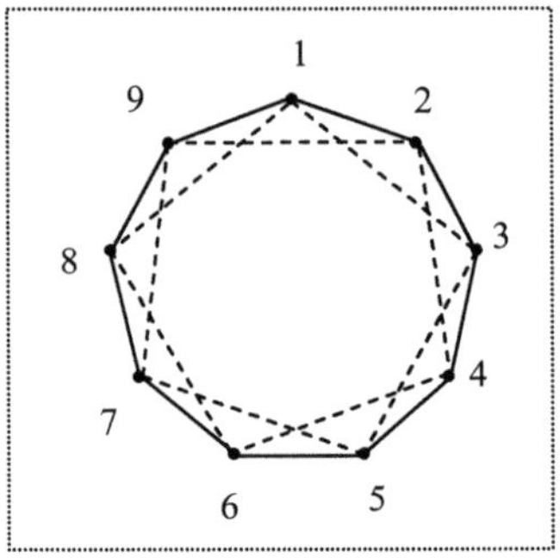

Fig. 7.29 Parents of NX

Then they use several simple curves, such as straight lines, triangles, rectangles, and ellipses, to divide the area into several parts. Parts covered by an odd number of curves become gray and those covered by an even number of curves remain white. An example of such curves is illustrated in Fig. 7.30.

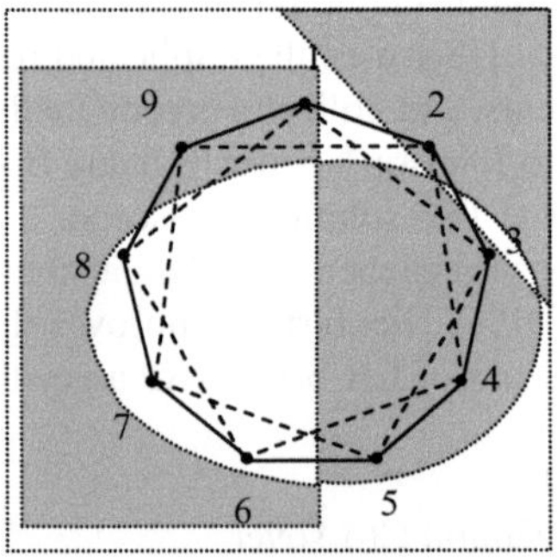

Fig. 7.30 Parts covered by several curves

For a road in parent 1, if both of its endpoints are in the gray part, it is kept. Otherwise it is discarded. In contrast, for a road in parent 2, if both of its endpoints are in the white part, it is kept. Otherwise it is discarded. The procedure generates several subtours, illustrated in Fig. 7.31.

A greedy strategy is taken to connect these subtours into a Hamiltonian cycle. A possible result is illustrated in Fig. 7.32, in which the solid lines are roads in parents and dashed lines are generated by a greedy method.

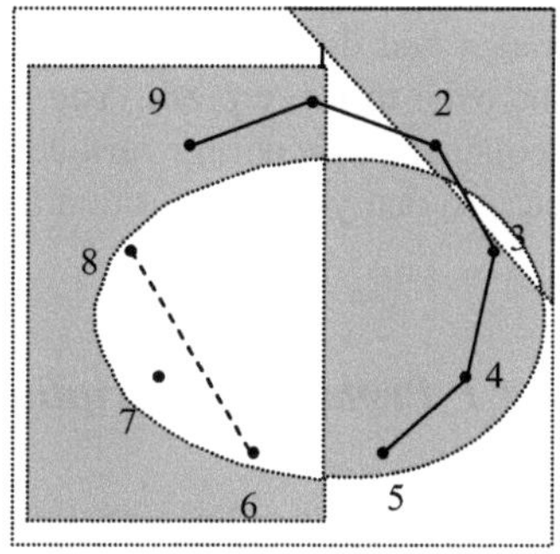

Fig. 7.31 Subtours generated by NX

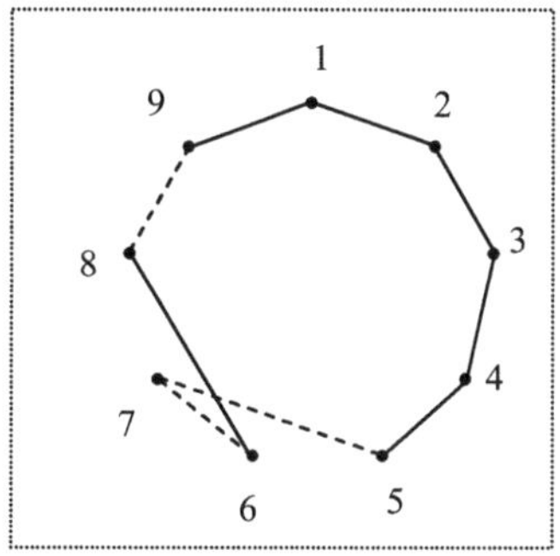

Fig. 7.32 Offspring of NX

As can be seen from the above process, NX keeps certain roads in the parents while introducing a certain perturbation from parents. After that, Jung and Moon utilize the Lin–Kernighan algorithm to improve the offspring. In Jung and Moon's philosophy, the perturbations provide promising initial solutions for local search methods.

They implemented the algorithm in a steady state way and compared NX, DPX, and EAX using benchmark TSPs from TSPLIB. According to their numerical experiments, the result of NX + Lin–Kernighan algorithm is slightly better than DPX + Lin–Kernighan algorithm but with a slightly longer time.

7.4 Job-shop Scheduling Problem

Scheduling is the process of generating a *schedule*. A schedule is a sequence of events subject to some predefined constraints. The objective of scheduling is to achieve some goals, such as the shortest time or the smallest expense to finish a job.

Let us suppose you want to have friends over for a meal. The menu has been made and there are several dishes to serve. Each dish will require different kitchen facilities with different sequences, e.g., dough should first be made in the bread maker and then, after covering it with cheese, tomato, and sausage, etc., baked in the oven to make pizza. You need to schedule the sequence of all events, e.g., the sequence of preparing various dishes on the stove and in the oven and bread maker, etc., so that your preparation time is minimized. What a mess!

7.4.1 Problem Description

In scholarly disciplines, these events that need to be scheduled are called *operations*. The resources to process these operations are called *machines*. Dishes you want to serve at the party are called *jobs*.

Without loss of generality, we suppose there are n *jobs* needing to be finished, each of which should be processed[35] with m different *machines* in a job shop. Different jobs might have different processing sequences on these machines.[36] We use o_{ijk} to represent that job i's operation j will be processed on machine k with *processing time* p_{ijk}.

For operation o_{ijk}, $o_{i,(j+1),h}$ is its *direct succeeding operation* or *direct successor*, $o_{i,(l>j),h}$ are its *succeeding operations* or *successors*, $o_{i,(j-1),h}$ is its *direct preceding operation* or *direct predecessor*, and $o_{i,(l<j),h}$ are its *preceding operations* or *predecessors*, where h is another machine.

An example of such predefined parameters is illustrated in Table 7.9. According to Table 7.9, job 2 should first be processed by machine 2 with time 2, then by machine 1 with time 3, and finally by machine 3 with time 3.[37]

Table 7.9 An example of job-shop scheduling with $n = 4$ jobs and $m = 3$ machines

	p_{ijk}			Machine		
Operation	1	2	3	1	2	3
Job 1	2	2	12	M_1	M_3	M_2
Job 2	2	3	3	M_2	M_1	M_3
Job 3	3	1	5	M_1	M_2	M_3
Job 4	2	2	3	M_3	M_2	M_1

The predefined job processing sequence on different machines is called the *precedence relationship* or *precedence constraint*. The processing time and precedence relationship together are called *technological requirements*. The *job-shop scheduling problem* (JSSP), which is the most famous scheduling problem, is to determine operations' processing sequences on each machine, subject to precedence con-

[35] Each process is an *operation*.

[36] It is different from flow-shop introduced in Sect. 7.1.1.

[37] Here we omit the unit of processing time for simplicity's sake.

straints and considering the processing time, so as to minimize the total processing time, i.e., the *makespan*.[38] The mathematical model of JSSP can be illustrated as follows:

$$\begin{aligned}\min\ t_M &= \max_{i,j,k}\{t_{ijk}\}\\ \text{s.t. } t_{i,j-1,h} + p_{ijk} &\le t_{ijk}, \quad \forall i,j,k,h \\ t_{ijk} &\ge 0, \quad \forall i,j,k\end{aligned} \tag{7.15}$$

where t_{ijk} is the *completion time* of operation o_{ijk} and t_M is the makespan. $t_{i,j-1,h} + p_{ijk} \le t_{ijk}$ means that for every operation, its direct predecessor's completion time plus its processing time might be smaller that its completion time because it might not be processed right after the completion of its direct predecessor.[39]

The solution space of JSSP is rather large. We can consider it in the following way. Suppose that all the precedence relationships of n jobs are $1 \to \cdots \to m$, which makes the JSSP a flow-shop scheduling problem.[40] There are $n!$ possible solutions for the first machine. The possible number of solutions for the second machine is also $n!$ because n jobs must be processed by machine 2 after machine 1. So in total there are $(n!)^m$ possible solutions for this problem. This is a rather large number.[41]

The above discussion is a simplification so that we can estimate the maximum feasible solution space. Actually, the precedence constraints will limit the number of feasible solutions but increase the difficulty. Generally, n permutations on each machine might generate an infeasible solution. Let us discuss this point with a very simple JSSP with $n = m = 2$. Job 1 needs to be processed by machine 2 first and then machine 1, but job 2 needs to be processed in the sequence $1 \to 2$. So the only possible solution for this problem is to let machine 1 process job 2 first and then job 1 and let machine 2 process job 1 first and then job 2. But permutation $(1,2)$ on machine 1 and permutation $(2,1)$ on machine 2 are infeasible solutions. This phenomenon is called *deadlock*.[42]

There are several ways to illustrate the solution of a JSSP. We will introduce two of them and focus on the second one later.

Because of the precedence relationship, the initial idea to represent a JSSP and its solution was to use a directed graph, as in Fig. 7.33. We have a virtual start point and a virtual end point. Each solid arrow represents the precedence relationship of one job. The number in the circle represents the machine number and the number above the circle is the processing time of that operation. For example, job 1 needs

---

[38] Makespan could be understood as the span, i.e., period, of making jobs. There are other objectives. For simplicity's sake, we only discuss makespan in this text. Interested readers are referred to [2, 35].

[39] Perhaps machine k, which needs to process job i, is processing other jobs at $t_{i,j-1,h}$.

[40] Not a permutation flow-shop scheduling problem.

[41] Do you remember the solution size of 50 jobs for a permutation flow-shop scheduling problem and how long it would take to solve that problem using a "crude" enumeration method with the fastest supercomputer? Try $n = m = 15$ for JSSP.

[42] Any permutation code for problems with precedence constraints might face deadlock. Special techniques, e.g., repair, are necessary for unlocking the solution. We do not discuss this in the text.

to be processed by machine 1 with time 2 first, then machine 3 with time 2, and finally machine 2 with time 12 according to the technological requirements given by Table 7.9. The solution of the JSSP is to find m Hamiltonian directed paths, each of which is within the vertices with the same machine number to represent the processing sequence of jobs on one machine. We use a dashed line to represent the Hamiltonian directed path on machine 1, i.e., job 1 first, then job 3, then job 4, and finally job 2, a dotted line to represent the processing sequence on machine 2, and a dashed-and-doted line to represent machine 3. These directed edges between the same machines are called *disjunctive arcs*.

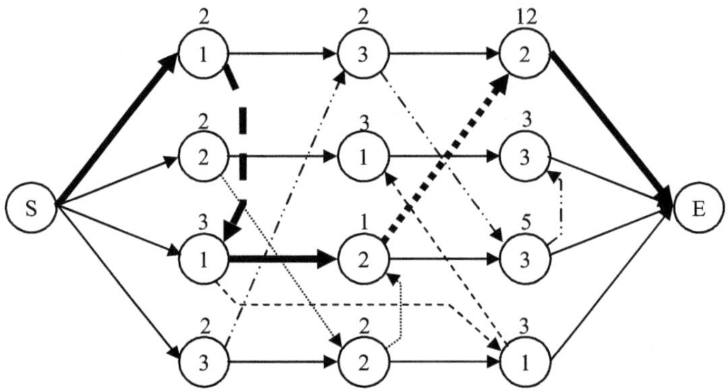

Fig. 7.33 An example directed graph for JSSP

Then from the start point to the end point we have many Hamiltonian directed paths, including both processing sequences of the jobs and disjunctive arcs of the machines. We define the length of a path as the sum of the processing times passing through in the path. The longest length in these Hamiltonian directed paths determines the makespan.[43] We use heavy lines in Fig. 7.33 to demonstrate it. The makespan is $2+3+1+12=18$. This path is called the *critical path*.

H. Gantt was an American mechanical engineer and management consultant. In the 1910s, he suggested a type of bar chart to illustrate a project schedule. An example Gantt chart for the JSSP of Table 7.9 is Fig. 7.34.

The horizontal axis of a Gantt chart is time and the vertical axis represents several machines with the sequence of operations processed on them.[44] For example, on machine 2, job 2's first operation is processed first, then job 4's second operation is processed. Job 3's second operation is processed until time 5 because its direct predecessor, o_{311}, is finished at time 5 by machine 1. Finally job 1's last operation is processed. We can also determine the critical path in Fig. 7.34 by finding the path

[43] Why?

[44] Another way a Gantt chart can represent the solution is to use the vertical axis for jobs.

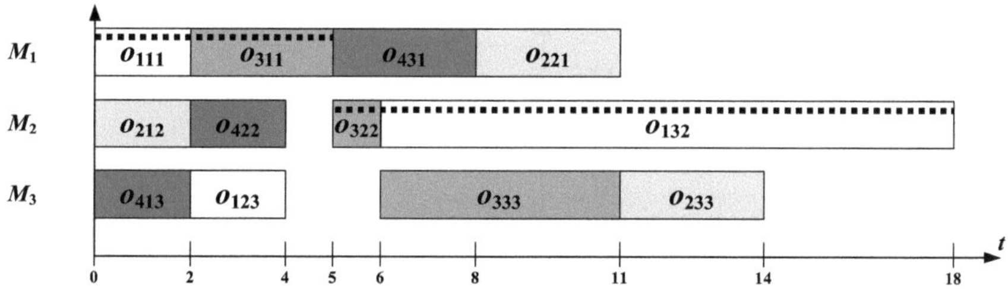

Fig. 7.34 An example Gantt chart for JSSP

from time 0 to t_M without *idle time*,[45] provided that we can jump seamlessly among machines. A heavy dotted line is used to represent this path in Fig. 7.34.

The alert reader might have noticed from Fig. 7.34 that we can actually generate an infinite number of feasible solutions for a JSSP because we can add artificially the idle time between every two operations on one machine. This is true. But the objective of our JSSP is to minimize t_M, so we want to eliminate any useless idle time. If we move an operation rectangle in the Gantt chart to the left and do not change any operation's sequence, this movement is called a *local left-shift*. If there exists a local left-shift, we can decrease the makespan by moving this operation to the left. A solution without a local left-shift is called a *semiactive schedule*. It is not difficult to understand that there is finite number of semiactive schedules for a JSSP.

If we change the sequence of some operations on a machine so that one operation is finished earlier but others are not delayed, this movement is called a *global left-shift*. A solution without a global left-shift is called an *active schedule*.[46]

If a machine is available, i.e., it is not processing any job, and a job is available, i.e., its previous operations are finished, but we do not process this job on this machine immediately, there is a *delay* for the job. A solution without a delay is called a *nondelay schedule*.

These concepts are clearly illustrated by Fig. 7.35.

There is no local left-shift in Fig. 7.35. So it is semiactive. o_{222} can be global left-shifted to start between time $[2,3]$ without delay o_{122}. So it is not active. At time 2, o_{222} could be started but we delay it. So it is not nondelay.

It has been proved that the relationship between these solution sets and the possible optimal solution sets for JSSP can be illustrated by Fig. 7.36. The optimal solution belongs to the active schedule set but might not belong to the nondelay schedule set. So we generally want to search the active schedule set.

Sometimes we can expand a JSSP into a parallel environment, i.e., there are c_1 identical machines with type 1 in work center 1, $\cdots$, c_m identical machines with type m in work center m. Every job still needs to be processed by m operations in

---

[45] A machine is idle if it is not processing jobs.

[46] To be clear, an active schedule means that for every operation in that schedule there is not enough idle time on its machine before its current location, or there is enough idle time but it cannot be scheduled at that time due to the precedence constraint.

Fig. 7.35 A semiactive, not active, not nondelay partial schedule

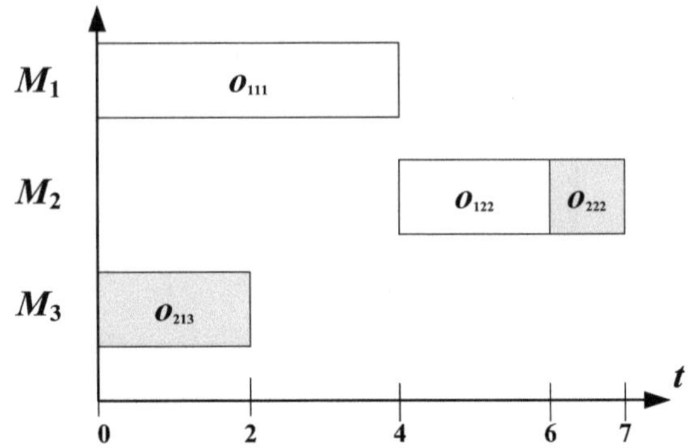

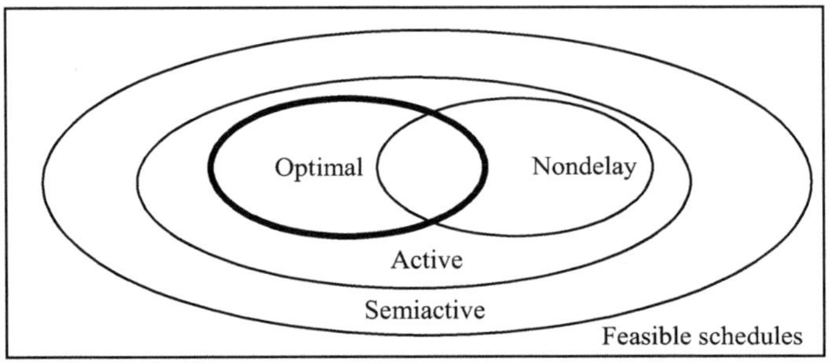

Fig. 7.36 The relationship between semiactive, active, and nondelay schedules and the possible optimal solution sets for JSSP

m work centers in a predefined sequence. But it can be processed by any machine in that work center. This expansion causes a *flexible job-shop scheduling problem* (FJSSP).

Another expansion from JSSP assumes that every machine has same capability, i.e., every machine can process every operation. We still want every job to be processed by every machine but without a predefined sequence. Then the problem is to determine both the sequence of processing machines for every job and the sequence of processed jobs for every machine. This expansion causes an *open-shop scheduling problem* (OSSP).

Like the knapsack problem and TSP, there are many benchmark JSSPs. Beasley published a paper advocating the establishment of an OR-Library [36]. Since then, he has maintained a Web site containing many of the famous combinatorial optimization problems, which includes scheduling, bin packing, knapsack, network flow, shortest path, TSP, vehicle routing, etc., and their current best solutions.[47] In 1993 Taillard proposed a method to randomly generate large scheduling problems [37].

[47] http://people.brunel.ac.uk/~mastjjb/jeb/info.html.

In the remainder of this section, we will only discuss the basic JSSP using the example given by Table 7.9.

7.4.2 *Heuristic Methods for Job-shop Scheduling*

JSSP is a rather famous and widely applicable combinatorial optimization problem. In addition, its search space is extremely large. So there are many heuristic methods for JSSP. We will introduce one construction method and one local search method.

7.4.2.1 Construction Methods

From Fig. 7.36 we know that the optimal solution of a JSSP is an active schedule. Then why not limit our search to the field of active schedules? In 1960 Giffler and Thompson proposed the concept of active schedules and suggested a way to generate active schedules [38]. Baker and Trietsch clearly restated the algorithm in their famous textbook published in 2009 [35].

The *Giffler–Thompson algorithm* generates an active schedule (or all active schedules) by selecting one operation at a time. For operation o_{ijk}, it is *schedulable* if operations $o_{i,(l<j),h}$ have all been scheduled. These scheduled operations are called a *partial schedule*. We start from an empty set of the partial schedule, add one schedulable operation one time, until all $n \times m$ operations are scheduled. Then the partial schedule becomes the active schedule. We use $PS(\tau)$ to represent the partial schedule containing τ scheduled operations and $SO(\tau)$ to represent the set of schedulable operations at stage τ. Because we schedule one operation at one stage, PS and SO have the same variable τ. For every operation $o_{ijk} \in SO(\tau)$, s_j is used to represent its earliest starting time and f_j is used to represent its earliest completion time. s_j is determined by the maximum value of the completion time of the last scheduled operation on machine k and the completion time of $o_{i,(j-1),h}$. And $f_j = s_j + p_{ijk}$. The Giffler–Thompson algorithm can be illustrated as follows.

Giffler–Thompson Algorithm for Generating an Active Schedule

Step 1: $\tau = 0$. $PS(\tau) = \emptyset$. $SO(\tau)$ includes all operations without predecessors.

Step 2: Determine $f^* = \min_{j \in SO(\tau)} \{f_j\}$ and the machine M^* by which f^* can be realized. If more than one machine has f^*, ties are split randomly.

Step 3: For the operations in $SO(\tau)$ that can be operated by machine M^*, find those where $s_j < f^*$. Select one operation, e.g., operation j, from among them and put in $PS(\tau)$ according to some rule.

Step 4: Remove operation j from $SO(\tau)$. $\tau = \tau + 1$. Add operation j's successor to $SO(\tau)$.

Step 5: Repeat step 2 to step 4 until all the operations have been added to PS.

In step 2 of the above algorithm, PS contains $(\tau - 1)$ scheduled operations. Step 3 determines the τth operation in PS.

The criterion $s_j < f^*$ in step 3 is very critical for generating an active schedule. We can use a partial mixed Gantt chart to illustrate this situation in Fig. 7.37. Suppose currently there are more than three schedulable operations in SO but f^* and M^* are determined by job 1. The previous finished operation on M^* is illustrated by the thick rectangle. There are three jobs with schedulable operations on M^*. The direct predecessors of these schedulable operations are represented by solid rectangles and the schedulable operations are represented by dotted rectangles. The earliest starting times and completion times are marked.

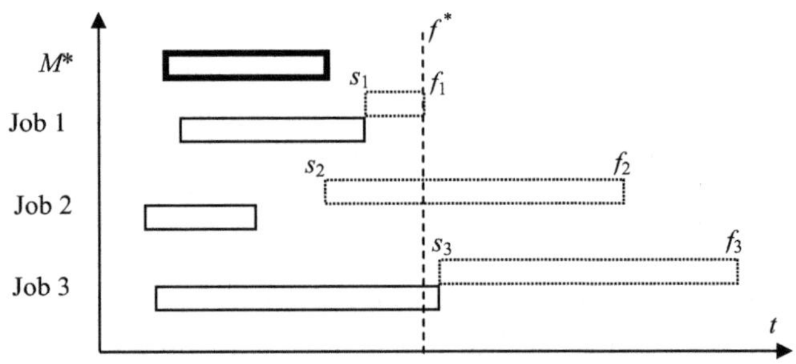

Fig. 7.37 The reason why criterion $s_j < f^*$ ensures active schedules

As can be seen from Fig. 7.37, the operations of jobs 1 and 2 can be scheduled in PS according to different rules. But if we schedule operation of job 3 in PS, later we could get a global left-shift operation of job 1 so that the operation of job 1 can be finished earlier without affecting other operations. So the operation of job 3 is forbidden. Criterion $s_j < f^*$ is to forbid the possibility of a future global left-shift so as to ensure that the generated schedule will be active.

We can generate all the active schedules of a JSSP if we add every possible schedulable operation to PS in step 3 and maintain a tree data structure to record all the partial schedules. Due to the complexity of JSSPs, even though the number of active schedules is smaller than that of feasible schedules, we will not proceed in this way. The branch and bound method will limit the search space in this tree data structure but is not applicable to moderate-scale JSSPs as well.

There are many rules, known as *priority dispatching rules*, to determine which schedulable operation is added to *PS* in step 3. We list some of them as follows [39].[48]

- **SPT (shortest processing time).** Select the operation with the shortest processing time.
- **LPT (longest processing time).** Select the operation with the longest processing time.
- **MWR (most work remaining).** Select the operation for the job with the most total remaining processing time.
- **LWR (least work remaining).** Select the operation for the job with the least total remaining processing time.
- **MOR (most operations remaining).** Select the operation for the job with the most remaining operations.
- **LOR (least operations remaining).** Select the operation for the job with the least remaining operations.
- **FCFS (first come first served).** Select the first operation in the machine's waiting queue.
- **R (random).** Select an operation at random.

It was suggested by Baker and Trietsch that the MWR rule often produces a good makespan [35]. So we just use the MWR rule as an example to illustrate the generating procedure of JSSPs illustrated by Table 7.9. The procedure can be illustrated by Table 7.10, where *wr* stands for the work remaining, i.e., total remaining processing time of the job. We only illustrate the newly added operation in column "$PS(\tau)$" for simplicity. The resulting schedule can be illustrated by Fig. 7.38, in which the critical path is illustrated by dotted lines.

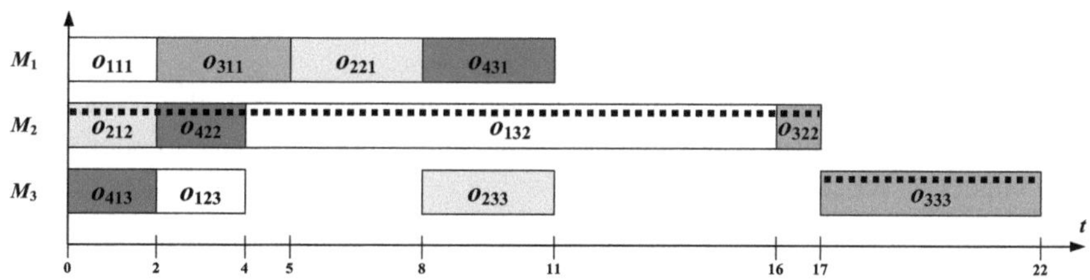

Fig. 7.38 The active schedule generated by the Giffler–Thompson algorithm with the MWR rule

[48] Panwalkar and Iskander summarized these rules in 1977 [40]; the rules were later developed by Blackstone *et al.* in 1982 [41] and Dorndorf and Pesch in 1995 [42]. A numerical comparison of these dispatching rules was carried out by Kim *et al.* in 2008 [43]. A survey on job-shop scheduling by local search was presented by Vaessens *et al.* in 1996 [44].

Table 7.10 The procedure for generating an active schedule using the Giffler–Thompson algorithm with the MWR rule

τ	$SO(\tau)$	f	f^*	M^*	s	wr	$PS(\tau)$
1	o_{111}, o_{212}, o_{311}, o_{413}	$f_{111}=2$, $f_{212}=2$, $f_{311}=3$, $f_{413}=2$	2	1	$s_{111}=0$, $s_{311}=0$	$wr_{111}=16$, $wr_{311}=9$	o_{111}
2	o_{123}, o_{212}, o_{311}, o_{413}	$f_{123}=4$, $f_{212}=2$, $f_{311}=5$, $f_{413}=2$	2	2	$s_{212}=0$	$wr_{212}=8$	o_{212}
3	o_{123}, o_{221}, o_{311}, o_{413}	$f_{123}=4$, $f_{221}=5$, $f_{311}=5$, $f_{413}=2$	2	3	$s_{413}=0$	$wr_{413}=7$	o_{413}
4	o_{123}, o_{221}, o_{311}, o_{422}	$f_{123}=4$, $f_{221}=5$, $f_{311}=5$, $f_{422}=4$	4	2	$s_{422}=2$	$wr_{422}=5$	o_{422}
5	o_{123}, o_{221}, o_{311}, o_{431}	$f_{123}=4$, $f_{221}=5$, $f_{311}=5$, $f_{431}=7$	4	3	$s_{123}=2$	$wr_{123}=14$	o_{123}
6	o_{132}, o_{221}, o_{311}, o_{431}	$f_{132}=16$, $f_{221}=5$, $f_{311}=5$, $f_{431}=7$	5	1	$s_{221}=2$, $s_{311}=2$, $s_{431}=4$	$wr_{221}=6$, $wr_{311}=9$, $wr_{431}=3$	o_{311}
7	o_{132}, o_{221}, o_{322}, o_{431}	$f_{132}=16$, $f_{221}=8$, $f_{322}=6$, $f_{431}=8$	6	2	$s_{132}=4$, $s_{322}=5$	$wr_{132}=12$, $wr_{322}=6$	o_{132}
8	o_{221}, o_{322}, o_{431}	$f_{221}=8$, $f_{322}=17$, $f_{431}=8$	8	1	$s_{221}=5$, $s_{431}=5$	$wr_{221}=6$, $wr_{431}=3$	o_{221}
9	o_{233}, o_{322}, o_{431}	$f_{233}=11$, $f_{322}=17$, $f_{431}=11$	11	1	$s_{431}=8$	$wr_{431}=3$	o_{431}
10	o_{233}, o_{322}	$f_{233}=11$, $f_{322}=17$	11	3	$s_{233}=8$	$wr_{233}=3$	o_{233}
11	o_{322}	$f_{322}=17$	17	2	$s_{322}=16$	$wr_{322}=6$	o_{322}
12	o_{333}	$f_{333}=22$	22	3	$s_{333}=17$	$wr_{333}=5$	o_{333}

The schedule in Fig. 7.38 is nondelay. By comparing it with the schedule in Fig. 7.34 we can confirm that a nondelay schedule is not necessarily better than a delay but active schedule.[49]

If we want to generate nondelay schedules, the only change in the above algorithm is in steps 2 and 3.[50]

Giffler–Thompson Algorithm for Generating a Nondelay Schedule

[49] The schedule in Fig. 7.34 is not nondelay because o_{132} could be started at time 4, but we deliberately delay it until time 6.

[50] Why?

Step 2: Determine $s^* = \min_{j \in SO(\tau)} \{s_j\}$ and the machine M^* by which s^* could be realized.

Step 3: For the operations in $SO(\tau)$ that should be operated by machine M^*, find those $s_j = s^*$. Select one operation, e.g., operation j, from among them and put in $PS(\tau)$ according to some rule.

Another famous construction method for JSSP is the *shifting bottleneck procedure* suggested by Adams *et al.* in 1988 [45].

7.4.2.2 Local Search Method

As for local searches, we need to define neighbors first.[51] Here we introduce the definition of neighbor according to the critical path [35].

As mentioned above, a critical path is a group of operations in a Gantt chart. Giffler and Thompson proved that every active schedule has at least one critical path [38]. The contiguous operations on the critical path is either belong to the same job and are processed by different machines consecutively or belong to the same machine and are processed without idle time. Operations in the latter situation form *blocks*. For example, o_{212}, o_{422}, o_{132}, and o_{322} form a block on machine 2 in Fig. 7.38.

The neighbor of the current schedule can be defined as the result of interchanging adjacent operations in the block of the critical path. This method is called *adjacent pairwise interchange*.

If we interchange operations o_{212} and o_{422}, the resulting schedule is illustrated by Fig. 7.39.[52]

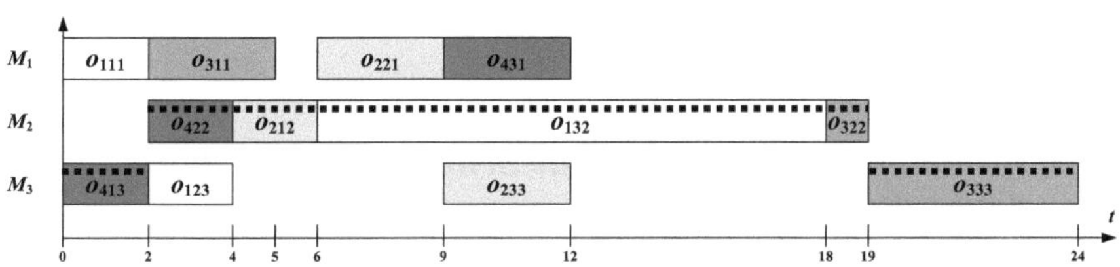

Fig. 7.39 The first neighbor of the schedule in Fig. 7.38

If we interchange operations o_{422} and o_{132}, the resulting schedule is illustrated by Fig. 7.40.

---

[51] Recall the definition of neighbor in the knapsack problem and in the TSP.

[52] Other operations might be changed accordingly due to precedence constraints.

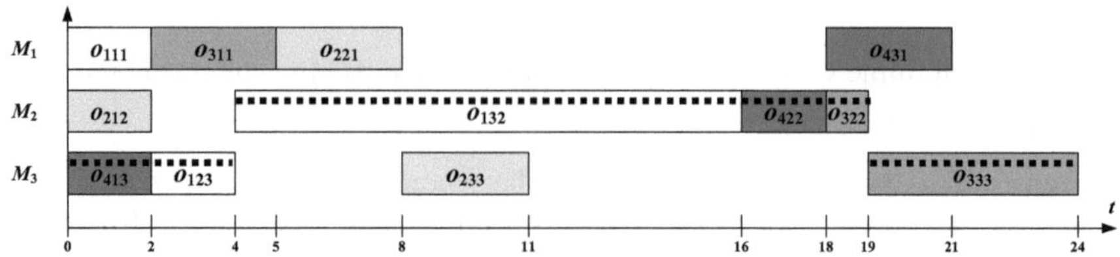

Fig. 7.40 The second neighbor of the schedule in Fig. 7.38

If we interchange operations o_{132} and o_{322}, the resulting schedule is illustrated by Fig. 7.41.

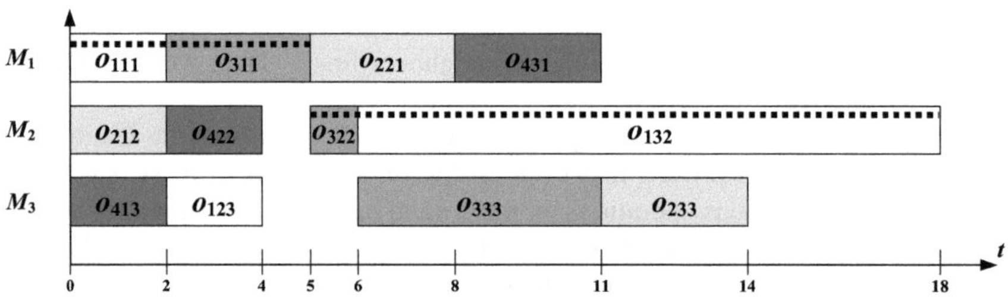

Fig. 7.41 The third neighbor of the schedule in Fig. 7.38

Thus the one-step local search by adjacent pairwise interchange finds a better solution, i.e., Fig. 7.41.

Every result of a local search has its critical path, illustrated in Figs. 7.39–7.41 with dotted lines. So we can continue the local search from them by adjacent pairwise interchange. Baker and Trietsch proved that we can find the optimal solution of a JSSP by adjacent pairwise interchange on active schedules [35].

7.4.3 Evolutionary Algorithm Code Schemes for Job-shop Scheduling

The code scheme for a JSSP can be roughly divided into two categories: *direct approaches* and *indirect approaches*. Every chromosome of a direct approach determines a schedule itself, but every chromosome of an indirect approach determines a way to generate a schedule.

We will use the example in Table 7.9 throughout this subsection to demonstrate the code schemes.

7.4.3.1 Direct Approaches

Operation-based Code

Since a JSSP is basically the schedule of operations, it is quite natural to number the $n \times m$ operations from 1 to $n \times m$. Every permutation of these $n \times m$ operations represents a (perhaps infeasible) schedule. This is *operation-based code*. In 1994 Gen *et al.* proposed an alternative idea to always generate a feasible schedule [46] (Fig. 7.42).

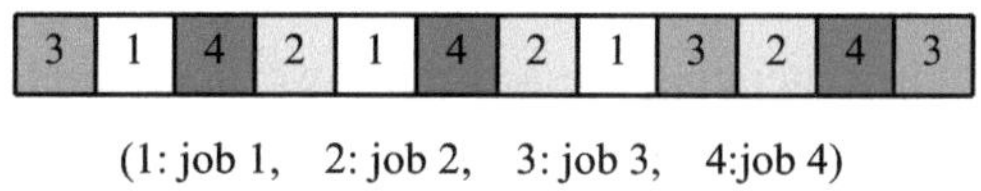

Fig. 7.42 A chromosome of operation-based code

Every gene represents an operation. The loci of the chromosome determine the sequence of the operations (from left to right) and the alleles of the chromosome are the job number. For example, according to the precedence constraint of Table 7.9, from left to right, the first allele 3 represents operation o_{311}, the second allele 3 represents operation o_{322}, and the third allele 3 represents operation o_{333}. The corresponding operation sequence is illustrated in Fig. 7.43.

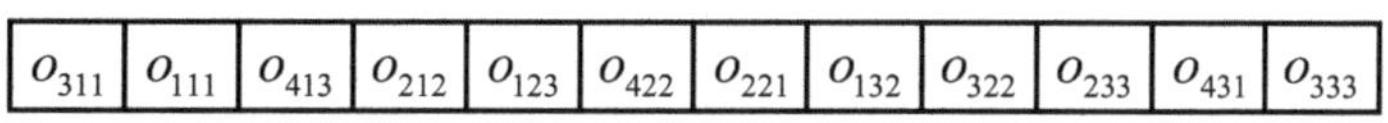

Fig. 7.43 Operation sequence of Fig. 7.42

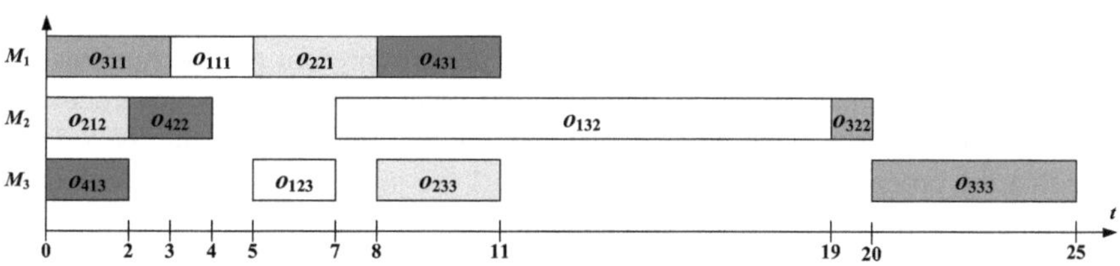

Fig. 7.44 Gantt chart of operation-based chromosome in Fig. 7.42

After we get the whole operation sequence, the Gantt chart of the schedule can be generated as Fig. 7.44.[53] In generating the Gantt chart, we start an operation whenever its predecessor has been finished and the machine to process it is available [54].

[53] Why we could always generate feasible operation sequence?

[54] This procedure will generate a nondelay schedule.

Job-based Code

Job-based code for JSSP was suggested by Holsapple *et al.* in 1993 [47]. It uses n genes to represent n jobs. The loci of the chromosome determines the sequence of scheduling jobs and the alleles of the chromosome are the job number. An example of a job-based chromosome is illustrated in Fig. 7.45.

Fig. 7.45 A chromosome of job-based code

3	1	2	4

In generating the schedule, we first simply arrange the operations of job 3 in their earliest possible time, then jobs 1, 2, and 4 sequentially. The procedure can be illustrated by Fig. 7.46. If there are enough empty time blocks in the early processing stage, the operations of the later arranged job could be inserted into the empty time block, such as operations o_{212}, o_{413}, and o_{422} in Fig. 7.46.

According to the procedure of decoding, we know that job-based code will generate an active schedule. But it is necessary to mention that not all active solutions can be represented by job-based code, i.e., the code is not surjective, which might lose the optimal solution. For example, both of the schedules represented by Figs. 7.33 and 7.41 cannot be represented by job-based code.

Random-key-based Code

We have already introduced the random-key-based code for TSPs in Sect. 7.3.3.3. *Random-key-based code* for JSSPs was suggested by Bean in 1994 [48].

We use $n \times m$ random numbers to represent a schedule. The chromosome is divided into m parts. The first n genes are with integer part 1, ..., and the last n genes are with integer part m (Fig. 7.47).

In each part, i.e., each machine, the loci of genes represent the job number. We rank the alleles of that part in ascending order so as to determine the processing sequences of jobs on a machine. For example, the first four random numbers determine that the processing sequence on machine 1 are jobs 3, 1, 4, and 2. With these job sequences on the machines, we can generate the scheduling illustrated in Fig. 7.48.

As can be seen from Fig. 7.48, random-key-based code might not guarantee generation of an active schedule, e.g., o_{322} can be global left-shifted.

Random-key-based code provides sequences of operations on each machine directly, which might conflict with precedence constraints, i.e., suffering deadlock. In 1995 Norman and Bean suggested a method to handle the conflict with precedence constraints [49].

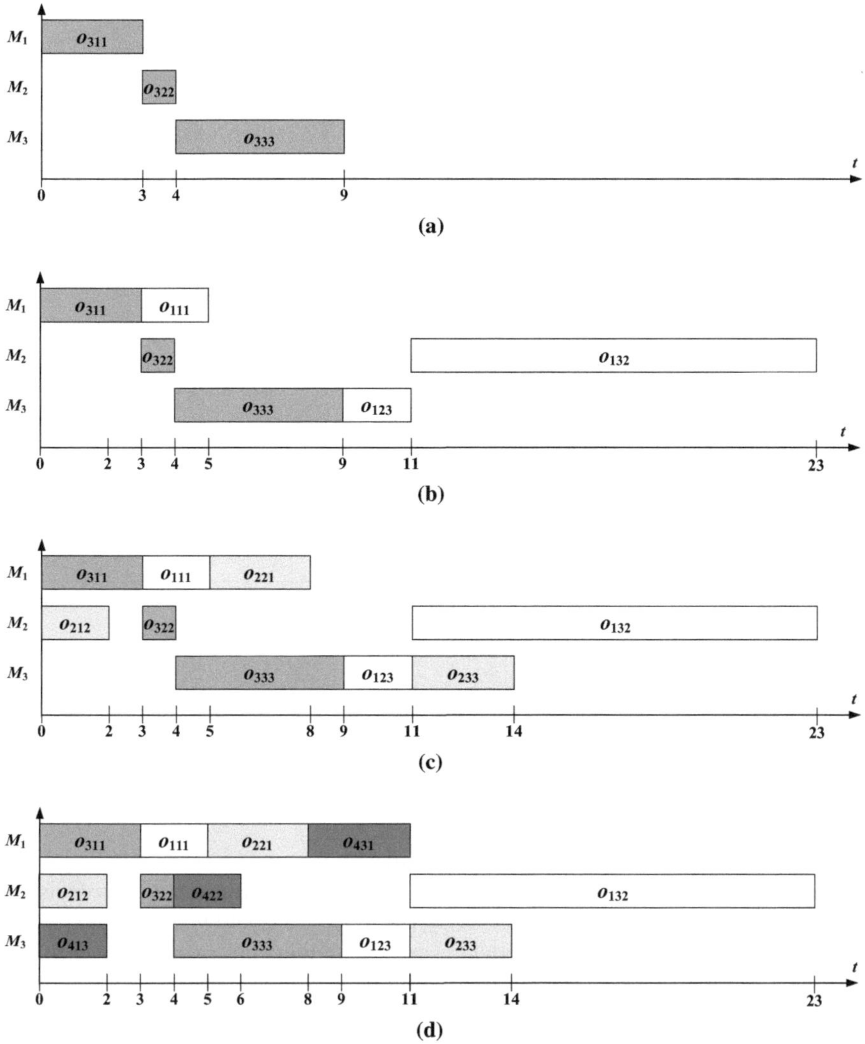

Fig. 7.46 The decoding process of the job-based chromosome in Fig. 7.45: (a) Gantt chart after scheduling job 3, (b) Gantt chart after scheduling job 1, (c) Gantt chart after scheduling job 2, and (d) Gantt chart after scheduling job 4

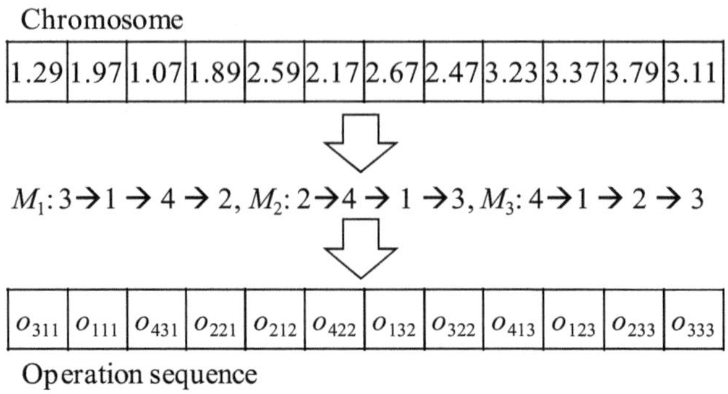

Fig. 7.47 A chromosome of random-key-based code

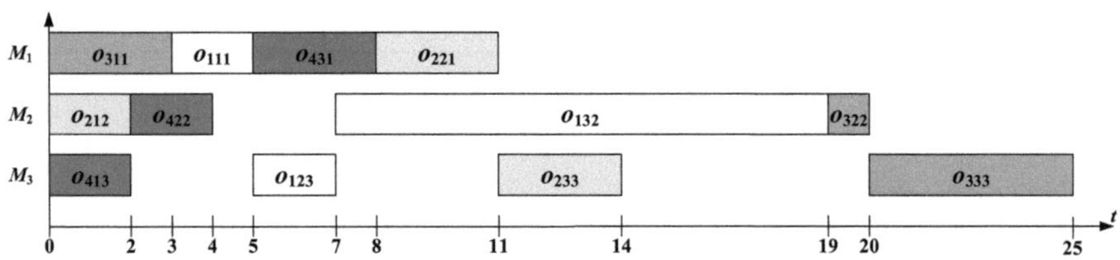

Fig. 7.48 Gantt chart of the random-key-based chromosome in Fig. 7.47

7.4.3.2 Indirect Approaches

Preference-list-based Code

In scheduling operations for JSSPs, sometimes the sequence is obvious, e.g., o_{212} and o_{413} of Table 7.9 should be processed at the beginning. But sometimes decisions should be made by schedulers, such as o_{311} and o_{111}. They all want to use machine 1 at the beginning. If we could provide some preference information for each machine on different jobs, these conflicts could be solved automatically. *Preference-list-based code* for JSSP was suggested by Davis in 1985 [50] and developed by Falkenauer and Croce in 1991 [51].

In preference-list-based code, we connect m subchromosomes to one chromosome. Each subchromosome is used to represent the preference of one machine for n jobs. This is illustrated by Fig. 7.49. Machine 1 favors job 3 the most, then job 2, then job 1, and dislikes job 4 the most. This is the preference list of machine 1. These rules are useful for solving conflicts while decoding a schedule.

To generate a schedule, we need to maintain a *waiting queue* for each machine. Whenever a machine finishes its current operation, it checks its waiting queue. If there is only one operation, the machine processes it. If there is more than one operation, the machine processes the preferred one according to its preference list. If no operations are in the waiting queue, it keeps idle until an operation comes. Figure 7.50 is the schedule generated by preference list in Fig. 7.49.

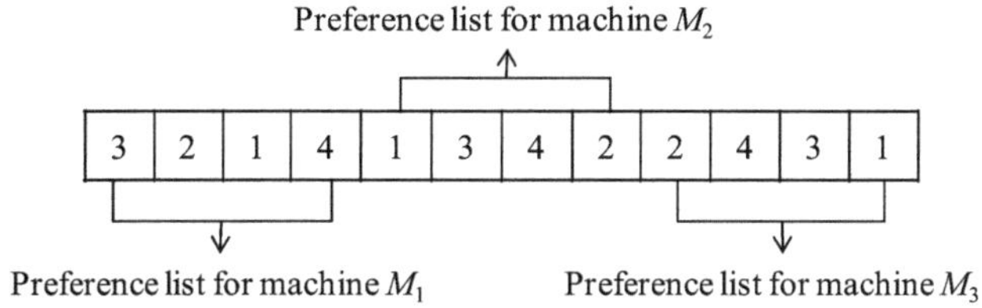

Fig. 7.49 Chromosome of preference-list-based code

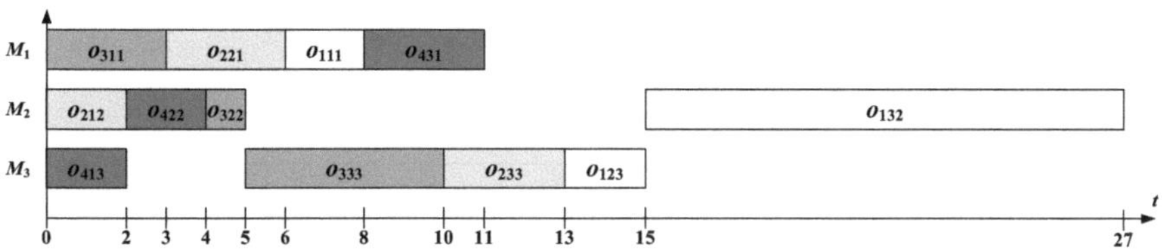

Fig. 7.50 Gantt chart of the preference-list-based chromosome in Fig. 7.49

The decoding procedure is illustrated by Table 7.11. We record the completion time for every selected operation so that we can check the waiting queue for the machine at that time. It basically involves using the precedence constraint in Table 7.9 and solving conflicts using the preference list in Fig. 7.49.

1. At the beginning, operations o_{111} and o_{311} are in machine 1's waiting queue, operation o_{212} is in machine 2's waiting queue, and operation o_{413} is in machine 3's waiting queue. Machine 1 selects o_{311} due to its preference list.
2. Then at time 2, o_{422} is the only selection for machine 2.[55]
3. At time 3, o_{311} is finished. The waiting queue for machine 1 contains o_{111} and o_{221}. Machine 1 selects o_{221} according to its preference list.
4. At time 4, o_{422} is finished. o_{322} is the only operation waiting for machine 2.
5. At time 5, o_{322} is finished. No operation is waiting for machine 2. But operation o_{333} is now the only one waiting for machine 3.
6. At time 6, o_{221} is finished. o_{111} and o_{431} are in machine 1's waiting queue. Machine 1 picks o_{111} according to its preference rule.
7. At time 8, o_{111} is finished. o_{431} is the only operation waiting for machine 1.
8. At time 10, o_{333} is finished. o_{123} and o_{233} are in machine 3's waiting queue. Machine 3 picks o_{233} according to its preference rule.
9. At time 13, o_{233} is finished. o_{123} is the only operation waiting for machine 2.
10. At time 15, o_{123} is finished, which makes o_{132} wait for machine 2.

According to the above procedure, operations will be processed whenever it is available. So preference-list-based code only generates nondelay schedules. In 1995 Croce *et al.* suggested a look-ahead simulation method to generate active schedules by preference-list-based code [52].

[55] o_{311} has not finished, so o_{322} is not in machine 2's waiting queue.

Table 7.11 Decoding procedure of the chromosome in Fig. 7.49

	Waiting queue			Operation selected Number in ( ) is the operation's completion time		
t	M_1	M_2	M_3	M_1	M_2	M_3
0	o_{111}, o_{311}	o_{212}	o_{413}	o_{311} (3)	o_{212} (2)	o_{413} (2)
2		o_{422}			o_{422} (4)	
3	o_{111}, o_{221}			o_{221} (6)		
4		o_{322}			o_{322} (5)	
5			o_{333}			o_{333} (10)
6	o_{111}, o_{431}			o_{111} (8)		
8	o_{431}			o_{431} (11)		
10			o_{123}, o_{233}			o_{233} (13)
13			o_{123}			o_{123} (15)
15		o_{132}			o_{132} (27)	

Priority-rule-based Code

In Sect. 7.4.2.1 we introduced how to generate an active schedule using the Giffler–Thompson algorithm. There we only used one priority dispatching rule, i.e., MWR, as an example. It is clear that every priority dispatching rule can generate one active schedule using the Giffler–Thompson algorithm. If we use different priority dispatching rules in different stages, i.e., τ, of the Giffler–Thompson algorithm, different schedules might be generated. That is the idea behind *priority-rule-based code*, which was suggested by Dorndorf and Pesch in 1995 [42].

The chromosome has $n \times m$ genes. The loci of genes represent the stage number. From left to right, the alleles of τ's gene represent the rule number in stage τ of the Giffler–Thompson algorithm.

Here we just use an example with four priority dispatching rules: 1 for SPT (shortest processing time), 2 for LPT (longest processing time), 3 for MWR (most work remaining), and 4 for LWR (least work remaining). One chromosome is illustrated in Fig. 7.51. The decoding procedure can be illustrated by Table 7.12, in which the "rule" column represents the priority dispatching rule determined by the corresponding gene and the "compare" column is the comparison basics according to various rules. While $\tau = 2$ and $\tau = 3$, there are situations where $s_j = f^*$. These operations cannot be scheduled.[56]

The Gantt chart of the schedule can be illustrated by Fig. 7.52.

Because most of the above code schemes are permutations or integers, we do not discuss their variation operators. But the selection of variation operators for TSPs and JSSPs might be different because the locus of chromosomes in JSSPs has specific meanings but not in TSP. We introduce a lot of consideration in maintaining

[56] Why?

1	4	2	1	3	4	2	1	3	2	4	1

(1: SPT, 2: LPT, 3: MWR, 4: LWR)

Fig. 7.51 Chromosome of priority-rule-based code

Table 7.12 Decoding procedure of the chromosome in Fig. 7.51

τ	$SO(\tau)$	f	f^*	M^*	s	Rule	Compare	$PS(\tau)$
1	o_{111}, o_{212}, o_{311}, o_{413}	$f_{111}=2$, $f_{212}=2$, $f_{311}=3$, $f_{413}=2$	2	1	$s_{111}=0$, $s_{311}=0$	SPT	$pt_{111}=2$, $pt_{311}=3$	o_{111}
2	o_{123}, o_{212}, o_{311}, o_{413}	$f_{123}=4$, $f_{212}=2$, $f_{311}=5$, $f_{413}=2$	2	3	$s_{123}=2$, $s_{413}=0$	LWR	-	o_{413}
3	o_{123}, o_{212}, o_{311}, o_{422}	$f_{123}=4$, $f_{212}=2$, $f_{311}=5$, $f_{422}=4$	2	2	$s_{212}=0$, $s_{422}=2$	LPT	-	o_{212}
4	o_{123}, o_{221}, o_{311}, o_{422}	$f_{123}=4$, $f_{221}=5$, $f_{311}=5$, $f_{422}=4$	4	2	$s_{422}=2$	SPT	-	o_{422}
5	o_{123}, o_{221}, o_{311}, o_{431}	$f_{123}=4$, $f_{221}=5$, $f_{311}=5$, $f_{431}=7$	4	3	$s_{123}=2$	MWR	-	o_{123}
6	o_{132}, o_{221}, o_{311}, o_{431}	$f_{132}=16$, $f_{221}=5$, $f_{311}=5$, $f_{431}=7$	5	1	$s_{221}=2$, $s_{311}=2$, $s_{431}=4$	LWR	$wr_{221}=6$, $wr_{311}=9$, $wr_{431}=3$	o_{431}
7	o_{132}, o_{221}, o_{311}	$f_{132}=16$, $f_{221}=10$, $f_{311}=10$	10	1	$s_{221}=7$, $s_{311}=7$	LPT	$pt_{221}=3$, $pt_{311}=3$	o_{311}
8	o_{132}, o_{221}, o_{322}	$f_{132}=16$, $f_{221}=13$, $f_{322}=11$	11	2	$s_{132}=4$, $s_{322}=10$	SPT	$pt_{132}=12$, $pt_{322}=1$	o_{322}
9	o_{132}, o_{221}, o_{333}	$f_{132}=23$, $f_{221}=13$, $f_{333}=16$	13	1	$s_{221}=10$	MWR	-	o_{221}
10	o_{132}, o_{233}, o_{333}	$f_{132}=23$, $f_{233}=16$, $f_{333}=16$	16	3	$s_{233}=13$, $s_{333}=11$	LPT	$pt_{233}=3$, $pt_{333}=5$	o_{333}
11	o_{132}, o_{233}	$f_{132}=23$, $f_{233}=19$	19	3	$s_{233}=16$	LWR	-	o_{233}
12	o_{132}	$f_{132}=23$	23	2	$s_{132}=11$	SPT	-	o_{132}

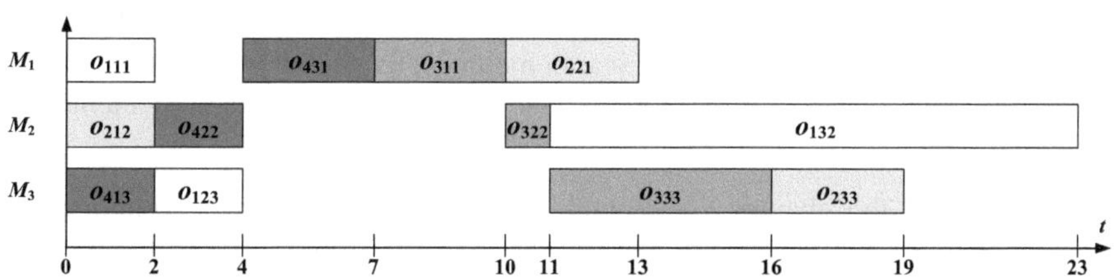

Fig. 7.52 Gantt chart of the priority-rule-based chromosome in Fig. 7.51

the connection of two successive genes. In JSSP, preserving the absolute location of genes in parents is also valuable for inheriting the merits of parents.

7.5 Summary

Combinatorial optimization has strong ties with economics and industry, so it has been intensively studied by mathematicians, economists, and industrial engineering researchers since the 1960s. Its solution space is generally extra large, which makes it intractable for exact algorithms, such as the "crude" enumeration method.[57] Heuristic algorithms are the only option before EAs.

In the beginning of applying EAs in combinatorial optimization, researchers tended to consider that EAs might be a substitution of heuristic algorithms. But comprehensive research reveals the fact that EAs should be considered complementary rather than competitive [16, 30, 53]. So MAs and hyper-heuristics, introduced in Sects. 3.7.2 and 3.7.3 respectively, are of great importance for combinatorial optimization.

In the three main sections for the knapsack problem, TSP, and JSSP, we can summarize the following features of EAs for combinatorial optimization.

- The code scheme for combinatorial optimization is flexible due to its hard constraints. The first priority of designing a successful EA for combinatorial optimization is to suggest an effective code scheme that limits the search space without losing the optimal solution and is easy to decode.
- Owing to the flexible code schemes, the variation operators are code-dependent. Good variation operators should (1) maintain the feasibility of offspring, (2) maintain the successful information of parents, and (3) explore the solution space effectively.
- Heuristic algorithms are very important for good EAs, or MAs. Construction algorithms can be used to generate the initial population and a local search algorithm can be used to improve the offspring. The advantage of MAs is obviously the fast convergence speed but it requires more involved decoding procedure and objective evaluation in one generation and thus might weaken the search ability of EAs when the total objective function evaluation number is the stop criterion. So the balance between genetic search and local search is critical for effective search [54].
- Provided the application of local search methods, the variation operators need to put more focus on the exploration because exploitation could be implemented by a local search method.

After reading this chapter, you should understand the difficulty of combinatorial optimization and the position of EAs in combinatorial optimization, know some

[57] Other techniques, such as cutting plane, branch and bound, dynamic programming, etc., could eliminate the search space and guarantee finding the optimal solution. But they also suffer from "combinatorial explosion."

code schemes for different problems, be acquainted with variation operators of permutation codes, and understand the idea and implementation of some heuristic methods for different problems.

In all, designing an EA for combinatorial optimization is the art of tradeoff between exploitation caused by local search methods and exploration caused by genetic operators.

Suggestions for Further Reading

Surveys on the application of metaheuristics in combinatorial optimization are given by Blum and Roli [55] and Gendreau and Potvin [56]. In 1999 Calégari *et al.* suggested a taxonomy of evolutionary algorithms in combinatorial optimization [57]. Bianchi *et al.* published a survey on metaheuristics for stochastic combinatorial optimization[58] in 2009 [58].

There are so many combinatorial optimization problems with distinct characteristics. Many real-world problems can be formulated as examples. We only introduce several examples in this chapter. Interested readers are referred to the textbook on combinatorial optimization [59].

There are several monographs on the three problems introduced in this chapter, such as Kellerer *et al.*'s book on *knapsack problems* in 2004 [60], Applegate *et al.*'s, Gutin and Punnen's, and Lawler *et al.*'s books on *TSP* in 2007, 2007, and 1985, respectively [61–63], and Pinedo's, Baker and Trietsch's, and Chakraborty's books on *scheduling* and *job-shop scheduling* in 2008, 2009, and 2009, respectively [2, 35, 64].

The topic of selecting proper *memes*, i.e., local search methods, in knapsack problems is critical for designing a successful algorithm. See two papers published in 2008 and 2009 [5, 8]. Laporte gave an intensive survey on the approximate algorithms for TSP in 1992 [65]. In 2009 Hasan published a paper to discuss in detail the procedure of MAs for JSSP [66]. In 2007 Tang *et al.* suggested a parallel MA including a diversity-based static/dynamic adaptive strategy to solve the large-scale NP-hard combinatorial problem, e.g., quadratic assignment problems (QAPs) [67].

The *multiple traveling salesman problem* (MTSP) seeks a partition of n cities into m salesmen and arranges the tour for each salesman so that (1) the total distance traveled by all salesmen is minimized and (2) the maximum distance traveled by any salesman is minimized to balance the load. MTSP has many real-world applications. Bektas gave a survey in 2006 [68] and Singh and Baghel suggested an improvement in 2009 [69].

In 1999 Larrañaga *et al.* published a comprehensive survey on the representations and variation operators for TSP [70] and Michalewicz and Fogel gave an interesting introduction on this topic in 2004 [71]. Poon and Carter's paper on crossover

[58] Stochastic combinatorial optimization means that part of the information about the problem data is only available in the probability distribution form.

operations might give the reader some fresh ideas [72]. In 2006 Raidl *et al.* suggested a biased edge exchange mutation operator with a preference on short distance roads [73].

JSSP might be the most intensively studied scheduling model in academia. Thus there are many reviews and comparisons on this topic. An early survey was provided by Blackstone *et al.* in 1982 [41]. In 1998 Jain and Meeran gave a comprehensive survey on job-shop scheduling techniques [53]. Gen and Cheng gave clear and comprehensive introductions in their books published in 1997 and 2008 [39, 74].

Evolutionary scheduling was reviewed by Hart *et al.* in 2005 [75]. Two earlier reviews specifically focused on representations and hybrids of EAs for job shop were published by Cheng *et al.* in 1996 and 1999, respectively [76, 77]. In 2000 Vázquez and Whitley compared several GAs and tabu search for static and dynamic job-shop scheduling in 2000, respectively [78, 79].

Optimization in dynamic environments always attracts the interest of researchers. In 2003 Jensen discussed the situation of rescheduling due to failures, e.g., machine breakdowns, sickness of employees, deliveries getting delayed, etc., to find robust and flexible schedules with a low makespan [80].

The must-read papers of this chapter are [81] for constructing effective MAs, [3] for an intensive survey on knapsack EAs, [70] for code schemes and variation operators for TSPs, and [76] for code schemes for JSSPs. If readers want to design effective MAs for TSPs and JSSPs, two old papers are highly recommended: [15, 38].

Exercises and Potential Research Projects

7.1. Search the SCI for the number of papers on knapsack problem, TSP, and JSSP to see how hot they are.

7.2. Prove that in 3-opt with one set of three edges which do not share the cities, seven new solutions can be generated, in which three solutions are actually 2-opt results.

7.3. What is the corresponding genetic variation operators in permutation code to 3-opt?

7.4. Could you provide another proof for Theorem 7.1 using path code?

7.5. Prove that an n-vertices complete graph has $\frac{n(n-1)}{2}$ edges.

7.6. Give an example to illustrate that the heritability of ordinal code is not good.

7.7. Generate offspring 2 in Figs. 7.20–7.22.

7.8. Could MTSP, introduced on page 319, belong to grouping problems? Why?

7.9. What is the solution size of $n = m = 15$ JSSP? How long would it take to solve this problem on your computer using the enumeration method?

7.10. Why does the longest length in Hamiltonian directed paths determine the makespan in Fig. 7.33?

7.11. Is Fig. 7.34 a semiactive, active, or nondelay schedule? Why?

7.12. Draw a Gantt chart whose vertical axis is for jobs according to Fig. 7.34.

7.13. How can the modification on page 308 generate a nondelay schedule?

7.14. Generate a nondelay schedule using the Giffler–Thompson algorithm for the problem in Table 7.9.

7.15. Implement at least one code for a knapsack problem, design your variation operators, and use the benchmark problem in Appendix and the techniques introduced in Sect. 3.6 to draw statistical conclusions according to your numerical experiments.

7.16. Use path code for a TSP, implement at least two crossover operators and two mutation operators, use the benchmark problem in Appendix and the techniques introduced in Sect. 3.6 to do a fair comparison, and draw statistical conclusions according to your numerical experiments.

7.17. Use the implementation of the above problem, combine it with one heuristic method for TSPs, use the benchmark problem in Appendix and the techniques introduced in Sect. 3.6 to do a fair comparison about your MA and the original EA, and draw statistical conclusions according to your numerical experiments.

7.18. Implement at least one code for JSSPs, design your variation operators, and use the benchmark problem in Appendix and the techniques introduced in Sect. 3.6 to draw statistical conclusions according to your numerical experiments.

7.19. Use the implementation of the above problem, combine it with one heuristic method for JSSPs, use the benchmark problem in Appendix and the techniques introduced in Sect. 3.6 to do a fair comparison about your MA and the original EA, and draw statistical conclusions according to your numerical experiments.

References

1. Falkenauer E (1998) Genetic algorithms and grouping problems. Wiley, New York
2. Pinedo ML (2008) Scheduling: theory, algorithms, and systems, 3rd edn. Springer, Berlin Heidelberg New York
3. Raidl GR, Gottlieb J (2005) Empirical analysis of locality, heritability and heuristic bias in evolutionary algorithms: a case study for the multidimensional knapsack problem. Evol Comput 13(4):441–475
4. Garey MR, Johnson DS (1979) Computers and intractability: a guide to the theory of NP-completeness. Freeman, San Francisco

5. Özcan E, Başaran C (2009) A case study of memetic algorithms for constraint optimization. Soft Comput 13(8):871–882
6. Zitzler E, Thiele L (1999) Multiobjective evolutionary algorithms: a comparative case study and the strength pareto approach. IEEE Trans Evol Comput 3(4):257–271
7. Jaszkiewicz A (2002) On the performance of multiple-objective genetic local search on the 0/1 knapsack problem - a comparative experiment. IEEE Trans Evol Comput 6(4):402–412
8. Zinflou A, Gagné C, Gravel M *et al* (2008) Pareto memetic algorithm for multiple objective optimization with an industrial application. J Heurist 14(4):313–333
9. Gordon VS, Whitley LD (1993) Serial and parallel genetic algorithms as function optimizers. In: Proceedings of the international conference on genetic algorithms, pp 177–183
10. Olsen A (1994) Penalty functions and the knapsack problem. In: Proceedings of the IEEE genetic and evolutionary computation conference, pp 554–558
11. Hinterding R (1994) Mapping, order-independent genes and the knapsack problem. In: Proceedings of the IEEE genetic and evolutionary computation conference, pp 13–17.
12. Hadley G (1964) Nonlinear and dynamic programming,. Addison-Wesley, Boston, MA
13. Croes GA (1958) A method for solving traveling-salesman problems. Oper Res 6(6):791–812
14. Lin S (1965) Computer solutions of the traveling salesman problem. Bell Syst Tech J 44:2245–2269
15. Lin S, Kernighan B (1973) An effective heuristic algorithm for the traveling-salesman problem. Oper Res 21(2):498–516
16. Johnson D, Mcgeoch L (1997) The traveling salesman problem: a case study in local optimization. In: Aarts EHL and Lenstra JK (eds) Local search in combinatorial optimization Wiley, New York, pp 215–310
17. Ronald S, Asenstorfer J, Vincent M (1995) Representational redundancy in evolutionary algorithms. In: Proceedings of the IEEE genetic and evolutionary computation conference, pp 631–636
18. Grefenstette JJ, Gopal R, Rosmaita BJ *et al* (1985) Genetic algorithms for the traveling salesman problem. In: Proceedings of the 1st international conference on genetic algorithms, pp 160–168
19. Fox B, McMahon M (1991) Genetic operators for sequencing problems. In: Proceedings of international workshop on foundations of genetic algorithms, pp 284–300
20. Seniw D (1991) A genetic algorithm for the traveling salesman problem. Master thesis, University of North Carolina at Charlotte
21. Banzhaf W (1990) The "molecular" traveling salesman. Biol Cybern 64(1):7–14
22. Michalewicz Z (1998) Genetic algorithms + data structures = evolution programs. Springer, Berlin Heidelberg New York
23. Fogel D (1990) A parallel processing approach to amultiple traveling salesman problem using evolutionary programming. In: Proceedings on the fourth annual parallel processing symposium, pp 318–326
24. Goldberg D, Lingle K (1985) Alleles, loci and the TSP. In: Proceedings of the first international conference on genetic algorithms and their applications, pp 154–159
25. Davis L (1991) Handbook of genetic algorithms. Van Nostrand Reinhold Company,New York
26. Syswerda G (1991) Schedule optimization using genetic algorithms. In: Davis L (ed) Handbook of genetic algorithms. Van Nostrand Reinhold Company, New York, pp 332–349
27. Oliver IM, Smith DJ, Holland JRC (1987) A study of permutation crossover operators on the traveling salesman problem. In: Proceedings of the second international conference on genetic algorithms and their application, pp 224–230
28. Whitley D, Starkweather T, Fuquay D (1989) Scheduling problems and traveling salesman: the genetic edge recombination. In: Proceedings of the third international conference on genetic algorithms, pp 133–140
29. Eiben AE, Smith JE (2003) Introduction to evolutionary computing. Springer, Berlin Heidelberg New York
30. Johnson D (1987) More approaches to the travelling salesman guide. Nature 330(6148):525

31. Nagata Y, Kobayashi S (1997) Edge assembly crossover: A high-power genetic algorithm for the traveling salesman problem. In: Proceedings of the 7th international conference on genetic algorithms, pp 450–457
32. Boese KD (1995) Cost versus distance in the traveling salesman problem. Tech. rep. TR-950018, UCLA
33. Freisleben B, Merz P (1996) New genetic local search operators for the traveling salesman problem. In: In: Proceedings of the international conference on parallel problem solving from nature, pp 890–899
34. Jung S, Moon B (2002) Toward minimal restriction of genetic encoding and crossovers for the two-dimensional euclidean TSP. IEEE Trans Evol Comput 6(6):557–565
35. Baker KR, Trietsch D (2009) Principles of sequencing and scheduling. Wiley, New York
36. Beasley J (1990) OR-Library: distributing test problems by electronic mail. J Oper Res Soc 41(11):1069–1072
37. Taillard E (1993) Benchmarks for basic scheduling problems. Eur J Oper Res 64(2):278–285
38. Giffler B, Thompson GL (1960) Algorithms for solving production-scheduling problems. Oper Re 8(4):487–503
39. Gen M, Cheng R (1997) Genetic algorithms and engineering design. Wiley-Interscience, New York
40. Panwalkar SS, Iskander W (1977) A survey of scheduling rules. Oper Res 25(1):45–61
41. Blackstone JH, Phillips DT, Hogg GL (1982) A state-of-the-art survey of dispatching rules for manufacturing job shop operations. Inter J Product Res 20(1):27–45
42. Dorndorf U, Pesch E (1995) Evolution based learning in a job shop scheduling environment. Comput Oper Res 22(1):25–40
43. Kim I, Watada J, Shigaki I (2008) A comparison of dispatching rules and genetic algorithms for job shop schedules of standard hydraulic cylinders. Soft Comput 12(2):121–128
44. Vaessens RJM, Aarts EH, Lenstra JK (1996) Job shop scheduling by local search. INFORMS J Comput 8:302–317
45. Adams J, Balas E, Zawack D (1988) The shifting bottleneck procedure for job shop scheduling. Manage Sci 34(3):391–401
46. Gen M, Tsujimura Y, Kubota E (1994) Solving job-shop scheduling problem using genetic algorithms. In: Proceedings of the 16th intertational conference on computer and industrial engineering, pp 576–579
47. Holsapple C, Jacob V, Pakath R *et al* (1993) A genetics-based hybrid scheduler for generating static schedules in flexible manufacturing contexts. IEEE Trans Syst Man Cybern 23(4):953–972
48. Bean JC (1994) Genetic algorithms and random keys for sequencing and optimization. ORSA J Comput 6(2):154–160
49. Norman B, Bean J (1995) Random keys genetic algorithm for job-shop scheduling. Tech. rep., University of Michigan
50. Davis L (1985) Job shop scheduling with genetic algorithms. In: Proceedings of the 1st international conference on genetic algorithms, pp 136–140
51. Falkenauer E, Bouffouix S (1991) A genetic algorithm for job shop. In: Proceedings of the IEEE international conference on robotics and automation, pp824–829
52. Croce FD, Tadei R, Volta G (1995) A genetic algorithm for the job shop problem. Comput Oper Res 22(1):15–24
53. Jain AS, Meeran S (1998) A state-of-the-art review of job-shop scheduling techniques. Tech. rep., University of Dundee
54. Ishibuchi H, Yoshida T, Murata T (2003) Balance between genetic search and local search in memetic algorithms for multiobjective permutation flowshop scheduling. IEEE Trans Evol Comput 7(2):204–223
55. Blum C, Roli A (2003) Metaheuristics in combinatorial optimization: Overview and conceptual comparison. ACM Comput Surv 35(3):268–308
56. Gendreau M, Potvin J (2005) Metaheuristics in combinatorial optimization. Ann Oper Res 140(1):189–213

57. Calégari P, Coray G, Hertz A *et al* (1999) A taxonomy of evolutionary algorithms in combinatorial optimization. J Heurist 5(2):145–158
58. Bianchi L, Dorigo M, Gambardella L *et al* (2009) A survey on metaheuristics for stochastic combinatorial optimization. Nat Comput 8(2):239–287
59. Papadimitriou CH, Steiglitz K (1998) Combinatorial optimization: algorithms and complexity. Dover, New York
60. Kellerer H, Pferschy U, Pisinger D (2004) Knapsack problems. Springer, Berlin Heidelberg New York
61. Applegate DL, Bixby RE, Chvatal V *et al* (2007) The traveling salesman problem: a computational study. Princeton University Press, Princeton, NJ
62. Gutin G, Punnen AP (2007) The traveling salesman problem and its variations. Springer, Berlin Heidelberg New York
63. Lawler EL, Lenstra JK, Kan AHGR *et al* (1985) The traveling salesman problem: a guided tour of combinatorial optimization. Wiley, New York
64. Chakraborty UK (2009) Computational intelligence in flow shop and hob shop scheduling. Springer, Berlin Heidelberg New York
65. Laporte G (1992) The traveling salesman problem: an overview of exact and approximate algorithms. Eur J Oper Res 59(2):231–247
66. Hasan S, Sarker R, Essam D *et al* (2009) Memetic algorithms for solving job-shop scheduling problems. Memet Comput 1(1):69–83
67. Tang J, Lim MH, Ong YS (2007) Diversity-adaptive parallel memetic algorithm for solving large scale combinatorial optimization problems. Soft Comput 11(9):873–888
68. Bektas T (2006) The multiple traveling salesman problem: an overview of formulations and solution procedures. Omega 34(3):209–219
69. Singh A, Baghel A (2009) A new grouping genetic algorithm approach to the multiple traveling salesperson problem. Soft Comput 13(1):95–101
70. Larranaga P, Kuijpers CMH, Murga RH *et al* (1999) Genetic algorithms for the travelling salesman problem: A review of representations and operators. Artif Intell Rev 13:129–170
71. Michalewicz Z, Fogel DB (2004) How to solve it: modern heuristics. Springer, Berlin Heidelberg New York
72. Poon PW, Carter JN (1995) Genetic algorithm crossover operators for ordering applications. Comput Oper Res 22(1):135–147
73. Raidl G, Koller G, Julstrom B (2006) Biased mutation operators for subgraph-selection problems. IEEE Trans Evol Comput 10(2):145–156
74. Gen M, Cheng R, Lin L (2008) Network models and optimization: multiobjective genetic algorithm approach. Springer, Berlin Heidelberg New York
75. Hart E, Ross P, Corne D (2005) Evolutionary scheduling: a review. Genet Programm Evolvable Mach 6(2):191–220
76. Cheng R, Gen M, Tsujimura Y (1996) A tutorial survey of job-shop scheduling problems using genetic algorithms. I: representation. Comput Ind Eng 30(4):983–997
77. Cheng R, Gen M, Tsujimura Y (1999) A tutorial survey of job-shop scheduling problems using genetic algorithms. II: hybrid genetic search strategies. Comput Ind Eng 37(1-2):51–55
78. Vázquez M, Whitley LD (2000) A comparison of genetic algorithms for the static job shop scheduling problem. In: Proceedings of the international conference on parallel problem solving from nature, pp 303–312
79. Vázquez M, Whitley LD (2000) A comparison of genetic algorithms for the dynamic job shop scheduling problem. In: Proceedings of the genetic and evolutionary computation conference, pp 1011–1018
80. Jensen M (2003) Generating robust and flexible job shop schedules using genetic algorithms. IEEE Trans Evol Comput 7(3):275–288
81. Krasnogor N, Smith J (2005) A tutorial for competent memetic algorithms: model, taxonomy, and design issues. IEEE Trans Evol Comput 9(5):474–488

Part III
Brief Introduction to Other Evolutionary Algorithms

Chapter 8
Swarm Intelligence

Abstract We look at the natural selection process as a learning or optimizing process and apply the survival of the fittest principle to designing the learning and optimizing algorithm. Then many EAs, e.g., GA, ES, EP, DE, etc., are suggested accordingly. There are other similar phenomena in nature. A swarm of "low-level" (not smart) insects sometimes surprises us with their amazing behaviors, such as foraging for food efficiently and constructing exquisite nests. We can also look at the process of foraging for food and constructing nest as a learning or optimizing process and learn to design corresponding algorithms. This swarm-level smart behavior generated by an agent-level, not smart property could enlighten us to suggest more robust algorithms for more complex problems in an uncertain environment.

8.1 Introduction

Swarm means a large group. It is generally used to describe social insects or social animals, e.g., ant colonies, bee colonies, fish schools, and bird flocks. We are often fascinated by the masterpiece of these "naive" small lives, such as sophisticated nests and highly efficient foraging behaviors. If you have ever considered how a swarm of low-level creatures could display high-level performance, you might have arrived at the following properties.

1. Every low-level creature can be regarded as an unsophisticated agent who acts according to simple rules.
2. The swarm fulfils tasks without central control.
3. Unsophisticated agents can be self-organized for coordinated/collective behaviors, through which they display complex problem solving skills. We say that intelligence emerges from these *collective behaviors*.
4. A collective behavior is implemented by local/indirect communications/ interactions between agents. These *indirect interactions* might be the local changes of the environment, which will affect the behaviors of the other creatures later.

The most spectacular part of swarm behaviors are that, through collective behavior, they represent $1+1>2$ effects. We can make these swarms a metaphor for implementing our learning and optimizing problems, i.e., we mimic the above properties to some extent. The artificial system that embodies some of the above properties of social creatures to solve learning and optimizing problems is called *swarm intelligence* (SI). SI is not an algorithm or system but a category of algorithms whose most famous examples are *ant colony optimization* (ACO) and *particle swarm optimization* (PSO).[1]

Compared with other EAs, SI's advantage lies in the fact that the intelligence-emergence process is self-organized, which means that SI is more suitable for problems in noisy and dynamic environments and for finding robust solutions.

Since their birth, ACO and PSO have attracted more and more focus and become the hottest algorithms in the EA field. Figure 8.1 illustrates the number of papers indexed by the SCI on ACO and PSO [2].

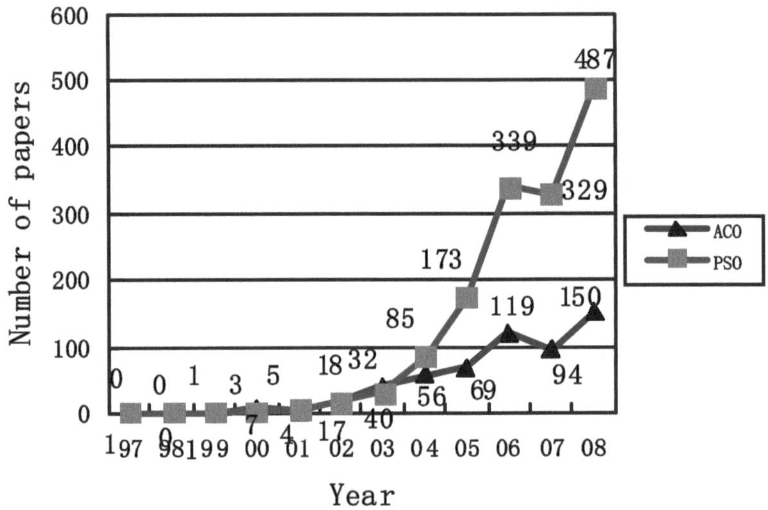

Fig. 8.1 Number of papers indexed by SCI on ACO and PSO

[1] They are the topics of Sects. 8.2 and 8.3, respectively.

[2] TS = ("ant colony optimization") and TS = ("particle swarm optimization" OR "swarm intelligence"). The SCI index "TS" is for the search topic in the title, the keywords, and the abstract.

8.2 Ant Colony Optimization

8.2.1 Rationale Behind Ant Colony Optimization

The foraging behavior of ant colonies has been intensively studied and the bionics corresponding to this behavior finds various applications. The indirect communication between ants is accomplished through *pheromones*. A pheromone is a chemical substance that triggers a natural response in another member of the same species. There are many types of pheromones deposited by ants, e.g., aggregation pheromones, alarm pheromones, trail pheromones, etc. Here we only care about *trail pheromones*, which are deposited by ants with food as they return to their nest; they attract other ants and get them to go along the pheromone trail to find food.

We can illustrate the simplified foraging behavior with the help of pheromones. Suppose there are only two ants with the same ability (speed and pheromone-depositing ability) and they have both found food at the same time. On their way to their nest, they deposit pheromones, which is illustrated by the dotted lines in Fig. 8.2a. Figure 8.2a is a snapshot of the time when these two ants are at the intersection of the two paths. The left one is shorter than the right one. This is the first time they need to make decisions. Because there is no any information about which path is shorter, they take one path with 50% probability. Suppose the triangle ant takes the shorter path and the square ant takes the longer path. So the triangle ant reaches the nest faster. After returning to its nest, the triangle ant selects the path according to the concentration of pheromones. Figure 8.2b is a snapshot of the triangle ant's second selection. We suppose at that time that the square ant is still on its way back home. So the triangle ant takes the shorter path with 100% probability. Figure 8.2c illustrates the situation where the triangle ant is taking the third selection with food and the square ant is taking the second selection without food. They take the paths with 50% probability because both paths have one pheromone trail. But as time elapses, more pheromones will be deposited along the shorter path because the ants taking this path will reach their nest faster. This situation will further stimulate more ants to take this path. Figure 8.2d is such a situation. Both the triangle ant and the square ant will take the shorter path with high probability.

In the above simple example, every ant, i.e., agent, only has two properties. The first one is to deposit a pheromone trail on the way back to the nest and the second one is to select the path containing a high *pheromone concentration* with high probability. These simple properties, together with the fact that the trail pheromone is an evaporable chemical substance, cause the ant colonies to emerge the intelligence for finding the shortest path.

In order to mimic the ants' foraging behavior, we need to consider two issues before designing an algorithm.

- The probability model for selecting different options, i.e., the way to form a solution.
- The way by which ants deposit pheromones and pheromones evaporate, i.e., the way to modify the solution-forming method.

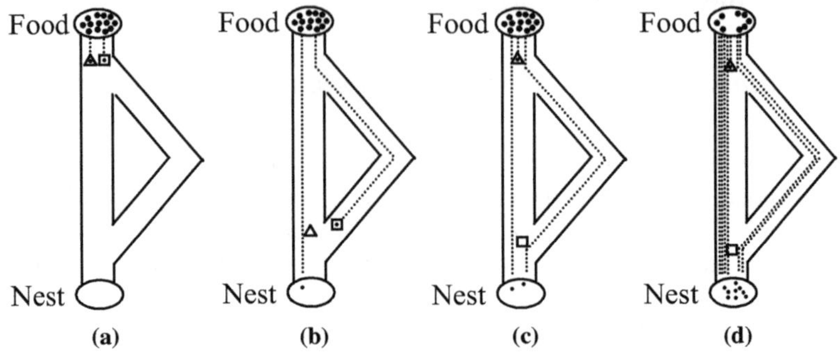

Fig. 8.2 Simplified example of ant foraging behavior: (a) first selection, (b) second selection, (c) third selection, and (d) latter selection

In the following section, we will focus on these two issues to introduce our algorithm.

8.2.2 Discrete Ant Colony Optimization

The emerging shortest path finding ability in ant colonies can be used directly in TSP (Sect. 7.3). We will take TSP as the example problem in most of this subsection and discuss JSSP, which is illustrated in Sect. 7.4, at the end.

We suppose that there are m ants to solve the TSP with n cities. At first, ants are located at different cities randomly. Every ant has $n-1$ options at first for which city to goto in the next step. They use the pheromone concentrations in the edge connecting their current city and the $n-1$ cities to make selections probabilistically, i.e., *transition probability*. Then all the ants move to their second city at the same pace. Now they probabilistically select one feasible city, among $n-2$ cities, according to the pheromone concentrations. This procedure continues until all m ants have visited n cities. Then they need to go back to their first city to form the solution of the TSP.

That is to say, we divide the process of forming the solution for the TSP with n cities into n time steps, with each ant going to a feasible city in each time step. This is different from the aforementioned example, where ants travel at the same speed. The different path length of account for the different pheromone concentrations. In the discretized model of TSP, all m ants finish their Hamiltonian cycles at the same time. So we need to modify the pheromone concentration artificially after the ants return to their first cities according to the length of their Hamiltonian cycles to mimic the effect of the aforementioned example.

The general solution process of ACO for TSP could be illustrated as follows.

ACO for TSP

Phase 1: Initialization. Assign the parameters for ACO, such as ant number m, initial pheromone concentration $\tau(0)$ for every link in the graph, local information importance factor β, evaporation factor ρ, stop criteria (such as *maxgen* and *stagnation*), etc. $gen = 0$. Stagnation is the condition that all m ants generate the same Hamiltonian cycle. It's the same meaning of the convergence in GA.

Phase 2: Main loop. Repeat the following steps until the stop criteria are satisfied.

Step 2.1: Generate a Hamiltonian cycle for each ant.

Substep 2.1.1: Place each ant in a randomly selected initial city.

Substep 2.1.2: For each ant, select the next feasible city to goto according to the transition probabilities in the links connecting the current city and the next feasible cities. After the movement, update the feasible city set. Whenever an ant goes through a link, it might modify the pheromone concentration of that link (optional).

Substep 2.1.3: Repeat substep 2.1.2 until all ants have visited all n cities. Then let them go back to their respective initial cities. Thus form the m Hamiltonian cycles and compute the lengths of the m Hamiltonian cycles. Determine the *best iteration solution* and update the *best solution so far*. The best iteration solution is the Hamiltonian cycle with the shortest length in the current generation and the best solution so far is the Hamiltonian cycle with the shortest length since the beginning of this run.

Step 2.2: Update the pheromone concentrations of the links according to the performance of these m Hamiltonian cycles. $gen = gen + 1$.

Phase 3: Submitting the best solution so far as a result of ACO for TSP.

The remaining part of this subsection will mainly discuss the steps 2.1.2 and 2.2.

8.2.2.1 Ant System for Traveling Salesman Problem

Dorigo *et al.* published the first paper on ACO in 1996 [1]. Their approach is called an *ant system* (AS).

In substep 2.1.2, the transition probability of the link (i, j) that connects city i and city j for ant k in generation gen is as follows:

$$p_{ij}^{k}(gen) = \begin{cases} \dfrac{\tau_{ij}(gen)\left(\eta_{ij}\right)^{\beta}}{\sum_{l \in allowed_k} \tau_{il}(gen)(\eta_{il})^{\beta}} & j \in allowed_k \\ 0 & \text{otherwise} \end{cases} \tag{8.1}$$

where $\tau_{ij}(gen)$ is the pheromone concentration, $\eta_{ij} = 1/d_{ij}$ is the *visibility* used to model the local perception ability of the ants,[3] β is a user-defined parameter to adjust the relative importance of $\tau_{ij}(gen)$ and η_{ij}, and $allowed_k$ is the set containing all the feasible cities of ant k currently. Equation 8.1 means that the ant will select the feasible link with a high pheromone concentration and short distance with high probability.

After m ants form the Hamiltonian cycles, they modify the pheromone concentration of the links they passed by in step 2.2. For the pheromone concentration τ_{ij} on the link between city i and city j, the update function is as follows:

$$\begin{aligned}\tau_{ij}(gen+1) &= (1-\rho)\,\tau_{ij}(gen) + \Delta\tau_{ij} \\ &= (1-\rho)\,\tau_{ij}(gen) + \sum_{k=1}^{m} \Delta\tau_{ij}^{k}\end{aligned} \tag{8.2}$$

where ρ is the evaporation factor[4] and $\Delta\tau_{ij}^{k}$ is the contribution of ant k to τ_{ij} as follows:

$$\Delta\tau_{ij}^{k} = \begin{cases} \frac{Q}{L_k} & \text{link}\,(i,j) \text{ in ant } k'\text{s tour} \\ 0 & \text{otherwise} \end{cases} \tag{8.3}$$

where L_k is the length of the Hamiltonian cycle formed by ant k and Q is a user-defined parameter affecting the contribution of the distance to the pheromone. Smaller L_k contributes a larger increase to τ. Equations 8.2 and 8.3 mean that: (1) the pheromone on every link evaporates with time, (2) ants only deposit pheromones on the links they pass, and (3) those ants with the better results deposit more pheromones.

Dorigo *et al.* used AS parameters $m = n$, $\beta = 5$, $\rho = 0.5$, and $Q = 100$ to successfully solve the TSP with 30 cities.

8.2.2.2 Ant Colony System for Traveling Salesman Problem

To further improve the search ability of AS, Dorigo and Gambardella suggested the *ant colony system* (ACS) in 1997 [2].

The main difference between ACS and AS lies in the following three points.

In substep 2.1.2, an ant selects the link with the largest value of $\tau_{il}(gen)(\eta_{il})^{\beta}$ in the feasible set with probability p_0. That is to say, if a uniform random number between $(0,1)$ is smaller than p_0, the ant will deterministically select the link with the largest value of $\tau_{il}(gen)(\eta_{il})^{\beta}$. Otherwise, it still uses Eq. 8.1 to probabilistically determine the next city to be visited.

[3] That is to say, an ant knows both the local information regarding the distance between its current city and any other feasible city, i.e., d_{ij}, and the history experience of the ant colony, i.e., $\tau_{ij}(gen)$.

[4] Thus $(1-\rho)$ is the remaining percentage.

In ACS, we have two kinds of pheromone updating rule. The global updating in step 2.2 considers only the best solution so far as follows, which is different from Eq. 8.2:

$$\tau_{ij}(gen+1) = (1-\alpha)\,\tau_{ij}(gen) + \alpha\Delta\tau_{ij} \tag{8.4}$$

where α is the global evaporation factor and $\Delta\tau_{ij}$ is defined as follows:

$$\Delta\tau_{ij} = \begin{cases} \frac{1}{L_{gb}} & \text{link}\,(i,j) \text{ in the best tour so far} \\ 0 & \text{otherwise} \end{cases} \tag{8.5}$$

where L_{gb} is the shortest length of the Hamiltonian cycle found thus far by all ants. Equations 8.4 and 8.5 mean that only the best solution so far can contribute to its corresponding pheromone concentration and the pheromone concentration in other links will evaporate with time.

The third improvement of ACS over AS is the local updating rule in substep 2.1.2. Whenever an ant passes a link, it will modify the pheromone concentration in that link as follows:

$$\tau_{ij} = (1-\rho)\,\tau_{ij} + \rho\Delta^{'}\tau_{ij} \tag{8.6}$$

where ρ is the local evaporation factor and $\Delta^{'}\tau_{ij}$ could be defined in two ways. The simplest way is as follows:

$$\Delta^{'}\tau_{ij} = \tau_0 \tag{8.7}$$

where τ_0 is a predefined factor.[5]

Another way is to modify the pheromone in a Q-learning way[6] as follows:

$$\Delta^{'}\tau_{ij} = \gamma \max_{l \in allowed_k(j)} \tau_{jl} \tag{8.8}$$

where γ is a control parameter and $allowed_k(j)$ is the feasible set while ant k is at city j. Equations. 8.6 and 8.8 mean that the contribution of ant k while passing link (i,j) is determined by the maximum pheromone concentration on the feasible links connecting city j.[7] The ACS with Eq. 8.8 is also called ant-Q.

All the above three modifications have a similar intuitive idea, i.e., reinforce the influence of the historically better solutions and thus accelerate the search speed.

---

[5] Dorigo and Gambardella suggested that $\tau_0 = (n \cdot L_{nn})^{-1}$. n is the city number and L_{nn} is the tour length generated by the closest neighbor construction heuristic, which is introduced in Sect. 7.3.2.1.

[6] Q-learning is an implementation of reinforcement learning [3, 4]. Interested readers can find a funny tutorial at `http://people.revoledu.com/kardi/tutorial/ReinforcementLearning/index.html`.

[7] Because the pheromone concentration embodies historical experiences during the evolving process, using the maximum pheromone concentration on the feasible links connecting city j to modify the pheromone concentration on line (i,j) might further accelerate the evolving process.

With the help of the above methods, Dorigo and Gambardella used ACS parameters $m = 10$, $\beta = 2$, $p_0 = 0.9$, and $\alpha = \rho = 0.1$ to successfully solve the TSP with 100 cities.

Apart from the above two implementations of ACO, there are several other variations, e.g., the rank-based ant system suggested by Bullnheimer *et al.* in 1999 [5], the MAX-MIN ant system suggested by Stüzel and Hoos in 2000 [6], and the hypercube framework suggested by Blum and Dorigo in 2004 and discussed by Birattari *et al.* in 2007 [7, 8].

8.2.2.3 Ant System for Job-shop Scheduling Problem

Because an ant colony has an innate ability to find the shortest path between the nest and food, it is quite natural to apply ACO to distance-related problems, e.g., TSP. For applying ACO to other combinatorial problems, several modifications should be made. Let us take JSSP, which was introduced in Sect. 7.4, for example.[8]

The two-machine and four-job JSSP example is illustrated by Table 8.1.

Table 8.1 The two-machine and four-job JSSP example

	p_{ijk}		Machine	
Operation	1	2	1	2
Job 1	2	2	M_1	M_2
Job 2	2	3	M_2	M_1
Job 3	3	1	M_1	M_2
Job 4	2	2	M_1	M_2

The first thing for solving JSSP with ACO is to model the JSSP as a half-directed graph, illustrated in Fig. 8.3.

There are nine nodes in this eight-operation JSSP. The 0 node in Fig. 8.3 represents the dummy starting point. Unlike Fig. 7.33, the number in the node does not represent the machine index, but only the index of operations.

There are both directed edges and undirected edges in Fig. 8.3, so we call it half-directed. The directed edge represents the precedence relationships in different jobs, e.g., the directed edge between nodes 1 and 5 means that the first operation, i.e., node 1, should be processed before the second operation, i.e., node 5, in job 1. The undirected edges between operations indicate the possible next state of the ants.

After that, we need special considerations for constructing a feasible solution with ants. All m ants start from node 0. We will only introduce the solution-generating procedure of one ant.

[8] There are many different implementation methods for ACO on JSSPs; we will only introduce the method of Dorigo *et al.* [1].

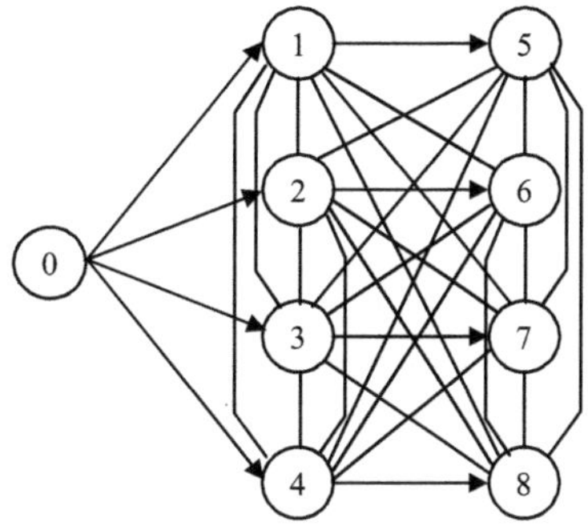

Fig. 8.3 A half-directed graph representing a JSSP in Table 8.1

We need to maintain two sets about ant k. G_k is the set of all the unvisited nodes and S_k represents the set of feasible nodes in the next step. Initially, for ant k, $G_k = \{1,2,3,4,5,6,7,8\}$ and $S_k = \{1,2,3,4\}$.

Then we need to know how to define η_{ij} in Eq. 8.1. We could use several priority dispatching rules, introduced on page 307, to define η_{ij}. Here we just take that $\eta_{ij} = 1/p_j$, where p_j is the processing time of operation j. Thus the next feasible operations with less processing time are preferred.

Suppose we take operation 2 as the first step, then node 2 should be deleted from G_k and its direct successor, if there is one, is added to S_k, which makes $G_k = \{1,3,4,5,6,7,8\}$ and $S_k = \{1,3,4,6\}$. We can continue this procedure until $G_k = \emptyset$. Thus a permutation of the eight operations is generated. For example, permutation $(2-3-7-1-6-4-8-5)$ represents the processing sequence of job $(3-1-2-4)$ for machine 1 and job $(2-3-4-1)$ for machine 2. The Gantt chart of this solution is illustrated by Fig. 8.4.

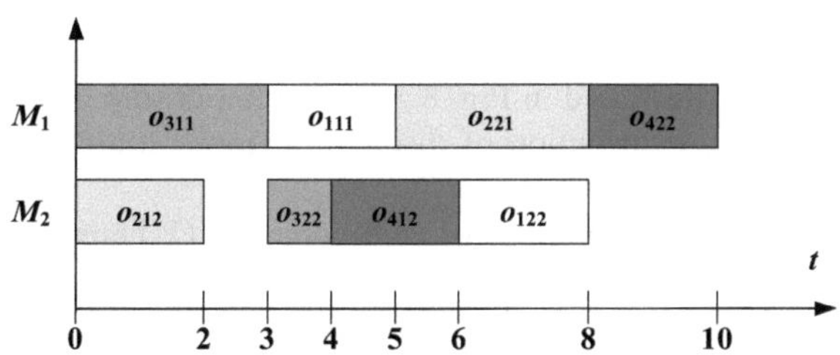

Fig. 8.4 The Gantt chart of the solution $(2-3-7-1-6-4-8-5)$

Then all the techniques introduced in AS and ACS could be used to optimize the JSSP. To sum up, at least three problems need to be resolved before applying ACO to a combinatorial optimization problem.

1. How does one use a graph to illustrate the problem?
2. How does one define the visibility η_{ij} in the problem?
3. How does one generate a feasible solution for an ant?

8.2.3 Continuous Ant Colony Optimization

As has been discussed above, the innate ability of an ant colony to find the shortest path can be easily applied to distance-related combinatorial optimization problems. By representing other combinatorial optimization problems to graphically, we can still use ACO. The basic idea is to use pheromone concentration information to construct new solutions and then use these newly generated solutions to modify the pheromone trails, as illustrated by Fig. 8.5a.

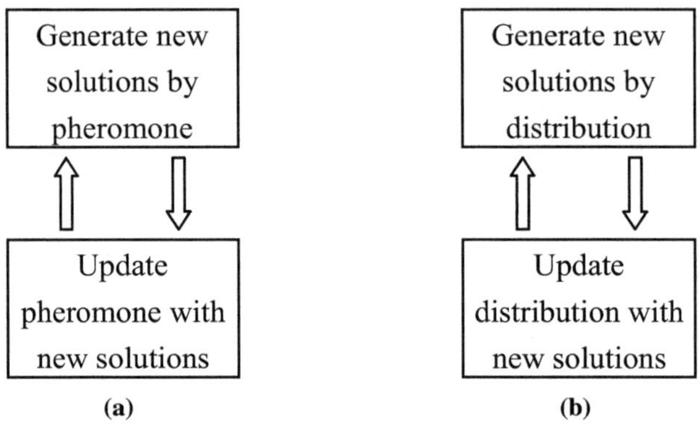

Fig. 8.5 Comparison of ACO for combinatorial and continuous optimization: (a) ACO for combinatorial optimization, and (b) ACO for continuous optimization

If we want to expand ACO into continuous optimization, i.e., topics introduced in Chaps. 4–6, a very intuitive idea is to change the discrete distributed pheromone on the edge into a continuous distributed probabilistic distribution function on the solution landscape, as illustrated in Fig. 8.5b. The relationship of these two figures is like the probabilistic distribution of discrete random variables and the density function of continuous random variables.

There are several implementations of continuous ACO; here we only introduce the $ACO_{\mathbb{R}}$ suggested by Socha and Dorigo in 2008 [9].

In $ACO_{\mathbb{R}}$, we maintain an archive with k good solutions with n variables and use them to generate normal distribution density functions, which are later used to generate m new solutions by ants. Then the m newly generated solutions replace the worst solutions in the archive. At first, an ant is used to generate a variable value, just like it is used to generate a step in TSP. For a problem with n variables, an ant needs n steps to generate a solution, just like it needs n steps to generate a Hamiltonian cycle in TSP. In $ACO_{\mathbb{R}}$, every variable is generated by a normal distribution density function, i.e., Eq. 2.2.

The general solution process of $ACO_{\mathbb{R}}$ for continuous optimization can be illustrated as follows.

ACO$_{\mathbb{R}}$ for Continuous Optimization

Phase 1: Initialization.

Step 1.1: Assign the parameters for ACO, such as ant number m, archive size k, weight factor q, deviation factor ξ, and stop criteria (such as *maxgen*), etc. $gen = 0$.

Step 1.2: Generate k initial solutions randomly and put them into the archive. Calculate their objective values and rank them so that the best solution has rank 1 and the worst solution has rank k. Calculate solution s^l's weight ω^l as follows:

$$\omega^l = \frac{1}{qk\sqrt{2\pi}} \mathrm{e}^{-\frac{(l-1)^2}{2q^2k^2}} \tag{8.9}$$

where l is the rank of solution s^l and q is the user-defined weight factor. We can draw two sets of weights with $q = 0.1$ and $q = 0.2$ ($k = 50$) in Fig 8.6. As can be seen from Fig. 8.6, smaller q favors a fitter solution.

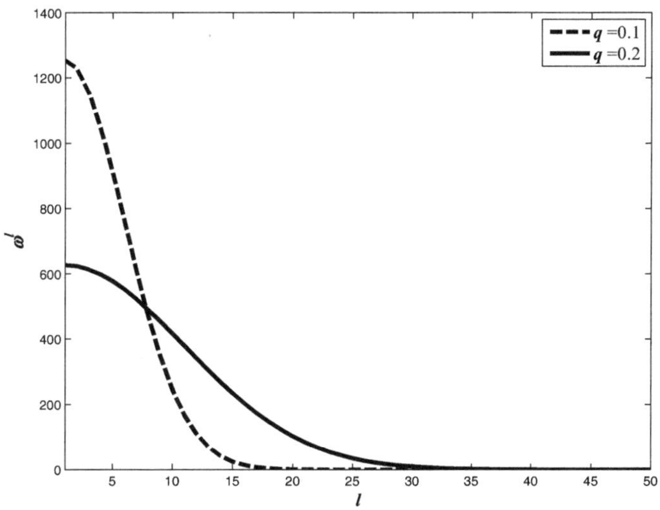

Fig. 8.6 Two sets of weights with different q

Then calculate the probability of being the expectation of the new solution as follows:

$$p^l = \frac{\omega^l}{\sum_{l=1}^{k} \omega^l} \tag{8.10}$$

According to Eqs. 8.9 and 8.10, fitter solutions have more chances of being selected.

Phase 2: Main loop. Repeat the following steps until the stop criteria are satisfied.

Step 2.1: Generate a solution by an ant. $i = 1$

Substep 2.1.1: Determine the mean of variable i. Select one solution in the archive with the probability illustrated by Eq. 8.10. Any proportional selection methods introduced in Sect. 3.3.2 could be used. This solution is denoted as s^j. The mean of variable i, i.e., μ_i, is the value x_i^j.

Substep 2.1.2: Determine the standard deviation of variable i as follows:

$$\sigma_i = \xi \sum_{e=1}^{k} \frac{\left|x_i^e - x_i^j\right|}{k-1} \tag{8.11}$$

where s^j is the selected solution in substep 2.1.1 and ξ is the user-defined deviation factor. It is easy to see that a larger value of ξ means a lower convergence speed of ACO$_{\mathbb{R}}$, similar to the effect of a larger value of evaporation factor ρ in AS and ACS.

Substep 2.1.3: Generate variable i with normal distribution $U(\mu_i, \sigma_i)$. $i = i + 1$. The ant moves to another variable.

Substep 2.1.4: Repeat substeps 2.1.1–2.1.3 until $i > n$. Different solutions in the archive might be selected for different variables.

Step 2.2: Go back to step 2.1 if the number of newly generated solution is less than m.

Step 2.3: Calculate the objective values of these m newly generated solutions.

Step 2.4: Replace the m worst solutions in the archive with the newly generated solutions. Recalculate the weight and the probability of being selected for each archive member using Eqs. 8.9 and 8.10, respectively.

Step 2.4: Update the best solution so far if necessary. $gen = gen + 1$.

Phase 3: Submitting the best solution so far as the result of ACO$_{\mathbb{R}}$ for continuous optimization.

ACO$_{\mathbb{R}}$ is quite similar to CMA and EDA, which were introduced in Sect. 3.5.2.6. Socha and Dorigo compared ACO$_{\mathbb{R}}$ with the most famous probabilistic model-based algorithms [10] with $k = 50$, $q = 10^{-4}$, $\xi = 0.85$, and $m = 2$ and came to the conclusion that $ACO_{\mathbb{R}}$ is a rather competitive continuous optimization algorithm.[9]

The innate ability to handle discrete variables in AS or ACS and the expanded ability to handle continuous variables in $ACO_{\mathbb{R}}$ make ACO a very promising method for mixed programming.

8.3 Particle Swarm Optimization

In 1995, Kennedy and Eberhart suggested a novel optimization method called *particle swarm optimization* (PSO) [11]. The intuitive idea behind its name is that they want to mimic the ability of a bird flock to fly synchronously, change directions suddenly, scatter, and regroup. The reason for such sophisticated behaviors can be summarized as the social sharing of information among these birds. Kennedy and Eberhart use velocity to describe the movement of birds and so gave the name "particle" (in the sense of physics) to their method. They first implemented the algorithm with a nearest neighbor velocity matching mechanism to mimic the behavior of bird flocks. But they later gave up this idea after obtaining numerical results and considered the more broader social information. Thus they also abandoned the name of "flock" and used "swarm."

Since the birth of PSO, there have been many variations extending in different directions. Due to the content limitation of this text, we can only illustrate the basic ideas and the interesting and landmark algorithms. The relationships between the variations introduced in this section can be summarized by Fig. 8.7.[10]

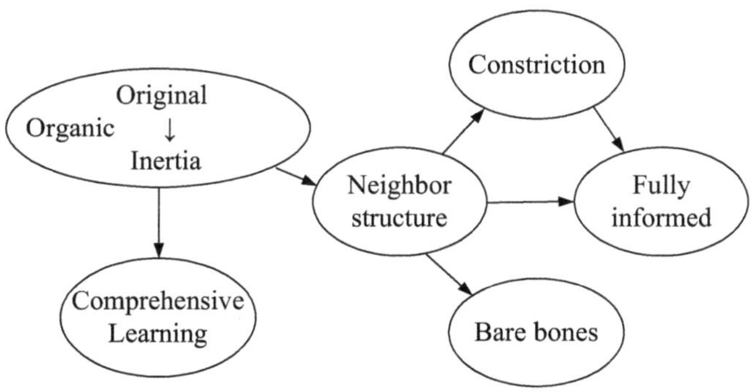

Fig. 8.7 The relationships between the PSO variations introduced in this section

[9] $ACO_{\mathbb{R}}$ cannot handle variable correlation, introduced in Sect. 3.2.2 as problems with separable and nonseparable variables. Socha and Dorigo suggested a pragmatic method to adaptively rotate the axes using solutions in the archive [9].

[10] The terms in the figure will be discussed later.

8.3.1 Organic Particle Swarm Optimization

In Kennedy and Eberhart's initial PSO, a particle i in n-dimensional real space "knows" three things as follows:

1. Its current location $\mathbf{x}^i = \{x_1^i, \cdots, x_n^i\}$;
2. Its historically best location $\mathbf{p}^i$; The "best" is evaluated by the objective value;
3. The overall best location found thus far in swarm $\mathbf{p}^g$, i.e., the historically best location in m particles.

We first calculate the next velocity as follows:

$$v_j^i(gen+1) = c_1 rand_1 \left(p_j^i - x_j^i(gen)\right) + c_2 rand_2 \left(p_j^g - x_j^i(gen)\right) \tag{8.12}$$

where gen is the iteration number of PSO, $rand_1$ and $rand_2 \sim U(0,1)$, $c_1 > 0$ and $c_2 > 0$ are *acceleration coefficients* controlling the influence of a particle's historical best location and the swarm's historical best location on its next velocity, respectively.[11] The first part of the right side of Eq. 8.12 is called the *cognitive component* and the second part is called the *social component*.[12] Then particle i moves to the next location with time step 1 as follows:

$$x_j^i(gen+1) = x_j^i(gen) + v_j^i(gen+1) \tag{8.13}$$

In 1999 Shi and Eberhart added the fourth thing a particle "knows" [12]:

4. Its current velocity $\mathbf{v}^i = v_1^i, \cdots, v_n^i$.

Then the velocity updating is developed as follows:

$$v_j^i(gen+1) = w(gen)\, v_j^i(gen) + c_1 rand_1 \left(p_j^i - x_j^i(gen)\right) + c_2 rand_2 \left(p_j^g - x_j^i(gen)\right) \tag{8.14}$$

where $0 < w < 1$ is the *inertia coefficient*[13] and the first part of the right side of Eq. 8.14 is called the *momentum component*.[14] In this book we call Eqs. 8.14 and 8.13 the *organic PSO*.

The general solution process of the PSO can be illustrated as follows.

---

[11] Kennedy and Eberhart suggested that $c_1 = c_2 = 2$.

[12] The cognitive component is for self-cognition and the social component is for social learning.

[13] Shi and Eberhart suggested that w decreases linearly from 0.9 at the beginning to 0.4 by the end of the algorithm, i.e., $w(gen) = 0.9 - (0.5 * gen/maxgen)$, for static problems [12] and $w = 0.5 + rand/2$ for dynamic problems [13].

[14] To consider the movement of particles in a more realistic physical environment.

Particle Swarm Optimization

Phase 1: Initialization.

Step 1.1: Assign the parameters for PSO, such as particle m, acceleration coefficients c_1 and c_2, and stop criteria (such as $maxgen$), etc. $gen = 0$.

Step 1.2: Generate m initial locations and velocities of the particles. Calculate their objective values and initialize the personal best locations and the overall best location accordingly.

Phase 2: Main loop. Repeat the following steps until the stop criteria are satisfied.

Step 2.1: $i = 1$.

Step 2.2: Determine the next location of particle i.

Substep 2.2.1: Determine the next velocity of particle i using Eq. 8.14 with all variables. Different $rand_1$ and $rang_2$ should be used for different variables.

Substep 2.2.2: Determine the next location of particle i using Eq. 8.13 with all variables.

Step 2.3: Calculate the objective value for particle i. Update $\mathbf{p}^i$ and $\mathbf{p}^g$ if necessary. $i = i+1$. If $i \leq m$, go back to step 2.2.

Step 2.4: Update w according to the deterministic control rule. $gen = gen + 1$.

Phase 3: Submitting $\mathbf{p}^g$ as the result.

To make the expression more concise, we can use the pointwise vector multiplication operator $\{.*\}$ to rewrite Eqs. 8.13 and 8.14 in vector form as follows.[15] We also neglect (gen) and $(gen+1)$ for clearer expressions:

$$\mathbf{v}^i = w\mathbf{v}^i + c_1\mathbf{rand}_1 .* \left(\mathbf{p}^i - \mathbf{x}^i\right) + c_2\mathbf{rand}_2 .* \left(\mathbf{p}^g - \mathbf{x}^i\right) \tag{8.15}$$

$$\mathbf{x}^i = \mathbf{x}^i + \mathbf{v}^i \tag{8.16}$$

where $\mathbf{rand}_1$ and $\mathbf{rand}_2$ are n-dimensional random vectors whose components are $\sim U(0,1)$. If all the elements of $\mathbf{rand}_1$ are the same, and likewise with $\mathbf{rand}_2$, Eqs. 8.15 and 8.16 become space vector operations, like DE introduced in Sects. 2.4.2.2 and 3.2.2.4.

According to Eq. 8.15, particle i will start from its current location and fly along the tradeoff direction to its personal best solution so far and the overall best solution so far in the swarm. When the swarm approaches convergence, i.e., $\mathbf{p}^i$ and $\mathbf{p}^g$ are both close to their current location, it will automatically decrease the velocity and thus decrease the step size according to Eq. 8.15. So we can say that PSO has the ability to adaptively control the step size.

[15] $(1,2,3).*(2,3,4) = (2,6,12)$

Too large a velocity may be generated by Eq. 8.15, which means the particles move too fast and the swarm blows up. Generally we can predefine an upper limitation $v_{\max}$. If the absolute value speed in any dimension exceeds $v_{\max}$, it will be assigned as $\pm v_{\max}$ considering its sign [14].

If the particle exceeds the definition domain in any dimension, then two methods can be adopted. The naive one is to just pull it back on the boundary. Liang *et al.* also suggested not to update its $\mathbf{p}^i$ so that it will fly back later [15].

The most attractive characteristics of PSO are its easy implementation (only two operations) and fast convergence speed (with the help of $\mathbf{p}^i$ and $\mathbf{p}^g$). But sometimes it suffers premature convergence due to too fast a convergence, especially in highly multimodal problems. So many studies have been done to control particle swarm diversity, which is the main topic of the following subsections.

8.3.2 Neighbor Structure and Related Extensions

The distinction of PSO lies in $\mathbf{p}^i$ and $\mathbf{p}^g$. Some researchers thought that letting $\mathbf{p}^g$ be the best solution in the swarm thus far might be too strong for multimodal problems. So they suggested a term *lbest*, corresponding to *gbest*. *gbest* is a concept used in organic PSO, i.e., the (global) best solution in the *swarm* thus far and *lbest* is for the (local) best solution in the *neighbor* thus far. Here the neighbor does not mean the real geometrical closeness relationship of the particles, but is the user-defined neighborhood relations among particles.

Many types of *neighbor structures* have been suggested [16, 17]. The *gbest* model can also be represented as an *all* structure, where all particles are the neighbors of every other particle. The all structure is the fastest way of propagating information because the newly found best solution in the swarm will be known by any particle in the next time step.

The simplest neighbor structure might be the *ring* structure. The nine-particle swarm with ring structure can be illustrated by Fig. 8.8. For example, only particles 2 and 9 are neighbors of particle 1.

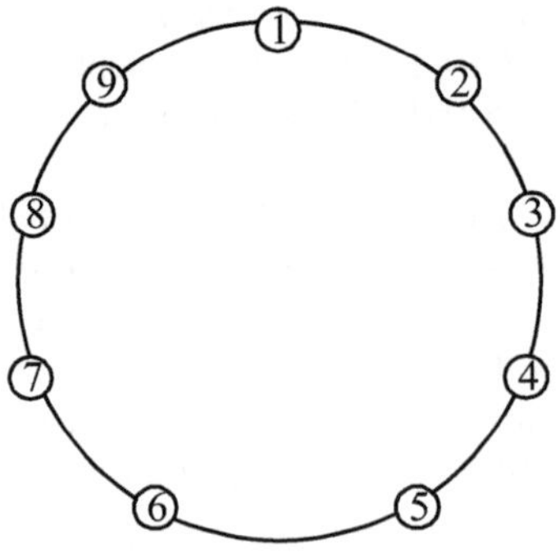

Fig. 8.8 The ring neighbor structure

Another typical structure is the *von Neumann* structure. Particles are placed in 2-D grids. Every particle has four neighbors. In order to make it so that the edge particles in 2-D space still have four neighbors, additional links should be added and the grids should be wrapped to expand the structure into 3-D space. A nine-particle swarm with the von Neumann structure can be illustrated by Fig. 8.9. For example, particles 2, 3, 4, 7 are neighbors of particle 1.

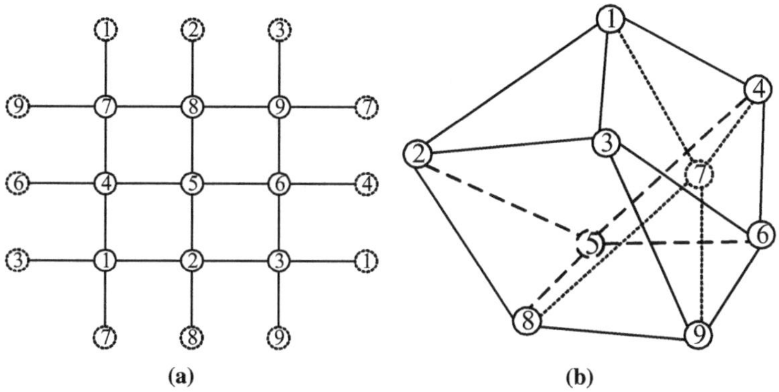

Fig. 8.9 The 2-D and 3-D representations of the von Neumann neighbor structure: (a) 2-D, and (b) 3-D

After defining the neighborhood, we can modify the velocity equation in Eq. 8.15 as follows:

$$\mathbf{v}^i = w\mathbf{v}^i + c_1\mathbf{rand}_1.*\left(\mathbf{p}^i - \mathbf{x}^i\right) + c_2\mathbf{rand}_2.*\left(\mathbf{p}^{l_i} - \mathbf{x}^i\right) \tag{8.17}$$

where $\mathbf{p}^{l_i}$ is the best solution so far (*lbest*) in particle *i*'s neighbors.

According to Eq. 8.17, the slowest way of information propagation is the ring structure. In Fig. 8.8, suppose particle 1 finds the global optimal solution; it will take four time steps to transfer this information to particles 5 and 6. But it might be an effective way to preserve the particle swarm diversity. Hence it might have a more powerful global search ability in highly multimodal space compared to the all neighbor structure. the von Neumann structure is in the middle of these two extremes.

We can also assign *rand* randomly selected particles as the neighbors of particle *i*, which is called *random* structure.

Selecting the proper neighbor structure for various problems might be a tough job.

8.3.2.1 Constriction Particle Swarm

In Sect. 8.3.1, we introduced the $v_{\max}$ method to limit particle speed. In 2002 Clerc and Kennedy introduced the *constriction coefficient* in order to cancel the requirement of setting $v_{\max}$ artificially [18].

The velocity equation considering the constriction coefficient can be illustrated as follows:

$$\mathbf{v}^i = \chi\left(\mathbf{v}^i + c_1\mathbf{rand}_1.*\left(\mathbf{p}^i - \mathbf{x}^i\right) + c_2\mathbf{rand}_2.*\left(\mathbf{p}^{l_i} - \mathbf{x}^i\right)\right) \tag{8.18}$$

where $\chi = \frac{2}{\left|2-\varphi-\sqrt{\varphi^2-4\varphi}\right|}$ is the constriction coefficient and $\varphi = c_1 + c_2$. Clerc and Kennedy proved that if $\varphi > 4$, Eq. 8.18 could limit the particle coefficient without $v_{\max}$.

The suggested values for the control parameter are $c_1 = c_2 = 2.05$, $\varphi = 4.10$, and thus $\chi = 0.7298$. Comparing Eq. 8.18 to Eq. 8.17, we could say that the constriction particle swarm is a special kind of organic particle swarm considering the neighbor structure. Clerc and Kennedy called the constriction particle swarm a *canonical particle swarm.*

In 2002 Kennedy and Mendes compared several neighbor structures using the canonical particle swarm and came to the conclusion that the von Neumann structure is the best structure for it [17]. In 2000 Eberhart and Shi also compared the canonical particle swarm and the organic particle swarm and suggested that the best approach is to utilize the constriction coefficient along with the velocity limit $v_{\max,i} = x_{\max,i}$, where $x_{\max,i}$ is the definition length in i's dimension and $v_{\max,i}$ is the absolute maximum velocity in i's dimension [19].

8.3.2.2 Fully Informed Particle Swarm

Both the canonical particle swarm and the organic particle swarm with a neighbor structure only consider the best other particle in the neighborhood. This characteristic might limit the search of PSO and therefore cause premature. In 2004 Mendes *et al.* suggested the *fully informed particle swarm* (FIPS) to consider the contribution of all neighbors [20].

We can define $\varphi_1 = c_1\mathbf{rand}_1$, $\varphi_2 = c_1\mathbf{rand}_2$, and $\varphi = \varphi_1 + \varphi_2$. Then Eq. 8.18 could be rewritten as follows:

$$\begin{aligned}\mathbf{v}^i &= \chi\left(\mathbf{v}^i + \varphi_1.*\left(\mathbf{p}^i - \mathbf{x}^i\right) + \varphi_2.*\left(\mathbf{p}^{l_i} - \mathbf{x}^i\right)\right)\\ &= \chi\left(\mathbf{v}^i + \varphi.*\left(\left(\varphi_1.*\mathbf{p}^i + \varphi_2.*\mathbf{p}^{l_i}\right)./\varphi - \mathbf{x}^i\right)\right)\end{aligned} \tag{8.19}$$

where ./ is the pointwise division operator.[16] Let $\mathbf{p}^m = \left(\varphi_1.*\mathbf{p}^i + \varphi_2.*\mathbf{p}^{l_i}\right)./\varphi$. Then we can further simplify the expression of the canonical particle swarm as follows:

$$\mathbf{v}^i = \chi\left(\mathbf{v}^i + \varphi\left(\mathbf{p}^m - \mathbf{x}^i\right)\right) \tag{8.20}$$

Equation 8.20 means that only the weight sum of the personal best location and the best neighbor location affects the velocity of particle i. FIPS considers all the best neighbor locations in the following way.

Suppose particle i's neighbor set is N_i and its neighbor number is $|N_i|$. For k's neighbor in N_i, its personal best location is $\mathbf{p}^k$, its acceleration random vector is $\varphi^k \sim \mathbf{U}\left(0, \frac{\varphi_{\max}}{|N_i|}\right)$, where $\varphi_{\max}$ is 4.10 according to Clerc and Kennedy's suggestion regarding the constriction coefficient, and $\mathbf{U}()$ is the vector uniform random function. Then the velocity updating equation in FIPS can be illustrated as follows:[17]

$$\mathbf{v}^i = \chi\left(\mathbf{v}^i + \sum_{k\in N_i}\varphi^k.*\left(\left(\sum_{k\in N_i}\varphi^k.*\mathbf{p}^k\right)./\left(\sum_{k\in N_i}\varphi^k\right) - \mathbf{x}^i\right)\right) \tag{8.21}$$

We can define $\varphi = \sum\limits_{k\in N_i}\varphi^k = \mathbf{U}(0, \varphi_{\max})$ to simplify Eq. 8.21 as follows:

$$\mathbf{v}^i = \chi\left(\mathbf{v}^i + \varphi.*\left(\left(\sum_{k\in N_i}\frac{\varphi.*\mathbf{p}^k}{|N_i|}\right)./\varphi - \frac{\sum\limits_{k\in N_i}\mathbf{x}^i}{|N_i|}\right)\right) \tag{8.22}$$

Then the final velocity updating rule in FIPS can be simplified as follows:

$$\mathbf{v}^i = \chi\left(\mathbf{v}^i + \left(\frac{\sum\limits_{k\in N_i}\varphi.*\left(\mathbf{p}^k - \mathbf{x}^i\right)}{|N_i|}\right)\right) \tag{8.23}$$

In Eq. 8.23, all of particle i's neighbors' personal best solutions will contribute to its velocity.[18]

In canonical particle swarm or organic particle swarm with a neighbor structure, the larger neighbor number means a faster convergence, but in FIPS this conclusion will be reversed.[19]

In 2006 Kennedy and Mendes compared the organic particle swarm with the neighbor structure and FIPS considering many types of structures [21]. They found that the organic particle swarm with a neighbor structure is less sensitive to the

---

[16] $(3,4,9)./(2,2,3) = (1.5,2,3)$

[17] Following Eq. 8.19 but considering the weights of the $|N_i|$ neighbors.

[18] It is necessary to note here that N_i in Eq. 8.19 does not contain particle i itself, i.e., Eq. 8.23 does not consider particle i's personal best location [21].

[19] Why?

neighbor structure than FIPS. But with a proper neighbor structure,[20] FIPS can generate better solutions.

8.3.2.3 Bare Bones Particle Swarm

In 2003 Kennedy revisited the canonical particle swarm and suggested a back-to-nature idea by eliminating the velocity updating rule, i.e., Eq. 8.18 [22].

Kennedy summarized the essence of PSO as the "sampling points on the problem space and using discovered knowledge to guide exploration".[21] Then he did a very interesting yet simple numerical experiment. Suppose we are facing a one-dimensional search space, $p^i \equiv -10$ and $p^{l_i} \equiv 10$ in Eq. 8.18. We only have one particle with $v(0) = 1$ and $x(0) = 0$. Its flying trajectory can be simplified as follows:

$$\begin{aligned} v' &= 0.7298\,(v + 2.05 * rand_1 * (10 - x) + 2.05 * rand_2 * (-10 - x)) \\ x' &= x + v' \end{aligned} \tag{8.24}$$

where v' and x' are simplified versions of $v(gen+1)$ and $x(gen+1)$, respectively, and v and x are simplified versions of $v(gen)$ and $x(gen)$, respectively. We record the location the particle has visited in the 100000 iteration steps and draw the histogram of the location in the scale of $(-40, 40)$ as in Fig. 8.10.[22] The horizontal axis is the small region of the particle location and the vertical axis is the amount of time the particle visited that region.

As can be seen from Fig. 8.10, the distribution presents a very clear bell shape. That means the most often visited region is determined by p^i and p^{l_i}. It seems that the particle flies in a normal distribution way with a mean $\mu = \frac{p^i + p^{l_i}}{2}$ and the standard deviation related to the distance between p^i and p^{l_i}.

So Kennedy discarded the velocity part and used a normal distribution to generate the new locations as follows:

$$\mathbf{x}^i\,(gen+1) \sim \mathbf{N}\left(\frac{\mathbf{p}^i + \mathbf{p}^{l_i}}{2}, \left|\mathbf{p}^i - \mathbf{p}^{l_i}\right|\right) \tag{8.25}$$

where $\mathbf{N}(\mu, \sigma)$ is the uncorrelated normal distribution vector with mean μ and the same standard deviation σ and $||$ is the norm function (distance).

At first glance, Eq. 8.25 has no sense of birds flying. Kennedy argued that the main characteristics of PSO lie in the mutual influence between particles, the self memory of the best location so far, and the neighbor structure. Equation 8.25 discards all the previous considerations about the velocity (related to "fly") but main-

[20] The average degree of the graph is 3 and the vertices in the graph are clustered naturally.

[21] Do these words invoke the continuous $ACO_{\mathbb{R}}$ and CMA?

[22] All the locations outside of the scale are neglected.

tains the above-discussed properties and embodies the essence of "sampling and modifying the distribution." So he named the algorithm the *bare bones particle swarm*.

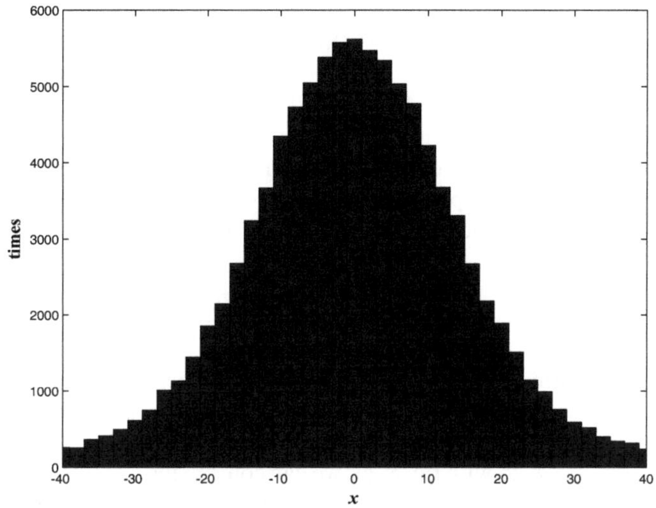

Fig. 8.10 Histogram of a simple example of the canonical particle swarm

Then Kennedy further introduced the interaction probability IP. Particles modify their location by Eq. 8.25 with probability IP.[23] Otherwise, $\mathbf{x}^i(gen+1) = \mathbf{p}^i$.

Kennedy compared the bare bones particle swarm with the canonical particle and FIPS using benchmark problems and different neighbor structures and found that the best algorithm is the bare bones particle swarm with $IP = 0.5$ and a random neighbor.

8.3.3 Extensions from Organic Particle Swarm Optimization

There are also many extensions of organic PSO, illustrated by Eq. 8.14, by introducing diversity in various ways. Here we introduce the *comprehensive learning particle swarm optimization* (CLPSO) suggested by Liang *et al.* in 2006 [15].

Liang *et al.* defined the *learning probability* p_c^i for particle i to describe its intention to learn from other particles. The velocity updating rule for dimension j of particle i is as follows:

$$v_j^i(gen+1) = w(gen)\,v_j^i(gen) + c * rand * \left(p_j^{l_i} - x_j^i(gen)\right) \tag{8.26}$$

[23] The suggested value of IP is 0.5.

where $p_j^{l_i}$ is called the *exemplar* of particle i. Other parameters in Eq. 8.26 have the same meaning in Eq. 8.14.

For variable j in particle i, we use tournament selection with probability p_c^i to pick its exemplar.[24] In binary tournament selection, we compare the objective values of two randomly selected particles from the swarm excluding the current particle and make the fitter one the exemplar.[25] Otherwise (with probability $1 - p_c^i$), its exemplar is itself.

The exemplar will remain unchanged in Eq. 8.26 until particle i has not been improved in the successive m generations.[26]

The learning probability p_c^i determines the frequency of a particle to learn from other particles. To further improve the swarm diversity, Liang *et al.* suggested that different learning probabilities should be assigned as follows:

$$p_c^i = 0.05 + 0.45 * \frac{\exp\left(\frac{10(i-1)}{m-1}\right) - 1}{\exp(10) - 1} \tag{8.27}$$

where m is the swarm number and i is the order number of the particles (from 1 to m). As can be seen from Eq. 8.27, various p_c^i could be generated in the range $(0.05, 0.5)$.[27]

By comparing Eq. 8.26 with Eq. 8.23, we can come to the conclusion that CLPSO introduces diversity into the swarm by probabilistically selecting the particle to learn, which is different from how it is done in FIPS, i.e., all the neighbor particles will have influences.

Liang compared CLPSO with nine PSO algorithms, including the above introduced organic particle swarm, canonical particle swarm, and FIPS, and came to the conclusion that CLPSO achieves the best results on most complex multimodal problems but is relatively weak for unimodal and simple multimodal problems.

8.4 Summary

Swarm intelligence has a lot of branches and we only briefly introduced the two most important of them, i.e., ACO and PSO.

Ant colonies can find the shortest length between food and the nest with the help of pheromones. So ACO is basically designed for shortest-length problems on a graph. Many techniques are suggested to adjust the pheromone concentration properly. In order to solve other combinatorial problems, we need to transfer the original problem to a graph and define the visibility and pheromone concentration accordingly. By the idea of sampling according to distribution and modifying the

---

[24] Different variables in particle i might select different particles as its exemplar.

[25] We could also say that CLPSO uses the "all" neighbor structure.

[26] Liang *et al.* suggested that $m = 7$ based on empirical numerical experiments on benchmark problems.

[27] In CLPSO, p_c^i remains unchanged during the evolving process.

distribution according to the result of the sampling, ACO can be expanded into continuous optimization and related to the estimation of distribution algorithm.

Bird flocks also emerge intelligence in foraging behavior. To mimic the coordinated flying property, PSO is recommended. PSO is basically designed for continuous optimization using a particle's historically best solution so far and its neighbors' historically best solution(s) so far. PSO is fast because it always uses the best information so far but suffers from premature thereby. So many variations have been proposed to introduce swarm diversity by the neighbor structure, by utilizing neighbor information, etc.

To some extent, ACO and PSO do not like to be EAs because they do not have a selection procedure and they do not use crossover operators to exchange information. But, according to the criteria given in Sect. 1.2, they are both population-based, fitness-oriented, and variation-driven. So we can still regard them as EAs.

After reading this chapter, you should be familiar with the procedure of applying ACO to combinatorial optimization and PSO to continuous optimization, understand the rationale of suggesting ACO and PSO behind the specific implementation steps, and learn some techniques for controlling parameters and adjusting the diversity.

Suggestions for Further Reading

On the topic of SI, there is a journal, *Swarm Intelligence* published by Springer since 2007, a biannual conference on SI, the International Conference on Ant Colony Optimization and Swarm Intelligence, whose proceedings are published by Springer, and an annual symposium, the IEEE Swarm Intelligence Symposium.

There are many recently published books about SI. Readers are encouraged to read Bonabeau *et al.*'s book published in 1999 [23], Weiss's book published in 2000 [24], Eberhart *et al.*'s book published in 2001 [25], Engelbrecht's book published in 2006 [26], Abraham *et al.*'s book published in 2006 [27], and Blum and Merkle's book published in 2008 [28].

Montes de Oca maintains a Web site on ACO containing plenty of information.[28] Two books about ACO are suggested for further reading, i.e., Dorigo and Stützle's book [29] and Solnon's book [30]. The paper published by Dorigo *et al.* in 1999 established the framework of ACO [31]. A tutorial for beginners on ACO was given by Dorigo *et al.* in 2006 [32].

At the theory level of ACO, in 2000 Dorigo *et al.* discuss the *stigmergy*, i.e., indirect coordination between agents, in ACO [33]; in 2005 Zecchin *et al.* discussed the parameter tuning guidelines for ACO [34]; and in 2005 Blum and Dorigo introduced the concept of a *competition-balanced system* to overcome the *second-order deception* in ACO [35].

---

[28] http://www.aco-metaheuristic.org/

At the application level, in 2002 Solnon used ACO to solve constrained optimization [36] and in 2009 Leguizamón and Coello Coello discussed the same topic in another way [37]. In 2002 Merkle used ACO in resource-constrained project scheduling [38]. In 2008 Martens *et al.* made ACO applicable to supervised learning problems, i.e., classification [39]. In 2006 Nezamabadi-pour *et al.* applied ACO to edge detection [40] and in 2008 Lim utilized ACO for path planning in sparse graphs [41].

A Web site Particle Swarm Central contains a lot of the latest information on PSO.[29] The books suggested on PSO included Clerc's book from 2006 [42], Poli *et al.*'s book from 2008 [43], Kishk's book from 2008 [44], and Parsopoulos and Vrahatis's book from 2009 [45]. The very intensive and up-to-date surveys on PSO given by Banks *et al.* in 2007 and 2008 are highly recommended [46, 47].

In 2004 Ratnaweera *et al.* suggested deterministic control mechanisms for various PSO control parameters [48]. In 2004 van den Bergh and Engelbrecht applied cooperative approach to PSO [49]. In 2006 Kadirkamanathan *et al.* discussed the stability analysis of PSO in a random environment [50]. Blackwell and Branke's paper and Janson and Middendorf's paper, both from 2006, discuss PSO in noisy and dynamic environments [51, 52]. Two comparison papers about PSO, DE, CMA, and GA are recommended, i.e., Langdon and Poli's paper from 2007 [53] and Vrugt *et al.*'s paper from 2009 [54].

In 2006 Liang and Suganthan applied PSO to constrained optimization [55]. Parsopoulos and Vrahatis's 2004 paper and Parrott and Li's 2006 paper discuss multimodal optimization [56, 57]. In 2004 Coello Coello *et al.* addressed MOP with PSO [58]. Ho *et al.* also discussed this topic in 2006 [59]. A comprehensive survey on PSO-based MOP solvers was given by Reyes-Sierra and Coello Coello in 2006 [60].

Many successful applications of PSO have been proposed. Some of them are recommended according to our interests: Hastings *et al.*'s paper on computer graphics and animation from 2009 [61], O'Neill and Brabazon's paper on unsupervised learning from 2008 [62], del Valle *et al.*'s paper on power systems from 2008 [63], Li *et al.*'s paper on gene selection from 2008 [64], and Rahimi-Vahed *et al.*'s paper on assembly line sequencing from 2007 [65].

The must-read papers of this chapter are [1] for AS, [2] for ACS, [9] for continuous ACO, [11] for the initial idea of PSO, [21] for FIPS and its related neighbor structure, and [15] for introducing diversity into a swarm.

Exercises and Potential Research Projects

8.1. Summarize the essences of [36] and [37] on a single sheet of paper separately and compare them using the benchmark problems in Appendix, PI in Sect. 5.5, and

---

[29] `http://www.particleswarm.info`

the techniques introduced in Sect. 3.6 to do a fair comparison and draw statistical conclusions according to your numerical experiments.

8.2. Implement $ACO_{\mathbb{R}}$ and compare it to the EA you finished in Chap. 3 using the benchmark problems in Appendix, PI in Sect. 5.5, and the techniques introduced in Sect. 3.6 to do a fair comparison and draw statistical conclusions according to your numerical experiments.

8.3. Implement AS or ACS for TSP or JSSP and compare it to the EA you finished in Chap. 7 using the benchmark problems in Appendix, PI in Sect. 5.5, and the techniques introduced in Sect. 3.6 to do a fair comparison and draw statistical conclusions according to your numerical experiments.

8.4. Draw the illustrative vector graph in 2-D real space for Eqs. 8.15 and 8.16 provided that all the elements of $\mathbf{rand}_1$ are the same, and likewise for $\mathbf{rand}_2$.

8.5. Why in canonical particle swarm or organic particle swarm with a neighbor structure does a larger neighbor number mean faster convergence but in FIPS the opposite is true?

8.6. Implement organic particle and CLPSO in your programming environment, use the benchmark problems in Appendix and the techniques introduced in Sect. 3.6 to do a fair comparison and draw statistical conclusions according to your numerical experiments.

8.7. Implement $ACO_{\mathbb{R}}$ and CLPSO, use the benchmark problems in Appendix and the techniques introduced in Sect. 3.6 to do a fair comparison and draw statistical conclusions according to your numerical experiments.

8.8. Try to improve CLPSO with techniques introduced in Chap. 3, use the benchmark problems in Appendix and the techniques introduced in Sect. 3.6 to do a fair comparison and draw statistical conclusions according to your numerical experiments.

8.9. Summarize the global search mechanism, convergence mechanism, and uphill mechanism for minimum optimization problems of ACO and PSO.

References

1. Dorigo M, Maniezzo V, Colorni A (1996) Ant system: optimization by a colony of cooperating agents. IEEE Trans Syst Man Cybern B Cybern 26(1):29–41
2. Dorigo M, Gambardella L (1997) Ant colony system: a cooperative learning approach to the traveling salesman problem. IEEE Trans Evol Comput 1(1):53–66
3. Watkins C (1989) Learning from delayed rewards. Ph.D. thesis, University of Cambridge, UK
4. Kaelbling LP, Littman M, Moore A (1996) Reinforcement learning: a survey. J Artif Intell Res 4:237–285

5. Bullnheimer B, Hartl RF, Strauß C (1999) A new rank based version of the ant system - a computational study. Central Eur J Oper Res and Econ 7:25–38
6. Stützle T, Hoos HH (2000) MAX-MIN ant system. Future Gener Comput Syst 16(8):889–914
7. Blum C, Dorigo M (2004) The hyper-cube framework for ant colony optimization. IEEE Trans Syst Man Cybern B Cybern 34(2):1161–1172
8. Birattari M, Pellegrini P, Dorigo M (2007) On the invariance of ant colony optimization. IEEE Trans Evol Comput 11(6):732–742
9. Socha K, Dorigo M (2008) Ant colony optimization for continuous domains. Eur J Oper Res 185(3):1155–1173
10. Kern S, Müller SD, Hansen N *et al* (2004) Learning probability distributions in continuous evolutionary algorithms - a comparative review. Nat Comput 3(3):355–356
11. Kennedy J, Eberhart R (1995) Particle swarm optimization. In: Proceedings of the IEEE international conference on neural networks, pp 1942–1948
12. Shi Y, Eberhart R (1999) Empirical study of particle swarm optimization. In: Proceedings of the IEEE congress on evolutionary computation, pp 1945–1950
13. Eberhart R, Shi Y (2001) Tracking and optimizing dynamic systems with particle swarms. In: Proceedings of the IEEE congress on evolutionary computation, pp 94–100
14. Eberhart RC, Dobbins R, Simpson PK (1996) Computational intelligence PC tools. Morgan Kaufmann, San Francisco
15. Liang J, Qin A, Suganthan P *et al* (2006) Comprehensive learning particle swarm optimizer for global optimization of multimodal functions. IEEE Trans Evol Comput 10(3):281–295
16. Kennedy J (1999) Small worlds and mega-minds: effects of neighborhood topology on particle swarm performance. In: Proceedings of the IEEE congress on evolutionary computation, pp 1931–1938
17. Kennedy J, Mendes R (2002) Population structure and particle swarm performance. In: Proceedings of the IEEE congress on evolutionary computation, pp 1671–1676
18. Clerc M, Kennedy J (2002) The particle swarm - explosion, stability, and convergence in a multidimensional complex space. IEEE Trans Evol Comput 6(1):58–73
19. Eberhart R, Shi Y (2000) Comparing inertia weights and constriction factors in particle swarm optimization. In: Proceedings of the IEEE congress on evolutionary computation, pp 84–88
20. Mendes R, Kennedy J, Neves J (2004) The fully informed particle swarm: simpler, maybe better. IEEE Trans Evol Comput 8(3):204–210
21. Kennedy J, Mendes R (2006) Neighborhood topologies in fully informed and best-of-neighborhood particle swarms. IEEE Trans Syst Man Cybern C Appl Rev 36(4):515–519
22. Kennedy J (2003) Bare bones particle swarms. In: Proceedings of the IEEE swarm intelligence symposium, pp 80–87
23. Bonabeau E, Dorigo M, Theraulaz G (1999) Swarm intelligence: from natural to artificial systems. Oxford University Press, Oxford, UK
24. Weiss G (2000) Multiagent systems: a modern approach to distributed artificial intelligence. MIT Press, Cambridge, MA
25. Eberhart RC, Shi Y, Kennedy J (2001) Swarm intelligence. Morgan Kaufmann, San Francisco
26. Engelbrecht AP (2006) Fundamentals of computational swarm intelligence. Wiley, New York
27. Abraham A, Grosan C, Ramos V (2006) Swarm intelligence in data mining. Springer, Berlin Heidelberg New York
28. Blum C, Merkle D (2008) Swarm intelligence: introduction and applications. Springer, Berlin Heidelberg New York
29. Dorigo M, Stützle T (2004) Ant colony optimization. MIT Press, Cambridge, MA
30. Solnon C (2010) Ant colony optimization and constraint programming. Wiley-ISTE, New York
31. Dorigo M, Caro GD, Gambardella LM (1999) Ant algorithms for discrete optimization. Artif Life 5(2):137–172
32. Dorigo M, Birattari M, Stutzle T (2006) Ant colony optimization. IEEE Comput Intell Mag 1(4):28–39
33. Dorigo M, Bonabeau E, Theraulaz G (2000) Ant algorithms and stigmergy. Future Gener Comput Syst 16(9):851–871

34. Zecchin A, Simpson A, Maier H *et al* (2005) Parametric study for an ant algorithm applied to water distribution system optimization. IEEE Trans Evol Comput 9(2):175–191
35. Blum C, Dorigo M (2005) Search bias in ant colony optimization: on the role of competition-balanced systems. IEEE Trans Evol Comput 9(2):159–174
36. Solnon C (2002) Ants can solve constraint satisfaction problems. IEEE Trans Evol Comput 6(4):347–357
37. Leguizamón G, Coello C (2009) Boundary search for constrained numerical optimization problems with an algorithm inspired by the ant colony metaphor. IEEE Trans Evol Comput 13(2):350–368
38. Merkle D, Middendorf M, Schmeck H (2002) Ant colony optimization for resource-constrained project scheduling. IEEE Trans Evol Comput 6(4):333–346
39. Martens D, Backer MD, Haesen R *et al* (2007) Classification with ant colony optimization. IEEE Trans Evol Comput 11(5):651–665
40. Nezamabadi-pour H, Saryazdi S, Rashedi E (2006) Edge detection using ant algorithms. Soft Comput 10(7):623–628
41. Lim KK, Ong Y, Lim MH *et al* (2008) Hybrid ant colony algorithms for path planning in sparse graphs. Soft Comput 12(10):981–994
42. Clerc M (2006) Particle swarm optimization. ISTE Publishing Company, London, UK
43. Poli R, Kennedy J, Blackwell T *et al* (2008) Particle swarms: the second decade. J Artif Evol Appl 2008:1–3
44. Mikki S, Kishk A (2008) Particle swarm optimizaton: a physics-based approach. Morgan and Claypool, San Rafael, CA
45. Parsopoulos KE, Vrahatis MN (2009) Particle swarm optimization and intelligence: advances and applications. Information Science, LinkHershey, PA
46. Banks A, Vincent J, Anyakoha C (2007) A review of particle swarm optimization. I: background and development. Nat Comput 6(4):467–484
47. Banks A, Vincent J, Anyakoha C (2008) A review of particle swarm optimization. II: hybridisation, combinatorial, multicriteria and constrained optimization, and indicative applications. Nat Comput 7(1):109–124
48. Ratnaweera A, Halgamuge S, Watson H (2004) Self-organizing hierarchical particle swarm optimizer with time-varying acceleration coefficients. IEEE Trans Evol Comput 8(3):240–255
49. van den Bergh F, Engelbrecht A (2004) A cooperative approach to particle swarm optimization. IEEE Trans Evol Comput 8(3):225–239
50. Kadirkamanathan V, Selvarajah K, Fleming P (2006) Stability analysis of the particle dynamics in particle swarm optimizer. IEEE Trans Evol Comput 10(3):245–255
51. Blackwell T, Branke J (2006) Multiswarms, exclusion, and anti-convergence in dynamic environments. IEEE Trans Evol Comput 10(4):459–472
52. Janson S, Middendorf M (2006) A hierarchical particle swarm optimizer for noisy and dynamic environments. Genet Programm Evolvable Mach 7(4):329–354
53. Langdon W, Poli R (2007) Evolving problems to learn about particle swarm optimizers and other search algorithms. IEEE Trans Evol Comput 11(5):561–578
54. Vrugt J, Robinson B, Hyman J (2009) Self-adaptive multimethod search for global optimization in Real-Parameter spaces. IEEE Trans Evol Comput 13(2):243–259
55. Liang J, Suganthan P (2006) Dynamic multi-swarm particle swarm optimizer with a novel constraint-handling mechanism. In: Proceedings of the IEEE congress on evolutionary computation, pp 9–16
56. Parsopoulos K, Vrahatis M (2004) On the computation of all global minimizers through particle swarm optimization. IEEE Trans Evol Comput 8(3):211–224
57. Parrott D, Li X (2006) Locating and tracking multiple dynamic optima by a particle swarm model using speciation. IEEE Trans Evol Comput 10(4):440–458
58. Coello C, Pulido G, Lechuga M (2004) Handling multiple objectives with particle swarm optimization. IEEE Trans Evol Comput 8(3):256–279

59. Ho S, Ku W, Jou J *et al* (2006) Intelligent particle swarm optimization in multi-objective problems, In: Ng WK, Kitsuregawa M, Li J *et al* (eds) Advances in knowledge discovery and data mining. Springer, Berlin Heidelberg New York, pp 790–800
60. Reyes-Sierra M, Coello Coello CA (2006) Multi-objective particle swarm optimizers: A survey of the-state-of-the-art. Tech. rep., CINVESTAV-IPN, Mexico
61. Hastings E, Guha R, Stanley K (2009) Interactive evolution of particle systems for computer graphics and animation. IEEE Trans Evol Comput 13(2):418–432
62. O'Neill M, Brabazon A (2008) Self-organising swarm (SOSwarm). Soft Comput 12:1073–1080
63. del Valle Y, Venayagamoorthy G, Mohagheghi S *et al* (2008) Particle swarm optimization: Basic concepts, variants and applications in power systems. IEEE Trans Evol Comput 12(2):171–195
64. Li S, Wu X, Tan M (2008) Gene selection using hybrid particle swarm optimization and genetic algorithm. Soft Comput 12(11):1039–1048
65. Rahimi-Vahed AR, Mirghorbani SM, Rabbani M (2007) A new particle swarm algorithm for a multi-objective mixed-model assembly line sequencing problem. Soft Comput 11(10):997–1012

Chapter 9
Artificial Immune Systems

Abstract In previous chapters, we introduced GAs, which mimic the natural evolving process, and ACO and PSO, which mimic the feeding behavior of an ant colony and the bird flock, respectively. Scientists in the EA field are used to taking these biological or physiological "intelligent" phenomena as metaphors to excite the imagination for generating algorithms for learning and optimizing problems. In this chapter, we will introduce a concept and a procedure borrowed from the human (vertebrate) immune system and focus on how to apply these ideas to implementing our algorithms for various applications.

9.1 Introduction

The human immune system is such a complicated system that not all cause-and-effect connections have been revealed. The basic function for the immune system is to identify and destroy foreign substances, denoted as *pathogens* or *invaders*, including viruses, bacteria, fungi, and parasites. Self-malfunction cells, such as cancer cells, are also targets of the immune system.

These pathogenic substances are recognized by special structure molecules, such as polypeptide or nucleotide segments, on their surfaces. These pathogenic substances are known as *antigen* (Ag).

The human body has two sets of immune systems: the innate immune system and the adaptive (acquired) immune system. The former is for general defense and has no preference for any specific antigen and the latter responds to specific antigens.[1] We will focus on the acquired immune system in this chapter.

We can regard the cells in our immune system as soldiers with various functions, e.g., scouters, logistics soldiers, advisors, and attackers. The two most important soldiers of them are *T cells* (with three types: TC, TH, TS) and *B cells*. Both are

[1] The adaptive immune system could adapt itself to various antigens and this adaptation is acquired after birth.

generated by the bone marrow. They are lymphatic cells, which are the white blood cells, circulating in the blood and lymphatic vessels.

Some pathogenic substances in the human body will be swallowed by phagocytes. Then their antigens are represented on these cells.[2] Cytotoxic T (TC) cells that match the antigen are then proliferated and differentiated to kill the pathogenic substances. This procedure is called *cell-mediated immunity*. Matched B cells are also proliferated and differentiated, and then secrete *antibodies* (Ab),[3] into body fluids. Helper T cells (TH) can help the procedure of B cell proliferation and differentiation. These antibodies bind with antigens to mark a tag. Then lymphocytes, phagocytic cells, and the complement system are responsible for eliminating the marked antigenic substances. This procedure is called *humoral immunity*. After the invader are eliminated, suppressor T (TS) Cells are activated to suppress the stimulated B cells.[4] Some of the stimulated B cells and T cells in both humoral immunity and cell-mediated immunity can exist inside the immune system for a certain period and be prepared for the second invasion by a similar substance. The procedure from the intrusion to the elimination of the pathogen is called the *immune response*.[5]

In the following three sections, we will illustrate three main ways of mimicking human immune systems, resulting in an *artificial immune system* (AIS).

According to de Castro and Timmis' suggestion, the AIS framework contains three basic elements, corresponding to three questions that need to be answered before designing an AIS [3].

- **Representation.** How does one represent antigens and antibodies in an AIS, which is equivalent to the code scheme in a GA?
- **Evaluation.** How does evaluate the binding effects of antibodies on antigens and the interactions between antibodies, which is equivalent to the fitness evaluation in GAs?
- **Immune principle.** What are the immune principles applied in AIS to direct learning and optimization, which is equivalent to the selection process and variation operators in GAs?[6]

We need to note here that AIS has by no means been an attempt to directly model all of the processes occurring within human (vertebrate) immune system. Instead, however, certain disciplines have been simplified and modeled in order to replicate some of the mechanisms of immune response that might best be suited to manipulating data, learning general patterns, or optimizing problems.

---

[2] So these cells are called *antigen-presenting cells* (APCs).

[3] Sometimes an antibody is also called a *receptor*, *sensor*, or *detector*.

[4] TC cells are also stimulated and suppressed by TH cells and TS cells.

[5] For more information on immunology knowledge, please refer to Kindt *et al.*'s book [1] and Murphy *et al.*'s book [2].

[6] In the subsequent sections, we will introduce three principles, i.e., clonal selection, immune network, and negative selection.

Figure 9.1 illustrates the number of papers indexed by the SCI on AIS,[7] which demonstrates that AIS attracts a lot of attention every year.

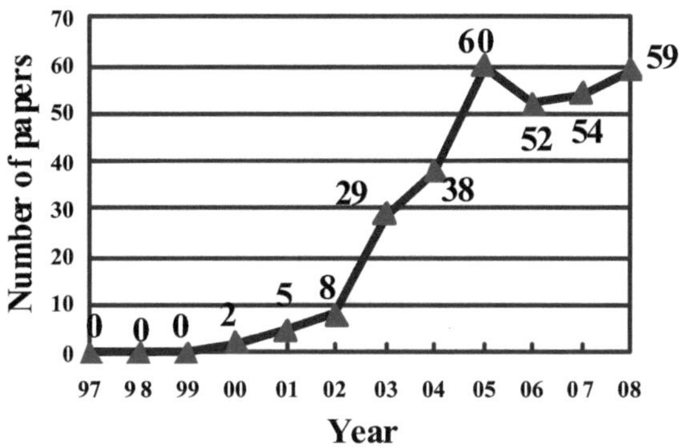

Fig. 9.1 Number of papers indexed by SCI on AIS

9.2 Artificial Immune System Based on Clonal Selection

9.2.1 Clonal Selection

We will only consider humoral immunity in this section. Every antigen has several characteristic protein segments, denoted as *epitopes*, which is illustrated as different geometrical shapes on the surface of the antigen in Fig. 9.2. This characteristic makes it possible for different B cells to recognize an antigen.

Every B cell has many receptors (antibodies), whose magnitude is about 10^5, on its surface with the same protein segment, i.e., one B cell only has one type of receptor. The part responsible for binding with the epitope of the antigen is called a *paratope*. Receptors also have a part, also called an *epitope*, that is possibly recognized and bound by other receptors.[8]

The human body has about 10^7 to 10^8 kinds of receptors, each of which has different match effects on antigens. The evaluation of the match is called an *affinity*.[9] As can be seen in Fig. 9.2, B cell_1 has a better match effect than B cell_2, i.e., the affinity measure of B cell_1 is higher than that of B cell_2.

---

[7] TS = ("artificial immune system" OR "artificial immune systems"). The SCI index "TS" is for the search topic in the title, the keywords, and the abstract.

[8] For simplicity, we neglect the Y-shaped protein structure of the receptor and suppose that every receptor has only one paratope and one epitope.

[9] According to Fig. 9.2, this match is a complementary form. For example, if we could use binary code to represent one antigen as (0010), then the receptor (1101) is the one with the highest affinity.

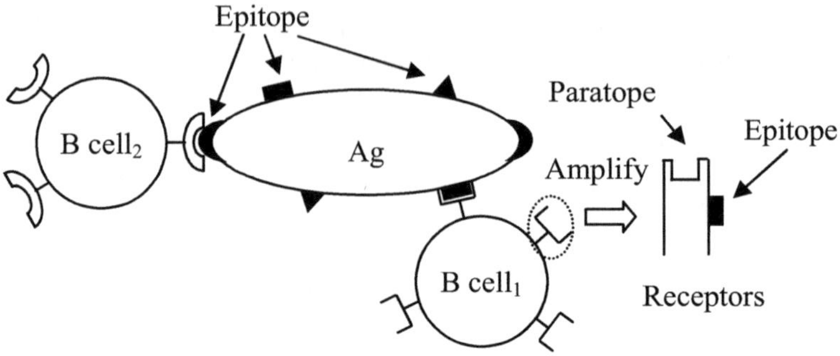

Fig. 9.2 Illustration of the binding of receptors to an antigen

The clone selection procedure can be roughly illustrated as follows:

1. The invader enters the human body and is swallowed by phagocytes.
2. The antigen of the invader is represented on the surface of the APC and then activates B cells.
3. B cells, with the help of TH cells, proliferate (clone) according to to their affinity measure, i.e., the clone number of one B cell is proportional to its affinity measure. The *concentration* of the fitter B cells increases faster.
4. The clones differentiate (mutate) according to their origin's affinity measure, i.e., the clones of the fitter origin will suffer less mutation. This procedure is called *somatic hypermutation*.[10]
5. The mutants are *plasma cells*, which is responsible for secreting a lot of specific antibodies. The antibodies will break away from the plasma cells, become free antibodies, and circulate in the body fluid. These B cells, and their secreted antibodies, constitute the *repertoire*.
6. antibodies mark antigens (thus mark the antigenic substances).
7. Lymphocytes, phagocytic cells, and the complement system eliminate the marked antigenic substances.
8. Some of the mutants with a high affinity measure will be memorized in the body.
9. The above procedure is called a *primary immune response*. After that, if a similar antigenic substance enters the human body again, the memory B cells will be quickly activated at a high stimulation level so that the human body can effectively respond to the recognized or similar pathogen. This procedure is called a *secondary immune response*.

Every memory B cell can monitor some types of antigens, so immunologists have suggested that a finite number of antibodies could protect us from infinite antigens. If we could imagine that memory B cells have different shapes corresponding to their ability to eliminate different antigens, The B cells construct the *shape space* that encompass (could respond to) all the possible antigens.

[10] The name comes from the fact that the mutation happened inside of the body and the mutation rate is higher than that of ordinary mutation.

9.2.2 Clonal Selection Algorithm

De Castro and Von Zuben utilized some parts of the above clonal selection principle and proposed the *clonal selection algorithm* (CSA) in 2002 [4]. CSA can be used in both unsupervised learning and optimization problems. We will only introduce the optimization version of CSA here.[11]

Before illustrating the CSA, it is better to explain the terms used in AIS and compare them to those of GA, as in Table 9.1.

Table 9.1 Comparison of the corresponding terms in AIS and GA

Terms in AIS	Corresponding terms in GA
Antigen	Objective function / Patterns needing to be recognized / Training data
Antibody	Individual (solution)
Affinity measure	Fitness value
Repertoire	Population
Somatic hypermutation	Mutation
Shape space	Solution landscape
Concentration	Crowdedness
Immune response	Optimizing/ learning process
Memory cell	Elitism
Regular update of B-cells	Maintain diversity by introducing new individuals

As shown in Table 9.1, the antigen could be an objective function, patterns needing to be recognized, or training data for optimizing, unsupervised learning, and supervised learning problems.[12]

Let us consider the maximum problem $g(\mathbf{x})$. De Castro and Von Zuben use binary code to represent antibodies, i.e., $l = 22$ binary genes for one variable. The solution process of the optimization version, *clonal selection algorithm* (CLONALG), can be illustrated as follows:

[11] Interested readers are referred to [4] for learning problems.

[12] These terms were introduced in Sect. 1.1.

Optimization Version of the Clonal Selection Algorithm (CLONALG)

Phase 1: Initialization.

Step 1.1: Assign the parameters for CLONALG, such as repertoire *popsize*, stop criteria (such as *maxgen*), clone factor β, etc.

Step 1.2: Generate *popsize* uniformly distributed antibodies randomly to form the initial repertoire and evaluate their affinity measures, i.e., fitness values. $gen = 0$.

Phase 2: Main loop. Repeat the following steps until the stop criteria are satisfied (such as $gen > maxgen$).

Step 2.1: Every antibody generates $round\,(\beta \cdot popsize)$ identical clones. Here *round*( ) is the round function toward the nearest integer. Thus make the repertoire size after cloning, N_c, as follows. It is necessary to mention that optimization version of CLONALG does not represent the affinity measure dependent characteristics but the learning version of CLONALG does.

$$N_c = \sum_{i=1}^{popsize} round\,(\beta \cdot popsize) + popsize \tag{9.1}$$

Step 2.2: The clones perform somatic hypermutation, i.e., the mutation extent of the fitter clone is smaller, while the *popsize* original antibodies are kept unchanged. This step is also called an *affinity maturation process*.

Step 2.3: Determine the affinity measures of the mutants.

Step 2.4: Select the *popsize* antibodies with the highest affinity measures among N_c antibodies and discard the others. $gen = gen + 1$.

Phase 3: Submit the final *popsize* antibodies as the results of CLONALG.

Compared to the solution process of GA on page 22, we can summarize the following characteristics of CLONALG.

- CLONALG does not have a crossover operator.
- The clone step is the most distinctive property of CLONALG.
- There is no parameter p_{m} to control the probability of mutation for every gene, like in GA. Instead, every clone suffers from mutation to a different extent proportional to its affinity measure.
- CLONALG has an elitism mechanism, i.e., *popsize* original antibodies are kept unchanged.

According to the above analysis, CLONALG might be roughly regarded as a parallel version of $(1 + round\,(\beta \cdot popsize))$-ES with adaptive mutation control. De Castro and Von Zuben applied CLONALG to multimodal optimization and TSP and got satisfactory results.

9.2.3 Artificial Immune System for Multiobjective Optimization Problems

In 2005 Coello Coello and Cortés developed a multiobjective immune system algorithm (MISA) based on a clonal selection principle [5]. For convenience, we rewrite the constrained MOP model as follows:

$$\begin{aligned} \min\ & \{z_1 = f_1(\mathbf{x}), z_2 = f_2(\mathbf{x}), \cdots, z_m = f_m(\mathbf{x})\} \\ \text{s.t.}\ & g_i(\mathbf{x}) \leqslant 0, \quad i = 1, \cdots, q \\ & h_j(\mathbf{x}) = 0, \quad j = q+1, \cdots, k \\ & \mathbf{x} \in \mathbb{R}^n \end{aligned} \tag{9.2}$$

Coello Coello and Cortés used binary strings to represent antibodies and also adopted the archive to store the nondominated solutions found thus far. The nondominated antibodies found thus far is stored in the hyperboxes illustrated in Sect. 6.6.2.1. We know the squeeze factor of each hyperbox.

In order to coordinate the multiple objectives and constraints, Coello Coello and Cortés assign preferences in the current repertoire by appointing feasible Pareto (FP) antibodies as rank 1, feasible dominated (FD) antibodies as rank 2, infeasible Pareto (IP) antibodies as rank 3, and infeasible dominated (ID) as rank 4.[13]

The solution process of MISA can be illustrated as follows:

Multiobjective Immune System Algorithm

Phase 1: Initialization.

Step 1.1: Assign the parameters for MISA, such as repertoire size *popsize*, archive size *asize*, stop criteria (such as *maxgen*), grid number div_i for variable i in the archive, binary length l for each variable, etc.

Step 1.2: Generate *popsize* uniformly distributed antibodies randomly to form the initial repertoire and evaluate their affinity measures and feasibilities. $gen = 0$.

Step 1.3: The initial archive is empty.

Phase 2: Main loop. Repeat the following steps until the stop criteria are satisfied (such as $gen > maxgen$).

Step 2.1: Assign the rank in the current repertoire. Pareto dominance is determined only among the same class, i.e., a feasible antibody is only compared with other feasible antibodies for Pareto dominance.

Step 2.2: Select the "best" antibodies. If the FP antibodies number is more than 5% *popsize*, just select all rank 1 antibodies. Otherwise select some rank 2 antibodies that are dominated by less other feasible antibodies until the "best" antibodies' size reaches 5% *popsize*. If the number of all FP

[13] Similar to the idea in Sect. 6.9.

and FD antibodies is still less than 5% *popsize*, select some rank 3 antibodies with a lesser constraint violation. If they still do not reach 5% *popsize*, rank 4 antibodies, which are dominated by less other antibodies, are selected. The result of this step is at least 5% *popsize* of the relatively "best" individuals in the current repertoire.

Step 2.3: Update the archive using the "best" antibodies.

Substep 2.3.1: Determine the Pareto dominance relationship between the archive members and the "best" antibodies. The dominated antibodies cannot enter the archive.

Substep 2.3.2: If the archive is not full, insert all the nondominated "best" antibodies.

Substep 2.3.3: Otherwise, only the "best" nondominated antibodies, which do not belong to the most crowded hyperbox, are allowed to enter the archive. Whenever an antibody is inserted into the archive, randomly eliminate a member in the most crowded hyperbox.

Substep 2.3.4: Adaptively maintain the hyperbox using techniques discussed in Sect. 6.6.2.1. Calculate the average squeeze factor of the archive after all the "best" antibodies have been inserted or rejected.

Step 2.4: Clone the "best" antibodies with the help of the archive.

Substep 2.4.1: The total estimated clone number is $6 \times popsize$. The clones are initially evenly distributed among these "best" antibodies.

Substep 2.4.2: If the archive is not full, determine the Euclidean distances between these "best" antibodies and get the average distance of every "best" antibody and the overall average distance among them. If the average distance of one antibody is smaller than the overall average distance, which means it belongs to the more crowded region, its clone number is reduced by half. Conversely, if the average distance of one antibody is larger than the overall average distance, which means it belongs to the less crowded region, its clone number is increased by 50%. Those whose average distance is equal to the overall average distance are kept unchanged.

Substep 2.4.3: If the archive is full, there are three possible cases.

Substep 2.4.3.1: If the "best" antibody is rejected in substeps 2.3.1 and 2.3.2 above, its clone number is zero.

Substep 2.4.3.2: If the "best" antibody belongs to the less crowded hyperbox, i.e., its squeeze factor is smaller than the average squeeze factor, double its clone number.

Substep 2.4.3.3: If the "best" antibody belongs to the more crowded hyperbox, i.e., its squeeze factor is larger than the average squeeze factor, its clone number is reduced by half.

Step 2.5: Perform a uniform somatic hypermutation for the clones. FP antibodies will suffer from n bit-flip mutations, where n is the number of variables in Eq. 9.2. FD antibodies will suffer from $n+1$ bit-flip mutations. The number of mutations for IP and ID antibodies is $n+2$ and $n+3$, respectively.

Step 2.6: Perform nonuniform mutation for other "not-so-good" antibodies. Here "nonuniform" has the different meaning to the one introduced in Sect. 3.2.2.3. Coello Coello and Cortés designed a deterministic control for the number of bit-flip mutations in an antibody, denoted as n_m, as follows:

$$n_m = 0.6L + \frac{gen}{maxgen}\left(\frac{1}{L} - 0.6L\right)$$

where $L = nl$ is the total binary chain length.

Step 2.7: Combine the mutants of steps 2.5 and 2.6. Rank them and select *popsize* antibodies using techniques introduced in Step 2.2.

Phase 3: Submit the final *popsize* antibodies as the results of MISA.

If we can understand rank as the affinity measure in AIS, MISA has a tendency toward proportional proliferation and differentiation in steps 2.2–2.6. Figure 9.3 illustrates the evolving process in one generation of MISA.

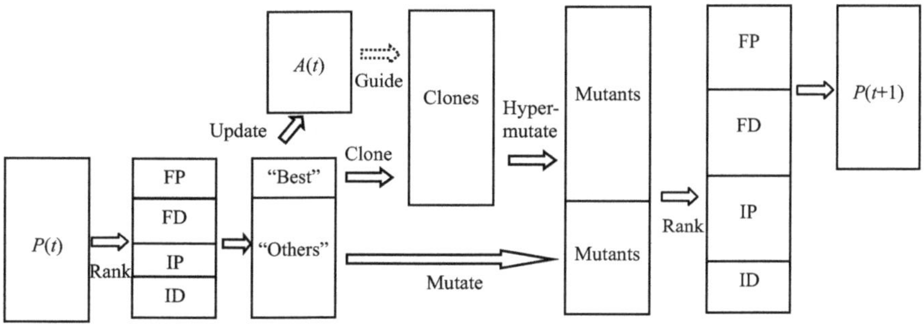

Fig. 9.3 The evolving process in one generation of MISA

It is not difficult to determine that MISA satisfies the three requirements (convergence, distribution, and elitism) in Sect. 6.4. So it possesses the basics of a good MOEA. Coello Coello and Cortés compared MISA with NSGA-II, PAES, and microGA[14] using performance indices *ER*, *SP*, and *GD*.[15] The comparison results suggested that MISA is "a viable alternative" to these MOEAs.

[14] Introduced in Sects. 6.5.1, 6.5.3, and 6.5.4, respectively.

[15] Introduced in Sect. 6.7.2.

9.3 Artificial Immune System Based on Immune Network

9.3.1 Immune Network Theory

In Sect. 9.2, we introduced the basic theory about the protective function of the human immune system and its metaphors for AIS. This and the next section will discuss the self-regulation of the immune system so that lymphatic cells can remember the invading antigens and avoid destroying the self cells.

After one type of antigen enters the human body, it causes an immune response and is eliminated by antibodies. Some of the matching B cells with a high affinity measure will go into a dormant state, waiting for the possibility of a second response. In a real immune system, antigens are sometimes remembered for long periods, in some cases equal to the lifespan of the human. why such a long time memory? Do the memory B cells have extra long life without being stimulated? Does bone marrow have the ability to generate the same memory B cells again and again? The *idiotypic network theory*,[16] suggested by Jerne in 1974 [6], gives us one way to explain this phenomenon.

As has been illustrated roughly in Fig. 9.2, every receptor, with its epitope, on B cell can also be matched by other receptors, with their paratopes. Without any antigen, those recognizing other B cells might be stimulated, proliferated, and differentiated, and those recognized by other B cells might be suppressed or even eliminated. Farmer *et al.* simplified Jerne's theory in 1986 [7] and Perelson further developed this theory in 1989 [8]. Since then, the *immune network theory* has became one of the most important principles in AIS.[17]

Suppose every antibody type has only one type of epitope and one type of idiotope, and these protein segments can be encoded by binary strings. We use l_e for the length of epitope, l_p for the length of paratope, and s for the threshold for stimulating proliferation and differentiation. Then we can compare the complementary matching situation for every two antibodies, i.e., the extent of antibody j matching antibody i, as follows:

$$m_{i,j} = \sum_k G\left(\sum_n e_i(n+k) \wedge p_j(n) - s + 1\right) \tag{9.3}$$

where n is for the comparing gene, k is for the left-shifting step, $p_j(n)$ is the nth gene of antibody j's paratope, $e_i(n+k)$ is the $(n+k)$th gene of antibody i's epitope, and $\wedge$ is the result of complementary matching.[18] $G(x) = x$ for $x > 0$ and $G(x) = 0$ otherwise. We can illustrate the simple example of $l_e = l_p = 6$ and $s = 3$ in Fig. 9.4. With

[16] Different antibody types may have epitopes in common. An epitope that is unique to a given antibody type is called an *idiotope*.

[17] The immune network theory is controversial among immunologists. However, the lack of acceptance of the immune network hypothesis by immunologists has not stopped AIS researchers from adapting it for their own use. The reason for this, we guess, lies in the fact that, as engineers and scientists, we like formulas, differential equations, and equilibrium points.

[18] $1 \wedge 0 = 1$, $0 \wedge 1 = 1$, $1 \wedge 1 = 0$, and $0 \wedge 0 = 0$.

these parameters and considering the definition of $G(x)$, we can easily determine that a possible value of k is $\{-3,-2,-1,0,1,2,3\}$.[19]

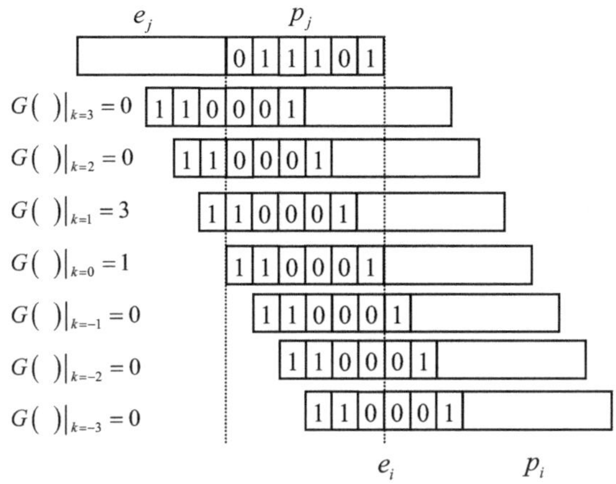

Fig. 9.4 Illustration of the antibody-antibody matching in an immune network. The total extent of antibody j matching antibody i is $m_{ij}=4$

Larger m_{ij} means that antibody j matches antibody i more and antibody j will be stimulated and antibody i will be suppressed. Farmer *et al.* suggested the following differential equation, i.e., the state function of state variable x_i, to illustrate the dynamic process among antibodies and antigens.

$$\frac{\mathrm{d}x_i}{\mathrm{d}t}=c\left(\sum_{j=1}^{N} m_{ji}x_ix_j-k_1\sum_{j=1}^{N} m_{ij}x_ix_j+\sum_{j=1}^{n} m_{ji}x_iy_j\right)-k_2x_i \tag{9.4}$$

where x_i is the concentration of antibody i, y_j is the concentration of antigen j. k_1, k_2, and c are all positive control parameters. According to Eq. 9.4, the number of antibody i will increase if it matches more other antibodies, is matched by fewer other antibodies, and matches more antigens.

Equation 9.4 suggests an explanation for the long life of memory B cells. If, under some conditions, the right part of Eq. 9.4 equals zero, the number of antibody i will be kept unchanged. That means antibody i will continue to exist in the dynamic stimulation and suppression environment.

Since the introduction of the immune network theory to AIS, many variants have been suggested and currently no one has been able to unify them into one framework. Some of them utilize both the clone selection principle and the immune network theory. In the following part of this section, we will introduce one type of continuous immune network and one type of discrete immune network. The general form of the dynamic process in the immune network can be illustrated as follows:

[19] $k<-3$ or $k>3$ will not provide $G()>0$.

$$\Delta x_i = (\text{Ab}\, i \text{ generated}) + (\text{Ab}\, i \text{ stimulated}) - (\text{Ab}\, i \text{ suppressed}) - (\text{Ab}\, i \text{ naturally died}) \tag{9.5}$$

9.3.2 Continuous Immune Network

In this subsection, we will introduce an interesting research field: *collaborative filtering* (CF) and how the immune network might contribute to CF.

CF is about giving predictions or recommendations according to other similar behaviors. For example, suppose that you are an Internet-based store owner. You have a lot of customers' profiles. If you could use one customer's historical consumption habits and those of others, who share similar tastes, to give appropriate purchase recommendations, your sales volume might increase significantly. These techniques are CF.

In 2002 Cayzer and Aickelin suggested a movie score prediction and recommendation system based on the continuous immune network [9, 10]. Suppose we need to predict viewer u's voting score for L unseen movies with the help of u's previous voting scores for other movies and N other viewers' voting scores for these movies.

The most common way to do CF is through a correlation coefficient among different viewers. Every viewer's voting scores for M total movies can be encoded as in Fig. 9.5.

Locus	Film_1	Film_2	Film_3		Film_M
Allele	Score_1	Score_2	Score_3	...	Score_M

Fig. 9.5 The code for movie recommendation system

where score_i might be $\{0, 0.2, 0.4, 0.6, 0.8, 1.0, -\}$. A larger voting score value means more favorable for a given movie. $-$ is for unseen movies.

Then the similarity measure, i.e., correlation coefficient, of viewer u and v can be calculated as follows:

$$r_{uv} = \frac{\sum_{i=1}^{n} (u_i - \bar{u})(v_i - \bar{v})}{\sqrt{\sum_{i=1}^{n} (u_i - \bar{u})^2 \sum_{i=1}^{n} (v_i - \bar{v})^2}} \tag{9.6}$$

where n is the number of movies both u and v have voted on, u_i and v_i are the voting scores of u and v for movie i, respectively, and $\bar{u}$ and $\bar{v}$ are the average voting scores of u and v for these n movies. The correlation coefficient between u and v defines their extent of linear correlation, i.e., the extent to which u and v are linearly correlated. The largest value $r = 1$ means that u and v share the very same taste in movies. The smallest value $r = -1$ means that u and v have complete opposite tastes in movies. If $r = 0$, then u and v have no common interests in movies. Obviously, in the N other viewers, we should prefer those whose similarity measures with u are large (absolute value).

After we get all the similarity measures for the N other viewers, the k ones with the highest absolute similarity measures are selected as the neighbors of u to predict the voting scores of the L unseen movies for u.[20] And then the predicted voting score of u for movie l is as follows:

$$p_l = \bar{u} + \frac{\sum\limits_{v \in \text{neighbor}} r_{uv} \left(v_l - \bar{v}\right)}{\sum\limits_{v \in neighbor} r_{uv}} \tag{9.7}$$

Equation 9.7 is a weighted average of k neighbors. The one with a larger correlation coefficient has more influence on the deviation from the average. In this way, the opinions of those with opposite tastes in movies with u can also be taken into consideration. After we get the L predicted voting scores for u, we can give recommendations for u from higher predicted voting scores.

In Cayzer and Aickelin' immune network, there is one antigen, i.e., u, and N antibodies, i.e., other viewers. After getting all the similarity measures for u by Eq. 9.6, we can use an immune network to generate neighbors.

We define the concentration of antibody i and the antigen as x_i and y, respectively. Then the state function for antibody i in the immune network can be modified, from Eq. 9.4, as follows:

$$\begin{aligned} \frac{\mathrm{d}x_i}{\mathrm{d}t} &= (\text{Ag stimulation}) - (\text{Ab suppression}) - (\text{death rate}) \\ &= k_1 m_i x_i y - \frac{k_2}{N} \sum_{j=1}^{N} m_{ij} x_i x_j - k_3 x_i \end{aligned} \tag{9.8}$$

where $m_{ij} = |r_{ij}|$, $m_i = |r_{ui}|$. k_1, k_2, k_3 are positive control factors representing stimulation, suppression, and death rate, respectively.[21] $\sum\limits_{j=1}^{N} m_{ij} x_i x_j / N$ is the average absolute similarity measure of antibody i. The larger, the more crowded. Equation 9.8 considers the suppression effect between antibodies, i.e., viewers with higher similarity measures to u and lower average absolute similarity measures are preferred.

The initial concentration of antibodies is set at 10. Then they evolve according to N differential equations. If the concentration value of one antibody is smaller than a predefined threshold, it is discarded from the immune network. The evolving process stops until the size of the immune network does not change for the succes-

---

[20] For movie l, we only select k highest absolute similarity measures among those who have seen l.

[21] Cayzer and Aickelin fixed $k_3 = 0.1$ and determined the best value for $k_1 = 0.5$ and $k_2 = 0.3$ with numerical experiments.

sive ten iterations, i.e., the immune network converges. Then these antibodies whose concentration values are larger than a threshold are selected as the neighbors of u.

We can imagine that u is a point in 3-D space and the neighbors found by the immune network might be distributed evenly within the ball surrounding u. The suppression part of Eq. 9.8 requires that these neighbors not be too close to each other, and so may contribute a more accurate prediction and recommendation.

The prediction equation could also consider the influence of the concentration value as follows, i.e., an antibody with a larger concentration value has more right to vote.

$$p_l = \bar{u} + \frac{\sum_{v \in \text{neighbor}} x_v r_{uv} (v_l - \bar{v})}{\sum_{v \in \text{neighbor}} x_v r_{uv}} \tag{9.9}$$

where x_v is the concentration value of neighbor v when the immune network stops evolving.

By comparing the immune-network-based CF and traditional CF, Cayzer and Aickelin concluded that these two methods have similar prediction accuracies, while the former has a better average accuracy of recommendations.

9.3.3 Discrete Immune Network

In this subsection, we will introduce an example of a discrete immune network for unsupervised learning (clustering). We talked about clustering several times in previous chapters. The most common methods are k-means clustering and hierarchical clustering [11–15]. These techniques classify data into groups according to the original data, which make the pattern characteristics in clusters hard to reveal. Also, users are generally required to assign the cluster number as a predefined parameter for these approaches.

Resource Limited Artificial Immune System

In 2000 Timmis *et al.* proposed an artificial immune network for data analysis [16] and then, in 2001, developed it into the *resource limited artificial immune system* (RLAIS) [17]. Other more recent reports are still available [18, 19]. These methods vary in their description. We will only introduce the RLAIS here [17], which uses real code representation.

In RLAIS, antibodies are not encoded and evolved directly. The entity to be evolved is the *artificial recognition ball* (ARB). ARBs form the immune network. Each ARB can represent a certain number of antibodies according to its stimulation level. Higher stimulated ARBs can acquire more resources, i.e., antibodies. But the total resource is limited, which means the requirement of some ARBs cannot be fulfilled. This situation causes lower stimulated ARBs to be removed from the repertoire. The concept of ARBs skilfully solves the problem of data compression, i.e., using a small number of ARBs to represent a large number of antigens, i.e., data needing to be clustered.

Before clustering, the data are normalized so that the Euclidean distances between every two points are within the scale $[0,1]$. The solution process of RLAIS can be illustrated as follows:

Resource Limited Artificial Immune System

Phase 1: Initialization.

Step 1.1: Assign the parameters for RLAIS, such as total resource number *mb*, clone and resource distribution parameter *k*, stop criteria (such as *maxgen*), mutation probability p_{m}, network affinity threshold *NAT*, etc.

Step 1.2: Randomly copy several distinct data into the immune network as initial ARBs to form the initial repertoire. Calculate their Euclidean distances. If the distance between two ARBs is smaller than *NAT*, a *link* is set up between them.

Phase 2: Main loop. Repeatedly train the immune network with one datum taken from the original data set, which are denoted as antigens, until the stop criteria are satisfied (such as $gen > maxgen$ or the size, connection, and location of ARBs have not changed for several generations). Every original datum, except the ones taken as the initial ARBs, could either be used once or more than once, depending on users' intention.

Step 2.1: Calculate the stimulation level for ARB j with the following consideration:

$$sl_j = \sum_{k=1}^{a} (1 - d_{Ag}(k,j)) + \sum_{i=1}^{n} (1 - d(i,j)) - \sum_{i=1}^{n} d(i,j) \tag{9.10}$$

where a is the number of antigens ARB j has been exposed to, $d_{Ag}(k,j)$ is the Euclidean distance between antigen k and ARB j, n is the number of links to ARB j, and $d(i,j)$ is the Euclidean distance between linked ARB i and ARB j. The first, second, and third parts of Eq. 9.10 represent, respectively, the primary stimulation, caused by antigen, second stimulation, caused by other ARBs, and suppression, caused by other ARBs. Equation 9.10 suggests that the ARBs that are close to an antigen and are closely linked get a high stimulation level.

Step 2.2: Clone every ARB to identical $k \times sl$ copies, where sl is its stimulation level. This step represents the proportional clone property of the clonal selection.

Step 2.3: Every clone undergoes mutation with probability p_{m}. Mutants are incorporated into the immune network as ARBs if their distances to the closest ARB are less than *NAT*. Then a *link* is set up between the mutant and its closest neighbor ARB. Those who do not mutate are discarded because the immune network already has this pattern.

Step 2.4: Recalculate the stimulation level for every ARB using Eq. 9.10.

Step 2.5: Distribute the resources, i.e., antibodies, according to the stimulation level. ARB j can apply for $k \times sl_j^2$ resources. But we only have mb resources for distribution. A pragmatic way to solve this is to fulfill the requirement of the ARB with the highest stimulation level, then the second highest stimulation level . . . until all the resources are distributed, i.e., the one with high stimulation level has more power to vote. This step is very critical for RLAIS because limited resources is used to restrict the number of the ARB and encourage those ARBs that are naturally closely existed in the AIS. Those ARBs without resource are discarded.

Phase 3: The ARBs with links represent a cluster. Then we can easily find out how many clusters are in the original data set and use the Euclidean distance between the data and the ARBs to determine which datum belongs to which cluster.

Among these control parameters, *NAT* is a very critical factor affecting the clustering results.[22] Timmis and Neal test RLAIS with "Fisher iris data set" to verify its capability.[23]

The last thing that needs to be mentioned is that the original complementary match, introduced by Eq. 9.3 and Fig. 9.4, is implemented inversely in some artificial immune networks. Equation 9.3 means roughly that a high stimulation level equals a large Hamming distance between two binary chains. But in Eq. 9.6, a high stimulation level, i.e., large m_{ij}, means that two viewers have the same/opposite taste in movies, i.e., they are close to each other. In Eq. 9.10, the stimulation level between ARBs, i.e., $(1 - d(i, j))$, is also inverse to distance.

9.4 Artificial Immune System Based on Negative Selection

T cells are a type of lymphatic cells that have the most important role in cell-mediated immunity. As was introduced in Sect. 9.1, TC cells that match the antigen are then proliferated and differentiated to kill the pathogenic substances. Anti-

[22] Small *NAT* values might correspond to large cluster numbers

[23] Asuncion and Newman maintain a Web site "Machine Learning Repository" containing hundreds of machine learning data sets at the University of California at Irvine (`http://archive.ics.uci.edu/ml/`). It might be the most cited benchmark problem source in the field of data mining, machine learning, or pattern recognition. Ultsch also provides a *fundamental clustering problems suite* via `http://www.uni-marburg.de/fb12/datenbionik/data?language_sync=1`.

gens are also proteins with gene structures. How does one differentiate human self proteins from the pathogenic nonself proteins for T cells? The negative selection principle gives an explanation.

After being generated by bone marrow, T cells need to undergo the maturation process in the thymus gland; where newly generated T cells meet various kinds (about 10^5) of self proteins. Those whose receptors complementarily match the self proteins are discarded. Other "matured" T cells are allowed to enter the lymphatic system. Then any matching between the receptors of T cells and the proteins will be recognized as foreign substances and thus stimulate an immune response. This maturation process is called *negative selection* (NS).

If we consider computer files and normal computer network communications as self cells, and changed files and abnormal communications as nonself cells. It is straightforward to use the NS principle in file protection and intrusion detection, which are the topics in the following two subsections.[24]

9.4.1 File Protection by Negative Selection

In 1994 Forrest *et al.* used NS in file protection events, caused by viruses or other accidents [20]. They gave a framework for applying NS to computer security that is still used by related researchers.

The protection requires two steps: training, similar to the maturation process inside the thymus gland, and protecting, similar to the immune response inside the body. These two steps can be illustrated in Fig. 9.6.

Let us take a simple descriptive example to illustrate these steps. Suppose we want to protect a very simple binary coded file (1 0 1 1 1 0 0 1 0 0 1 1 0 0 0 0). We divide this 16-bit file into 4 segments artificially, i.e., (1 0 1 1), (1 0 0 1), (0 0 1 1), and (0 0 0 0). These segments contain the self sting set (S).

We then randomly generate 4-bit strings to match S. As for the concept of "match," we need to define a threshold. If r contiguous bits are the same for the two strings being compared, we call them "matched".[25] For example, (1 0 1 1) matches (1 1 1 1) for $r = 2$ but does not match for $r \geq 3$.

Due to the simplicity of the example, we can enumerate all 4-bit strings for the training step, illustrated in Fig. 9.6a, and finally get the detector set (R) as (0 1 1 0) and (1 1 1 0) for $r = 2$. This R is like the matured T cells in the human immune system, which cannot match self proteins (S).

Then we enter the protecting step where the protected self strings are regularly compared to R. Whenever a match is detected, that means that R has found a modification to the protected file. Then an alarm can be generated for users.

---

[24] It should be mentioned here that NS-based file protection and intrusion detection is an alternative solution for the problem, not a substitute for others.

[25] We need to mention that the original complementary match in the human body is implemented inversely here.

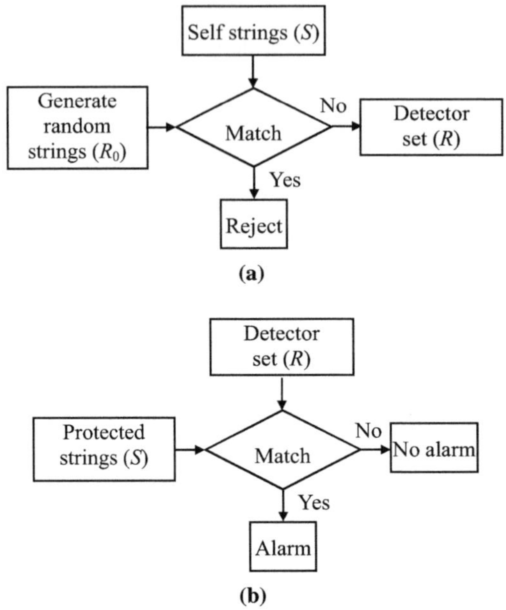

Fig. 9.6 Steps of NS-based file protection: (a) training, and (b) protecting

We can examine the effectiveness of NS-based file protection by single bit-flip mutations on S. In all 16 possible mutations, our R could match 6.[26]

Perhaps the reader is disappointed at the protection results of the example because we can only monitor 37.5% of file changes. We can explain it in several ways. Firstly, bit-flip mutation is a very strict condition for file protection. Generally, file changes, due to a virus or other reasons, will cause a larger number of bit changes, which will alleviate the problem. Another reason lies in the fact that the file size is to small. Forrest *et al.* tested a file with 512 bytes. They cut it into 128 self 4-byte strings S and got 46 4-byte strings in R using the training steps in Fig. 9.6a. Then they generated random changes in S, over 1000 trials, and the 46 detectors could recognize 84.3% of them ($r = 8$). Apart from that, viruses generally attach their codes at the end of a file, which makes NS-based file protection easier. Forrest *et al.* tested a virus that modified the first 5 bytes of a file and appended 300 bytes at the end of the file, with a small COM file with 2620 bytes. The NS-based system could detect the virus with probability 98% using ten detectors ($r = 10$).

[26] The matching detail is left as an exercise.

9.4.2 Intrusion Detection by Negative Selection

Intrusion detection in a computer network is an extremely hot yet difficult problem because there are many types of intrusions and many parameters to monitor. In 2002 Dasgupta and Gonzalez utilized NS-based GA to detect abnormal communications [21]. Due to space constraints, we only introduce the essence of their method and referr interested readers to the original paper.

Suppose we want to detect the intrusion with two parameters: number of bytes per second (S1) and number of ICMP packets per second (S2).[27] We need samples of normal communication and can illustrate these self samples on the 2-D space of S1-S2, like the points in Fig. 9.7, where we have already normalized the data in the scale $[0, 1]$. The dotted line forms the self region we considered as normal situations, i.e., like the self cells in the human body.

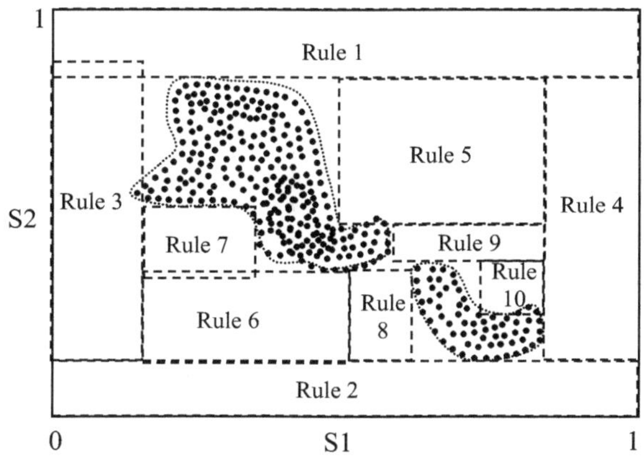

Fig. 9.7 illustrative example of NS-based intrusion detection

What we need to do is generate a set of rules to express abnormal network communications. One such rule can be illustrated as follows:

IF S1 is within [0,1] AND S2 is within [0.8,1], THEN nonself.

With these rules, we can monitor the network, and whenever a rule is activated, the algorithm will notify the administrator for further confirmation. One possible rule set for intrusion detection is illustrated by the dashed line forming rectangles in Fig. 9.7.

As can be seen from Fig. 9.7, due to the irregularity of the self data set, some of these rules might contain self samples, i.e., diagnose the normal communication as abnormal, and some of the nonself regions are not covered, i.e., no response to abnormal communications. And some rules might have duplicated covering regions.

[27] The specific meanings of these two parameters are irrelevant to the following discussion.

So Dasgupta and Gonzalez first transformed the intrusion detection problem into a detection rule generation problem and then transformed the latter problem into an optimization problem.

We actually have several objectives. Firstly, we want fewer rules. This requirement is straightforward and intuitive for fast response and small storage requirements. So we want a rule to cover as large an area as possible. Also, we want a less duplicated area between rules for the same reason. The last, but not least, objective is that we want the normal samples to be covered by the rules as little as possible. How do we effectively solve the multiple-objective problem?

Dasgupta and Gonzalez skillfully use the sequential niche, introduced in Sect. 5.2, to address the second requirement and combine the first and the third ones.

Actually, we only need to evolve the "IF" part of a rule because all the "THEN" parts are "nonself." So the representation scheme is clear. For the above example rule, its chromosome could be $(0,1,0.8,1)$.[28]

The first objective can be illustrated as follows:

$$volume(R) = \prod_{i=1}^{n} (high_i - low_i) \tag{9.11}$$

where R is a rule, i.e., a chromosome, n is the dimension of the space, i.e., $n = 2$ in our example, and $high_i$ and low_i are the higher and lower bounds of dimension i, respectively. Then $high_i - low_i$ is the (hyper-)rectangle's edge length in dimension i. So Eq. 9.11 determines the (hyper-)area of the (hyper-)rectangle, i.e., the covering area of that rule. Surely the larger the better.

The third objective can be illustrated as follows:

$$error\_number(R) = \left\{ \mathbf{x}^i \in S \,\middle|\, \mathbf{x}^i \in R \right\} \tag{9.12}$$

where $\mathbf{x}^i$ is the ith normal sample for training, S is the normal communication set, and R is a rule. Equation 9.12 counts the number of normal samples covered by the rule.[29] The smaller the better.

The raw fitness of individual R can be illustrated as follows:

$$raw\_fitness(R) = volume(R) - C \cdot error\_number(R) \tag{9.13}$$

where C is a user-defined parameter for evaluating the penalizing effects of covering the normal samples. The larger $raw\_fitness$, the better.

Then we run the GA for the first time. It will generate a rule, just like rule 1 in Fig. 9.7. Then we run the GA again and again to find other rules. How does one avoid finding the same rule again?

In a sequential niche, we need to modify the objective function after every run of the GA. Dasgupta and Gonzalez suggested the following modification:

---

[28] The first two alleles are the lower and higher bounds of the dimension 1, and likewise the next two alleles for dimension 2.

[29] This is where the NS principle is applied. Those rules that cover more normal data will be discarded; similarly those T cells that match more self proteins will be discarded.

$$fitness\,(R) = raw\_fitness\,(R) - \sum_{R^j \in \, rule\ set} volume\left(R \cap R^j\right) \tag{9.14}$$

where R^j is the previously found jth rule and $volume\left(R \cap R^j\right)$ is the common (hyper-)area shared by rule R and rule R^j. This simple yet powerful expression will achieve the second objective.[30]

From the second run of the GA, we use Eq. 9.14 as the fitness objective until the stop criteria, such as the maximum number of rules or maximum number of GA runs to generate a rule, are satisfied.

Dasgupta and Gonzalez compared their NS-based intrusion detection with the positive-characterization-based method[31] and found that they have a similar accuracy but the former requires much less memory.

For the benchmark problem in intrusion detection, MIT's Lincoln Laboratory, in collaboration with the Defense Advanced Research Projects Agency and the Air Force Research Laboratory, maintains a Web site of several intrusion detection data sets. [32]

9.5 Summary

AIS is a system, not an algorithm, so it has broader content than other chapters. We'd like to repeat that it is not our concern to reproduce immune phenomena by AIS, but to show that immune concepts can be used to develop powerful computational tools for data processing and optimization.

To some extent, we can regard the immune response process in our body as the fast version of a natural evolving process. The fitter lymphatic cells with similarity measures close to those of nonself cells and different from those of self cells will be proliferated and differentiated.

Clonal selection might be the most often used immune principle in AIS. It can be utilized in both optimizing and learning problems. We only focus on optimization in Sect. 9.2 owing to space constraints.

Immune networks are fascinating, even though the concept is disputed among immunologists. Thus there are many application variations of this principle to unsupervised learning.

NS is simple and has thoughts directly applicable to discrimination problems that are not limited to file protection and intrusion detections discussed in Sect. 9.4. Any situation with the requirement of discriminating normal from abnormal, i.e., self from nonself, might utilize these concepts.

[30] Here you might understand why in this textbook we introduce so many different types of considerations for one kind of problem.

[31] If the shortest distance of a real-time sample to the trained normal samples is less than the predefined threshold, it is an abnormal communication.

[32] http://www.ll.mit.edu/mission/communications/ist/corpora/ideval/index.html

After reading this chapter, you should be familiar with applications of ideas about the human immune response process to AIS design, understand the intuitive considerations of clonal selection for optimization, immune network for clustering, and NS for nonself recognition, and enjoy the various successful applications of these simple ideas.

Suggestions for Further Reading

Since the introduction of AIS, many researchers have published monographs and textbooks on the topic. Interested readers are referred to Dasgupta's book from 1998 [22], de Castro and Timmis's book from 2002 [3], Tarakanov *et al.*'s book from 2003 [23], and Ishida's book in 2004 [24]. A tutorial on AIS was given by Dasgupta in 2006 [25].

There is an annual conference on AIS, the International Conference on Artificial Immune Systems, and its proceedings are published by Springer. Aitkin *et al.* maintain a Web site containing plenty of information on AIS.[33]

In 2005 Garrett gave the criteria "distinctiveness" and "effectiveness" for evaluating AIS and surveyed the ordinary AIS models [26]. Timmis and colleagues recently published several review papers [27–29].

The AIS introduced in this chapter only contains mutation operators. Bersini incorporated chemical crossover operator into AIS in 2002 [30].

We only introduced the optimization version of CLONALG in Sect. 9.2.2. Those interested in exploratory data analysis are referred to the papers published by Wu and Fyfe in 2008 [31] and Do *et al.* in 2009 [32], in addition to [4]. In 2008 Lau *et al.* provided a parallel immune optimization algorithm [33].

In 1997, entropy-based concentration was suggested and applied by Chun *et al.* to AIS to further improve its optimization ability [34]. CLONALG has been successfully applied in protein structure prediction, e.g., work done by Anile *et al.* and Cutello in 2007 [35, 36], and in the optimal computation of finite field exponentiation, e.g., work done by Cruz-Cortés *et al.* in 2008 [37].

Immune-network-based learning is an attractive research field. Apart from the two examples introduced in Sect. 9.3, in 2001 de Castro and Zuben also included the stimulation and suppression idea among antibodies into CLONALG and suggested the artificial immune network for data analysis (aiNet) [38]. Like CLONALG, aiNet also has a learning version for clustering [39] and an optimization version for multimodal optimization [40, 41]. In 2004 a branch RLAIS, introduced in Sect. 9.3.3, for supervised learning was published by Watkins and Timmis [42]. For further discussions on immune networks, readers can refer to Ishida's previous work on active noise control and diagnosis from 1996 and 1997, respectively [43, 44].

For those interested in intrusion detection, we suggest two surveys published by Aickelin *et al.* in 2004 [45] and by Kim *et al.* in 2007 [46]. Apart from the theory

[33] `http://www.artificial-immune-systems.org/`.

introduced in Sect. 9.4.2, Hofmeyr and Forrest, Harmer *et al.*, and Kim and Bentley used negative selection principle in 2000, 2002, and 2002, respectively [47–49]. In 2003 Aickelin *et al.* applied danger theory in intrusion detection [50]. Esponda *et al.* provided a formal framework for positive and negative detection schemes, which is quite useful for intrusion detection [51].

As to the NS principle, Dasgupta and Forrest gave an industrial application example in 1999 [52]. In 2007 Ji and Dasgupta published an intensive survey [53].

We only introduce three immune methods and their applications in learning and optimization problems. In 2008 Lay and Bate used innate immune techniques for improving the reliability of real-time embedded systems [54]. In 2009 Berg used adaptive cellular immunity for designing AIS [55].

The must-read papers of this chapter are [28] for general introduction, [4] for clonal selection, [7] for immune networks, and [20] for NS.

Exercises and Potential Research Projects

9.1. Summarize the essence of aiNet [38] on a single sheet of paper and compare it with $(\mu + \lambda)$-ES, introduced in Sect. 2.3.1, for optimization.

9.2. Compare qualitatively aiNet, [38], and RLAIS, introduced in Sect. 9.3.3, for unsupervised learning, e.g., clustering.

9.3. Using qualitative and quantitative comparisons for multimodal optimization among standard fitness sharing algorithms, introduced in Sect. 5.3.1, CLONALG, introduced in Sect. 9.2.2, and aiNet [40, 41], use the benchmark problems in [4], PI in Sect. 5.5, and the techniques introduced in Sect. 3.6 to do a fair comparison and draw statistical conclusions according to your numerical experiments.

9.4. Summarize the global search mechanism, convergence mechanism, and uphill mechanism for minimum optimization problems in CLONALG.

9.5. List the six matches in the example of bit-flip mutation in Sect. 9.4.1.

9.6. Summarize [48] on a single sheet of paper and compare its methods with those of [21], introduced in Sect. 9.4.2.

References

1. Kindt TJ, Osborne BA, Goldsby RA (2006) Kuby immunology, 6th edn. Freeman, New York
2. Murphy KM, Travers P, Walport M (2007) Immunobiology: the immune system, 7th edn. Garland Science, London, UK
3. de Castro LN, Timmis J (2002) Artificial immune systems: a new computational intelligence approach. Springer, Berlin Heidelberg New York

4. de Castro L, Zuben FV (2002) Learning and optimization using the clonal selection principle. IEEE Trans Evol Comput 6(3):239–251
5. Coello Coello CA, Cortés NC (2005) Solving multiobjective optimization problems using an artificial immune system. Genet Programm Evolvable Mach 6(2):163–190
6. Jerne NK (1974) Towards a network theory of the immune system. Ann Immunol 125C(1–2):373–389
7. Farmer JD, Packard NH, Perelson AS (1986) The immune system, adaptation, and machine learning. Phys D 2(1–3):187–204
8. Perelson A (1989) Immune network theory. Immunol Rev 110(1):5–36
9. Cayzer S, Aickelin U (2002) A recommender system based on the immune network. In: Proceedings of the congress on evolutionary computation, pp 807–813
10. Cayzer S, Aickelin U (2005) A recommender system based on idiotypic artificial immune networks. J Math Modell Algorithms 4(2):181–198
11. Duda RO, Hart PE, Stork DG (2000) Pattern classification, 2nd edn. Wiley-Interscience, New York
12. Alpaydin E (2004) Introduction to machine learning. MIT Press, Cambridge, MA
13. Bishop CM (2007) Pattern recognition and machine learning. Springer, Berlin Heidelberg New York
14. Xu R, Wunsch D (2008) Clustering. Wiley-IEEE, New York
15. Xu R, Wunsch D (2005) Survey of clustering algorithms. IEEE Trans Neural Netw 16(3):645–678
16. Timmis J, Neal M, Hunt J (2000) An artificial immune system for data analysis. Biosystems 55(1–3):143–150
17. Timmis J, Neal M (2001) A resource limited artificial immune system for data analysis. Knowledge-Based Syst 14(3-4):121–130
18. Neal M (2002) An artificial immune system for continuous analysis of time-varying data. In: Proceedings of the 1st international conference on artificial immune systems, pp 76–85
19. Neal M (2003) Meta-stable memory in an artificial immune network. In: Proceedings of the 2nd international conference on artificial immune systems, pp 168–180
20. Forrest S, Perelson A, Allen L *et al* (1994) Self-nonself discrimination in a computer. In: Proceedings of the IEEE computer society symposium on research in security and privacy, pp 202–212
21. Dasgupta D, Gonzalez F (2002) An immunity-based technique to characterize intrusions in computer networks. IEEE Trans Evol Comput 6(3):281–291
22. Dasgupta D (1998) Artificial immune systems and their applications. Springer, Berlin Heidelberg New York
23. Tarakanov AO, Skormin VA, Sokolova SP (2003) Immunocomputing: principles and applications. Springer, Berlin Heidelberg New York
24. Ishida Y (2004) Immunity-based systems. Springer, Berlin Heidelberg New York
25. Dasgupta D (2006) Advances in artificial immune systems. IEEE Comput Intell Mag 1(4):40–49
26. Garrett SM (2005) How do we evaluate artificial immune systems? Evol Comput 13(2):145–177
27. Freitas A, Timmis J (2007) Revisiting the foundations of artificial immune systems for data mining. IEEE Trans Evol Comput 11(4):521–540
28. Timmis J (2007) Artificial immune systems—today and tomorrow. Nat Comput 6(1):1–18
29. Timmis J, Andrews P, Owens N *et al* (2008) An interdisciplinary perspective on artificial immune systems. Evol Intell 1(1):5–26
30. Bersini H (2002) The immune and the chemical crossover. IEEE Trans Evol Comput 6(3):306–313
31. Wu Y, Fyfe C (2008) Exploratory data analysis with artificial immune systems. Evol Intell 1(2):159–169
32. Do TD, Hui SC, Fong A (2009) Associative classification with artificial immune system. IEEE Trans Evol Comput 13(2):217–228

33. Lau H, Tsang W (2008) A parallel immune optimization algorithm for numeric function optimization. Evol Intell 1(3):171–185
34. Chun J, Lim J, Jung H (1997) Optimal design of synchronous motor with parameter correction using immune algorithm. In: Proceedings of the IEEE international electric machines and drives conference, pp 610–615
35. Anile AM, Cutello V, Narzisi G *et al* (2007) Determination of protein structure and dynamics combining immune algorithms and pattern search methods. Nat Comput 6(1):55–72
36. Cutello V, Nicosia G, Pavone M *et al* (2007) An immune algorithm for protein structure prediction on lattice models. IEEE Trans Evol Comput 11(1):101–117
37. Cruz-Cortés N, Rodriuez-Henriquez F, Coello CC (2008) An artificial immune system heuristic for generating short addition chains. IEEE Trans Evol Comput 12(1):1–24
38. de Castro LN, Zuben FJV (2002) aiNet: an artificial immune network for data analysis. Abbass HA, Sarker RA, Newton cs (eds) Data mining: a heuristic approach. Idea Group Publishing, Hershey, PA, pp 231–260
39. de Casto LN, Zuben FV (2000) An evolutionary immune network for data clustering. In: Proceedings of the sixth Brazilian symposium on neural networks, pp 84–89
40. de Castro L, Timmis J (2002) An artificial immune network for multimodal function optimization. In: Proceedings of the IEEE congress on evolutionary computation, pp 699–704
41. Timmis J, Edmonds C (2004) A comment on Opt-AiNET: an immune network algorithm for optimisation. In: Proceedings of the genetic and evolutionary computation conference, pp 308–317
42. Watkins A, Timmis J, Boggess L (2004) Artificial immune recognition system (AIRS): an immune-inspired supervised learning algorithm. Genet Programm Evolvable Mach 5(3):291–317
43. Ishida Y, Adachi N (1996) Active noise control by an immune algorithm: adaptation in immune system as an evolution. In: Proceedings of the IEEE international conference on evolutionary computation, pp 150–153
44. Ishida Y (1997) Active diagnosis by self-organization: an approach by the immune network metaphor. In: Proceedings of international joint conferences on artificial intelligence, pp 1084–1091
45. Aickelin U, Greensmith J, Twycross J (2004) Immune system approaches to intrusion detection - a review. In: Proceedings of the 3rd international conference on artificial immune systems, pp 316–329
46. Kim J, Bentley PJ, Aickelin U *et al* (2007) Immune system approaches to intrusion detection — a review. Nat Comput 6(4):413–466
47. Hofmeyr SA, Forrest S (2000) Architecture for an artificial immune system. Evol Comput 8(4):443–473
48. Harmer P, Williams P, Gunsch G *et al* (2002) An artificial immune system architecture for computer security applications. IEEE Trans Evol Comput 6(3):252–280
49. Kim J, Bentley PJ (2002) Towards an artificial immune system for network intrusion detection: An investigation of dynamic clonal selection. In: Proceedings of the IEEE world congress on computational intelligence, pp 1244–1252
50. Aickelin U, Bentley P, Cayzer S *et al* (2003) Danger theory: the link between AIS and IDS? In: Proceedings of the 2nd international conference on artificial immune systems, pp 147–155
51. Esponda F, Forrest S, Helman P (2004) A formal framework for positive and negative detection schemes. IEEE Trans Syst Man Cybern B Cybern 34(1):357–373
52. Dasgupta D, Forrest S (1999) Artificial immune systems in industrial applications. In: Proceedings of the second international conference on intelligent processing and manufacturing of materials, pp 257–267
53. Ji Z, Dasgupta D (2007) Revisiting negative selection algorithms. Evol Comput 15(2):223–251
54. Lay N, Bate I (2008) Improving the reliability of real-time embedded systems using innate immune techniques. Evol Intell 1(2):113–132
55. van den Berg HA (2009) Design principles of adaptive cellular immunity for artificial immune systems. Soft Comput 13(11):1073–1080

Chapter 10
Genetic Programming

Abstract Genetic programming is a very famous branch of EAs. The departure point of genetic programming is to automatically generate functional programs in the computer, whose elementary form could be an algebraic expression, logic expression, or a small program fragment. This idea can be expanded to generate artificial intelligence by computer. The reason for a separate chapter, instead of integrating it into Chap. 2, lies in its special representation, evaluation, and variation methods caused by the requirement of representing structure in the chromosome.

10.1 Introduction to Genetic Programming

Genetic programming (GP) was suggested by Koza in 1992 as a tool for computers to solve problems automatically [1]. The intuitive idea of GP is to generate computer programs automatically by genetic operations.

10.1.1 The Difference Between Genetic Programming and Genetic Algorithms

In GAs, we are generally facing an optimization problem.[1] What we need to do is to find the optimal variable values so that their objective function has the maximum/minimum value. This kind of optimization problem is called *parameter optimization*, i.e., to find the optimal problem parameters of the problem.

But sometimes problems are more complicated. We face not only parameter optimization but also structural optimization. In designing an artificial neural network, we need to determine not only the thresholds for each nerve cell but also the number of the internal level nerve cells. In the example of designing a digital filter, we

[1] We often transform a learning problem into an optimization problem in GAs.

are interested in both finding the optimal parameters of the filter and determining the order of the filter. Generally such designing problems have an underlying requirement for structure optimization. But traditional GAs cannot handle structure optimization, i.e., GAs cannot express the structure of a solution in chromosomes.

The above discussion comes from the operations research community, i.e., mathematics point of view. Let us discuss the difference between GP and GAs from the viewpoint of artificial intelligence, i.e., computer science point of view. There are a lot of data. Some of them are the input of the system, others are outputs. The requirement is to set up a nonlinear relationship[2] between the input and the output so that this relationship can map correctly the input data to the output data. If so, we say the relationship has some "intelligence." In the future, if new data come, we can hope that this intelligent relationship will suggest the correct output, that is a forecast, classification, pattern recognition, clustering, etc. This kind of nonlinear relationship setup problem is called a *learning problem*, and the research field is called *data mining* or *knowledge discovery*. Although we can transform learning problems into optimization problems and use GAs to solve them, GP can handle them directly.

After understanding the above two demands for GP, we need to point out several things for learning GP using the basics of GAs, provided that you have grasped the previous chapters.[3]

- How does one represent structure in GP, i.e., what is the meaning of genes and chromosomes in GP?
- How does one generate randomly distributed individuals in the initialization procedure?
- How does one evaluate the fitness value of an individual in the environment of a learning problem?
- How does one design variation operators for GP chromosomes?

In the next subsection, we will use a curve fitting problem to illustrate the basic concepts of GP.

Since its inception, GP has been a hot research field. Figure 10.1 illustrates the number of papers indexed by the SCI on GP.[4]

10.1.2 Genetic Programming for Curve Fitting

Suppose we are given a set of numbers as in Table 10.1, i.e., there are 17 observation data with the input and the output of the system. What we need to do is to set up the nonlinear relationship between the input and the output so that next time we can

---

[2] This relationship can be regarded as a machine, algorithm, or even a brain.

[3] The solution process of GP is the same as the one introduced on page 77.

[4] TS = ("genetic programming"). The SCI index "TS" is for the search topic in the title, the keywords, and the abstract.

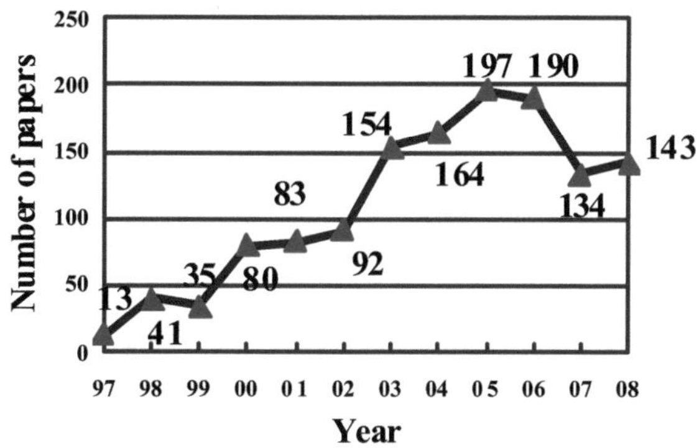

Fig. 10.1 Number of papers indexed by SCI on GP

feed back a "best fit" output with an input between, or out of, the range $[1,17]$.[5] This problem is called *curve fitting* or *symbolic regression* or *system identification*.[6]

Table 10.1 An example of curve fitting

Input	1	2	3	4	5	6
Output	1.841	2.324	1.873	1.243	1.277	2.170
Input	7	8	9	10	11	12
Output	3.303	3.818	3.412	2.618	2.317	2.928
Input	13	14	15	16	17	
Output	4.026	4.733	4.523	3.712	3.162	

The nonlinear relationship between the input and the output can be written as a nonlinear function between the input variable x and the output variable y. Actually, these observation data are generated with the function $y = f(x) = x^{0.5} + \sin(x)$, i.e., we need to identify the form and the parameter of function f.[7] The ideal function and its sampling points are illustrated by Fig. 10.2.

In the following part of this subsection, we will introduce the four points mentioned in Sect. 10.1.1 and omit the other details of GP.

[5] The so-called "best fit" can be understood as either exactly the same value or as close as possible. The former situation is called *interpolation* and the latter one is called *approximation*. If we want to forecast the value of the output outside the given range, this problem is called *extrapolation*.

[6] The first term comes from mathematics, the second one from computer science, and the last one from control theory.

[7] Here we regard "computer program" as an algebraic function expression. Other "programs," e.g., logic expressions, will be introduced in the following sections.

10.1.2.1 Syntax Tree Code

The key part of GP is the ability to represent a variable, e.g., x, and algebraic operators, e.g., $+$, $-$, $\times$, and $\div$, in genes. We can use a tree to represent an algebraic expression. For example, the ideal function to be identified, e.g., $y = f(x) = x^{0.5} + \sin(x)$, can be expressed by the tree in Fig. 10.3.[8]

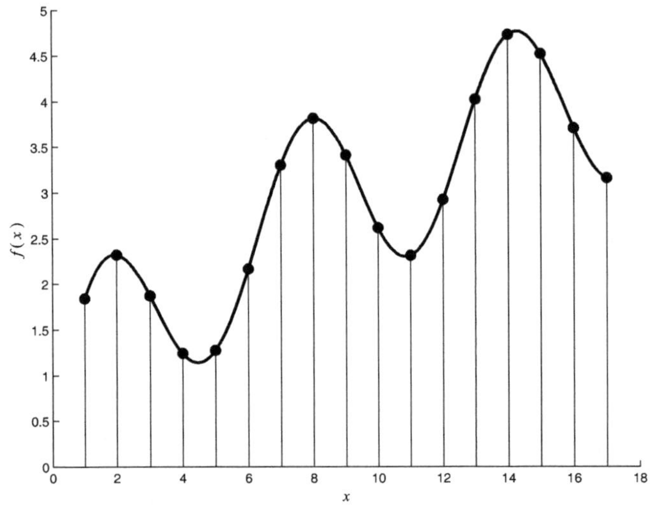

Fig. 10.2 The function and its sampling points for generating the data set of Table 10.1

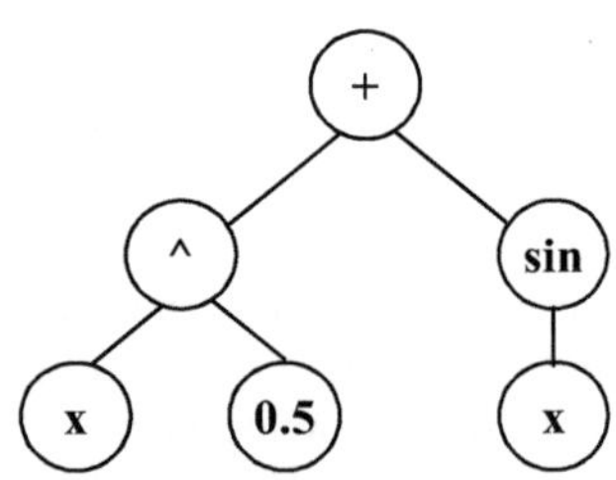

Fig. 10.3 Syntax tree representation for ideal function

The tree in Fig. 10.3 is called a *syntax tree*. Some programming languages, e.g., Lisp, can represent the expression in a prefix notation way, so that the syntax tree in Fig. 10.3 can be written as `(+(^(x 0.5) sin(x)))`, i.e., there is a bijective map between the syntax tree and its prefix expression. The real chromosome in the

[8] It is an upside down tree.

memory of the computer is `(+(^(x 0.5) sin(x)))`, which can be understood as Fig. 10.4.[9]

Fig. 10.4 The representation of a GP chromosome using the form of GAs

+	^	x	0.5	sin	x

A tree is a special form of graph. In a tree, every circle is called a *node*. The lines between nodes are called *edges* or *links*. The uppermost node is called the *root*. If the node has only one link but is not the root, it is called a *leaf*. Other nodes besides the root and leaves are called *internal nodes*.

As can be seen from Fig. 10.3, if we want to use a tree to represent an algebraic expression, internal nodes can be functions, variables, and constants, but leaves can only be variables and constants.[10]

10.1.2.2 Initialization

As can be seen from Fig. 10.4, the genes of GP could be variables, constants, functions, or something else. But they cannot be connected as you wish. The chromosomes need to have the legal corresponding syntax tree, i.e., they should have meanings. So randomly generating genes will likely produce illegal individuals.

In order to randomly generate a syntax tree from scratch, we need to know all the available functions, variables, and constants. This set is called a *primitive set*, which is predefined by users. In determining the primitive set, we need to make sure that the real solution of the problem could be expressed by the elements of the primitive set.[11]

Suppose we have defined the function set as sin, cos, tan, $\frac{1}{x}$, $+$, $-$, $\times$, $\div$, ^, square root, ln, *rand*(). Some of them require one variable while others require two.[12] The variable set is only x provided that we know in advance that this is a one-input and one-output system. The constant set is π and e.

[9] But for simplicity, in the following part of this section, we will only discuss the syntax tree.

[10] In GP, these variables and constants are called *terminals*, which means that the expression terminates here.

[11] In mathematics, given a set of $n+1$ distinct data of input and output, we can prove that there exists an $n$order polynomial that goes exactly through these points, i.e., interpolation. So if we only want to interpolate these given data, normal functions to generate polynomials are enough. But sometimes we want more than that, i.e., we want the forecast output to be as close as possible to the real output but abandon the requirement of the exact right output at the given data points, i.e., approximation. Then Fourier approximation and Chebyshev polynomial approximation are possible options.

[12] The *rand*() function does not require parameters and return a uniformly distributed random number in the range $(0,1)$. In the following part, for simplicity, we suppose that *rand*() can generate uniform random numbers between (a,b), where a and b are userdefined parameters. There are other functions that require three inputs, e.g., IF function in logic expression.

We can start the initialization procedure from the root. The root of an individual should be a function. After that, we know the variable number of that function, and then could randomly select elements from the primitive set. For every initialized node, if it's a function, we need to generate its variables (might be any element in the primitive set). If a generated node is a terminal, no follow-up nodes are necessary.

For example, in this way, we generated 4 individuals: $-4.3+1.21x$, $0.667e^{0.071x}$, $0.127+1.77\ln(x)$, and $1.242x^{0.76}$.[13] The syntax trees of them could be illustrated by Fig. 10.5.

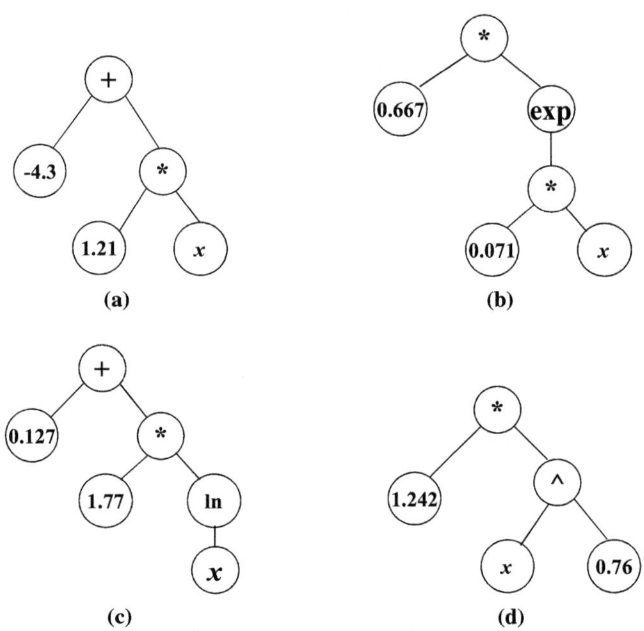

Fig. 10.5 Syntax tree of the four individuals after initialization: (a) individual 1, (b) individual 2, (c) individual 3, and (d) individual 4

The alert readers might have noticed that the above procedure could generate individuals with a lot of nodes if we always select function for node, which is not realistic for pragmatic applications. So special considerations should be taken to limit the node number of the initial individuals. We can define the *depth* of a node as the number of links from the root to reach that node and the depth of the tree as the maximum depth of its leaves. The depths of the four trees in Fig. 10.5 are 2, 3, 3, and 2, respectively. One pragmatic way to limit the node number is to predefine the maximum depth of the tree, denoted as D. Apart from that, the *size* of a tree is defined as its node number. The size of the four trees in Fig. 10.5 are 5, 6, 6, and 5, respectively. So another way is to define the maximum size.

[13] Suppose that $popsize = 4$ and the real numbers in these individuals are generated by $rand()$.

In order to increase the diversity of the initial individuals, Koza suggested three methods as follows:

1. **Full.** All leaves in the individual have the same depth D.
2. **Grow.** The depth of the individuals should be no more than D.
3. **Ramped half-and-half.** The depth of the initial individuals are randomly distributed between $[2, D]$. For example, $D = 5$, which means that $0.25 \times popsize$ individuals are generated with maximum depth $2, \ldots, 0.25 \times popsize$ individuals are generated with maximum depth 5.[14] In each maximum depth, the individual is generated by a full method or grow method with the same probability 0.5.

10.1.2.3 Evaluation

The initial four individuals and the original data are drawn in Fig. 10.6.

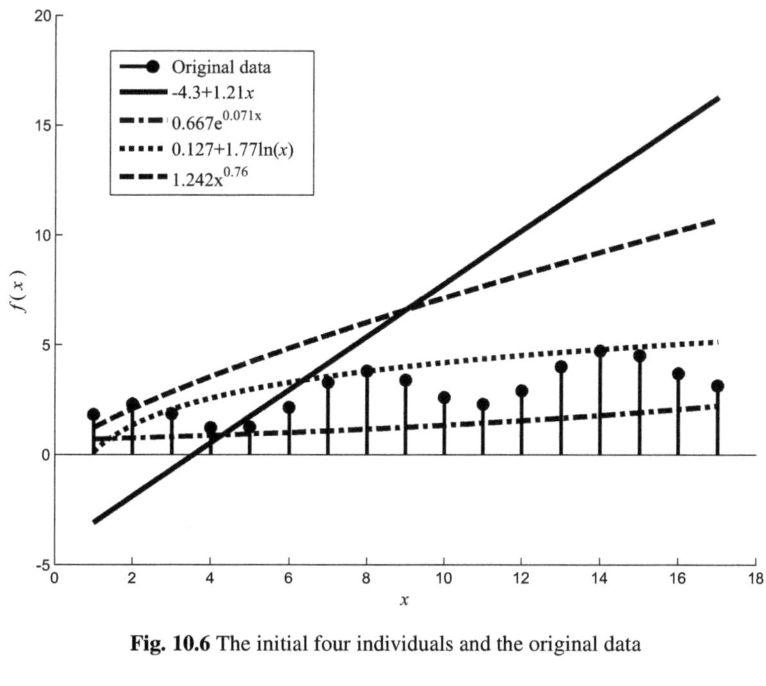

Fig. 10.6 The initial four individuals and the original data

As can be seen from Fig. 10.6, individuals 2 and 3 are closer to the original data compared to individuals 1 and 4. How does one evaluate the closeness quantitatively? As an option, we can use the *sum of the absolute errors* on the sampling points.

[14] Hence the meaning of ramped. Suppose we rank the new individuals with their depths. Then there is a ramp shape of the individuals' depths.

We sample 17 points in these individuals, i.e., curves, at $x = 1, \ldots, x = 17$, calculate their absolute errors to the original data at these points, and sum them up to represent the distance of these curves to the original data and thus assign these errors as their fitness values, i.e., Eq. 10.1

$$fitness_i = \sum_{k=1}^{17} |y(k) - \mathrm{ind}_i(k)| \tag{10.1}$$

where ind_i and $fitness_i$ are the curve value and the fitness value of individual i, respectively, and $y(k)$ is the original data at sampling point k. The smaller the error the fitter the individual. By this method, we can calculate the errors of these curves as 87.5140, 26.4728, 17.5729, and 60.9095, respectively, which corroborates the above qualitative guess for their fitness.

Here is another obvious difference between GP and GAs. In GAs, the fitness value of an individual can be obtained by only one run of the objective function. But here, we need to evaluate the individual function many times to get the sum of the absolute error, i.e., its fitness value. These evaluations are called *fitness cases*.

It is necessary to mention that although the constants in these four individuals are generated by a $rand()$ function in the initialization procedure, their values remain unchanged in the following genetic evolving processes, i.e., evaluation, selection, and variation operators. So these constants are called *ephemeral random constants*.[15]

After we get the fitness values of the initial population, selection can be processed with the techniques discussed in Chap. 3.

10.1.2.4 Variation Operators

The crossover operator and the mutation operator of GP are carried out with a tree representation. We only give a limited number of examples of the variation operators here.

Subtree Crossover

If two individuals need to perform crossover according to genetic rules, we can randomly pick one node in each individual and exchange their subtree. Thus two new individuals can be generated.

For example, individuals 1 and 4 need to perform crossover and the x node and the ^ node are selected, respectively. Figure 10.7 illustrates one offspring.[16]

---

[15] They are random numbers for a very short time, i.e., initialization, and then become constants.

[16] The other one is left as an exercise.

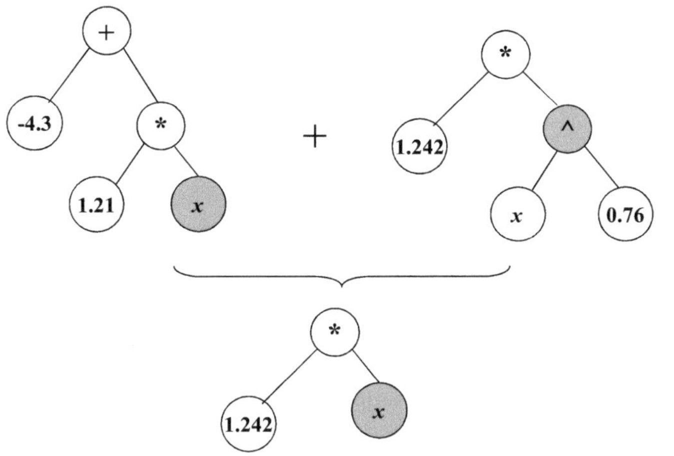

Fig. 10.7 Subtree crossover to generate one offspring

Point Mutation

If one individual requires mutation. We randomly pick one node and generate randomly another possible node to replace it. Then a mutant is generated.

For example, individual 4 needs to mutate and the ^ node is randomly selected. Because the ^ function requires two variables, we need to select a function with two variables randomly from the primitive set, e.g., $*$. Figure 10.8 illustrates the mutant.

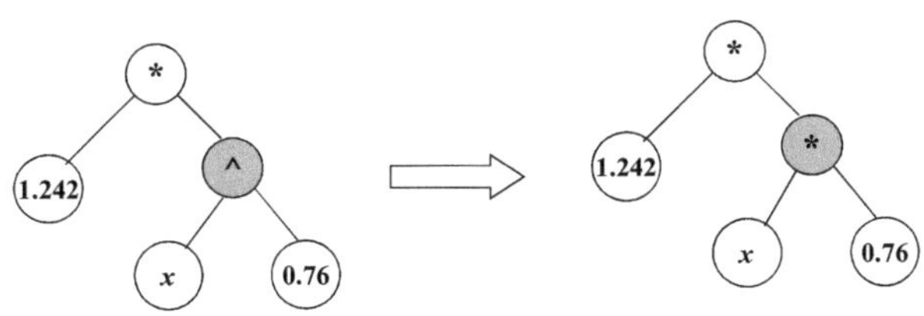

Fig. 10.8 Point mutation to generate mutant

Note: there should be some "safeguard" rules to ensure that the offspring and the mutant are still legal individuals. The violations might be divide-by-zero, negative variable in logarithmic function, etc.

10.2 Other Code Methods for Genetic Programming

The basic code method based on Lisp's prefix notation language has the following shortcomings.

- Chromosomes have variable length, which creates some inconveniences in designing variation operators.
- It is relatively hard to directly determine the crossover and mutation nodes according to a prefix expression. We need to transform it into syntax tree first.
- Simple crossover and mutation, e.g., single-point crossover and bit-flip mutation, might generate illegal offspring.

Several methods with fixed chromosome length have been proposed since the inception of GP that can undergo simple and standard variation operators. We just give two examples here.

10.2.1 Gene Expression Programming

Gene expression programming (GEP) was proposed by Ferreira in 2001 [2, 3]. The essence of GEP is to divide a chromosome into two parts: *head* and *tail*. The head has h genes, which are composed of functions and terminals, and the tail has t genes, which are only composed of terminals. h is the maximum number of internal nodes provided by the user. Then t is determined by the following equation:

$$t = h(n-1)+1 \tag{10.2}$$

where n is the maximum number of variables required by functions. Let us prove Eq. 10.2 by mathematical induction. For simplicity, we suppose that all the functions require n variables. Then if $h = 1$, i.e., only one root in an internal node set, the leaf number is $n = 1 \times (n-1)+1 = n$, which is equal to Eq. 10.2. Suppose the extreme condition that all the internal nodes are h functions and the leaf number is $t = h(n-1)+1$. The $h+1$ internal nodes means one leaf needs to be changed into a function with n variables. So the current leaf number is $t = h(n-1)+1-1+n = (h+1)(n-1)+1$,[17] which is Eq. 10.2 in the situation of $(h+1)$.

All the above discussions make the assumption that the internal nodes are all functions with n variables. So Eq. 10.2 gives the maximum length of the tail, which might not be used up completely in the decoding procedure.

Let us suppose that the function set includes square root, $\times$, $\div$, $+$, and $-$. The maximum number of variables of these functions is two. The terminal set includes variables a and b. After we set $h = 10$, we can get that the length of the chromosome is $h+t = 10+11 = 21$. One chromosome is illustrated by Fig. 10.9.
where Q is for square root. We use italic bold fonts in Fig. 10.9 to represent the tail. Others genes belong to the head.

[17] -1 means that one internal node is deleted and $+n$ means that n leaves are added.

0	1	2	3	4	5	6	7	8	9	0	1	2	3	4	5	6	7	8	9	0
+	Q	-	/	*b*	*	+	/	Q	*b*	*a*	*a*	*b*	*a*	*a*	*b*	*b*	*a*	*a*	*a*	*b*

Fig. 10.9 The chromosome of GEP

As can be see from Fig. 10.9, genes in the head can be any element in the primitive set but genes in the tail can only contain elements in the terminal set.

The decoding procedure is very straightforward. We start from the first gene in the chromosome. It is a $+$, which means that the root of the syntax tree is $+$. The $+$ function requires two variables, so the second and the third genes, i.e., Q and $-$, represent the second level in the syntax tree. The square root only requires one variable, while $-$ requires two variables. So the fourth to the sixth genes are arranged in the third level of the syntax tree from left to right. Execute the above procedure until all the nodes are terminals. Then we can get the syntax tree in Fig. 10.10. [18]

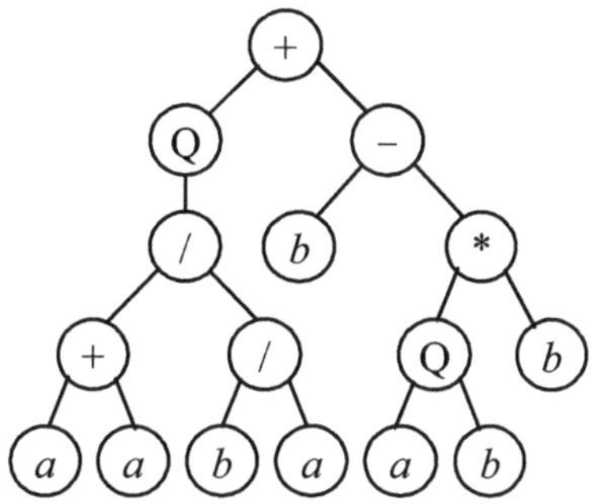

Fig. 10.10 The syntax tree of the chromosome in Fig. 10.9

According to the definition of the head and the tail and the decoding procedure, this code scheme will always represent legal syntax trees if we stick to the rule that the tail can only contain terminals.[19] So ordinary crossover and mutation can be used freely.

GEP uses fixed-length chromosomes with ordinary variation operators, which improves the shortcomings mentioned at the beginning of this subsection. Ferreira discussed other topics, e.g., the assignment of h, powerful variation operators, and the use of several subchromosomes in a chromosome to further improve the search ability of GEP. Interested readers are referred to [2, 3]. In 2003 Zhou *et al.* published a paper to apply GEP in classification [4]. In 2008 Karakasis combined GEP with AIS to get more accurate classification rules [5]. Readers interested in GEP might refer to these papers as examples of its applications.

---

[18] This decoding procedure is called *gene expression*. *baaab* in the tail is not used in the decoding procedure. So this code method has redundancy.

[19] Readers are encouraged to randomly change a gene in Fig. 10.9 and decode it to a syntax tree.

10.2.2 Grammatical Evolution for Solving Differential Equations

In computer science, the *Backus–Naur form* (BNF) is a syntax to illustrate a programming language. One sentence of the program source can be written in the form of BNF as follows:

```
<symbol> ::= __expression__
```

where <symbol> can be regarded as the expression needing to be determined here and __expression__ contains expressions, operators, functions, and digits, denoted as <expr>, <op>, <func>, and <digit>, respectively. It is quite like the primitive set discussed above. One example of the possible __expression__ to indicate algebraic expressions with one variable is illustrated in Table 10.2. Here we just list the set for algebraic expression for simplicity. A possible __expression__ for other programming language is also available. There are four types of rules in Table 10.2, i.e., <expr>, <op>, <func>, and <digit>, in which the rule <expr> can be recursively called.

Table 10.2 An example of the possible rules

Rule		Expression	Value	Rule		Expression	Value
<expr>	::=	<expr> <op> <expr>	0	<digit>	::=	0	0
		(<expr> <op> <expr>)	1			1	1
		<func> (<expr>)	2			2	2
		<digit>	3			3	3
		x	4			4	4
<op>	::=	+	0			5	5
		-	1			6	6
		*	2			7	7
		/	3			8	8
<func>	::=	sin	0			9	9
		cos	1				
		exp	2				
		ln	3				

In 2001 O'Neill *et al.* suggested a way to transfer a series of integer numbers into BNF so as to represent a programming language and use genetic evolution to automatically generate programs [6, 7]. They call this method *grammatical evolution*. Let us take Table 10.2 as a basic set and illustrate the decoding process for the chromosome illustrated in Fig. 10.11.

Fig. 10.11 An example chromosome of grammatical evolution

10	11	2	7	9	8	8	22	6	13	5	7

The genes of the chromosome are all integer numbers. We can denote the allele as V. The first expression must be <expr>. Then we can start the decoding procedure from the leftmost gene, i.e., 10, using the following equation:

$$\text{Rule} = V \bmod NR \tag{10.3}$$

where NR is the current number of rules in the leftmost undetermined expression, i.e., $NR = 5$ in Table 10.2 for rule <expr>, and Rule is the selected rule number. For example, the first gene in Fig. 10.11 is 10. Using Eq. 10.3, we know that allele 10 means Rule = 10 mod 5 = 0, i.e., rule <expr> <op> <expr>, which is used to replace the current leftmost undetermined expression <expr>. Then the leftmost undetermined expression is <expr>, which needs the second gene 11 to decode. 11 mod 5 = 1, which means we need to replace <expr> with (<expr> <op> <expr>). It is the expression (<expr> <op> <expr>) <op> <expr>. Continue the above procedure until there is no undetermined expression. The decoding procedure can be illustrated as in Table 10.3.

Table 10.3 The decoding procedure of the chromosome in Fig. 10.11

Expression	Current gene	Operation
<expr>	10	10 mod 5 = 0
<expr> <op> <expr>	11	11 mod 5 = 1
(<expr> <op> <expr>) <op> <expr>	2	2 mod 5 = 2
(<func> (<expr>) <op> <expr>) <op> <expr>	7	7 mod 4 = 3
(ln (<expr>) <op> <expr>) <op> <expr>	9	9 mod 5 = 4
(ln (x) <op> <expr>) <op> <expr>	8	8 mod 4 = 0
(ln (x) + <expr>) <op> <expr>	8	8 mod 5 = 3
(ln (x) + <digit>) <op> <expr>	22	22 mod 10 = 2
(ln (x) + 2) <op> <expr>	6	6 mod 4 = 2
(ln (x) + 2) * <expr>	13	13 mod 5 = 3
(ln (x) + 2) * <digit>	5	5 mod 10 = 5
(ln (x) + 2) * 5		

As can be seen from Table 10.3, sometimes we do not use all 12 genes to get a full expression, i.e., the final gene 7 is useless. If more than 12 genes are needed to determine the clear expression, Tsoulos and Lagaris further suggested starting over from the first gene, i.e., to wrap [8]. Then the fixed-length chromosome could be used to indicate the algebraic expressions that facilitate the common variation operators.[20]

In 2006 Tsoulos and Lagaris used grammatical evolution to find an analytical closed form solution for differential equations [8].[21]

An n-order ordinary differential equation (ODE) can be illustrated as follows:

$$f\left(x,y,y^{(1)},\cdots,y^{(n)}\right)=0,\quad x\in[a,b] \tag{10.4}$$

where $y^{(n)}$ is the n-order derivation of y over x and $[a,b]$ is the definition domain of variable x. To determine function $y(x)$, we still need n boundary conditions as follows:

$$g_k\left(x,y,y^{(1)},\cdots,y^{(n)}\right)|_{x=a\text{ or }b}=0,\quad k=1,\cdots,n \tag{10.5}$$

If $x=a$, Eq. 10.5 is called the initial value condition; if $x=b$, Eq. 10.5 is called the final value condition. What we need to do for solving ODE is to find the function $y(x)$ that satisfies Eqs. 10.4 and 10.5.

We can use an integer value chromosome, e.g., Fig. 10.11, to represent an algebraic expression, which might be the solution to the ODE problem. To determine the fitness value of the chromosome i, we should first decode it as algebraic expression M_i in the way illustrated in Table 10.3.[22] Then we can use some algebraic methods to get its deviations $M_i^{(1)}$, ..., $M_i^{(n)}$ [9]. After that, we can evaluate the closeness of M_i to the real y. The evaluation procedure is similar to that discussed in the previous section. We sample N points between $[a,b]$ and take the following expression as the sum square of the function errors in the N sampling points:

$$E(M_i)=\sum_{j=1}^{N}f^2\left(x_j,M_i(x_j),M_i^{(1)}(x_j),\cdots,M_i^{(n)}(x_j)\right) \tag{10.6}$$

Then for n boundary conditions, we can also get their sum square errors as follows:

$$P(M_i)=\lambda\sum_{k=1}^{n}g_k^2\left(x,M_i(x),M_i^{(1)}(x),\cdots,M_i^{(n)}(x)\right)|_{x=a\text{ or }b} \tag{10.7}$$

where λ is a user-defined parameter to scale these two errors on same magnitude level.

Finally, the fitness value of individual i is as follows:[23]

---

[20] Sometimes many wrappings still cannot parse a legal expression. So Tsoulos and Lagaris suggested allowing at most two wrappings. After that, if we still cannot get a clear expression, this individual is discarded.

[21] Their paper discusses high order ODEs, a system of first-order ODE, and elliptic partial differential equation. Here we just introduce the first case.

[22] Here we regard "computer program" as the algebraic function expression that satisfies the differential equation and boundary conditions.

[23] The smaller the fitter.

$$fitness_i = E(M_i) + P(M_i) \tag{10.8}$$

Thus, all the technical obstacles are removed for genetic evolving, and we can select, cross over, mutate, and replace individuals until the stop criteria are satisfied. After that, the best analytical closed form solution for Eqs. 10.4 and 10.5 can be suggested.

10.3 Example of Genetic Programming for Knowledge Discovery

In the previous two sections we only discussed how to use GP to generate algebraic expressions so as to implement the requirement of curve fitting or differential function solving. GP has been used to evolve other "computer programs" for various tasks. Here we will introduce the work published by Bojarczuk *et al.* in 2000 on applying GP to aid the diagnosis of chest-pain [10].[24]

Chest pain is a symptom related to several diseases. There are 165 *predicting attributes* related to chest pain. Every predicting attribute for the patient has a logic value of `TRUE` or `FALSE`. For example, "pain irradiates to the upper right region of the abdomen" is a predicting attribute. If the patient has such a symptom, his logic value for that attribute is `TRUE`, and conversely, if the patient does not have such a symptom, then his logic value for that attribute is `FALSE`.

There are 12 diseases related to chest pain, e.g., angina, esophageal pain, etc. The knowledge discovery problem here is to determine the disease given the 165 logic values for the predicting attributes of a patient. It can be expressed as the logic expression "IF - THEN." One of these logic expressions can be written as follows:[25]

IF (starting factor is emotion = TRUE) AND ((the pain lasts no more than seconds = TRUE) OR ((the pain begins gradually = TRUE) AND (the pain irradiates towards the upper left limb = TRUE))) THEN disease is stable angina.

In the above diagnostic logic expression, "starting factor is emotion," "the pain lasts no more than seconds," etc. are all predicting attributes, and "stable angina" is one diagnosed disease.

This is a classification problem. We need to classify one case, with 165 predicting attributes, into one of the 12 diseases. Bojarczuk *et al.* divide the problem into 12 subproblems. Subproblem *i* answers one question: could this case be diagnosed as disease *i*?[26]

So what we need to do is to evolve 12 logic expressions of the "IF" part in the above "IF - THEN" rule for 12 diseases. The "THEN" part is not necessary because

[24] Here we regard "computer program" as a logic expression that can generate correct decisions for chest-pain-related diseases.

[25] This rule is only for illustration. Readers should not diagnose themselves without consulting a doctor.

[26] Another way to illustrate their technique is to transform a multiclass classification problem into multiple two-class problems.

for disease i, we will only evolve the diagnostic rule for it. Any case that satisfies the logic expression will be diagnosed as disease i.[27] In the following part, we will only discuss the evolving procedure of logic rule i.

Bojarczuk *et al.* have 138 diagnosed chest-pain-related disease cases, each with 165 attributes. They use 90 cases to train rule i and 48 cases to test the generalization performance of that rule.[28] They further divide the 90 training cases into two groups: those with disease i and those without disease i.

Then they use GP to evolve the logic expressions. The function set contains only three elements: "AND," "OR," and "NOT." The terminal set contains the symbol of the 165 attributes, expressed as a, b, etc. Each symbol can have one of two values: "TRUE" or "FALSE." One syntax tree example of the rule for disease i is illustrated by Fig. 10.12.

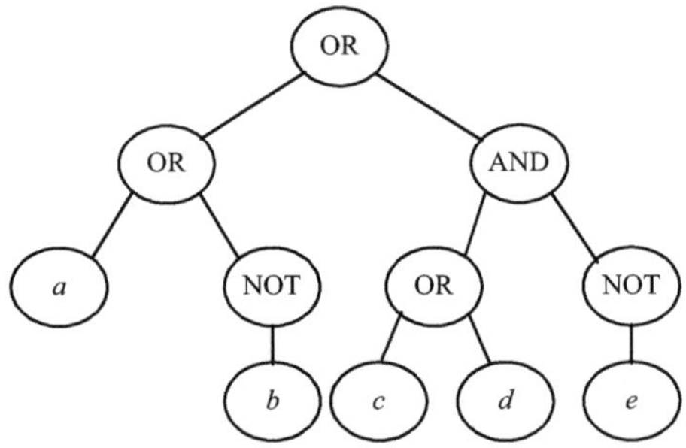

Fig. 10.12 An example syntax tree for disease i

The initialization and variation procedures are all similar techniques introduced in Sect. 10.1.2. Here we just focus on the evaluation of the individual.

For every training case, we know its 165 attributes and whether it has disease i or not. For an individual, illustrated by Fig.10.12, we can deduce whether these 165 attributes will cause disease i or not. So there are four situations illustrated as follows.[29]

- **True positive.** The rule predicts that the case has disease i and the case really has it. We use tp to represent the case numbers belonging to this situation.
- **False positive.** The rule predicts that the case has disease i, but the case does not actually have it. We use fp to represent the case numbers belonging to this situation.
- **True negative.** The rule predicts that the case does not have disease i, and in fact the case does not have it. We use tn to represent the case numbers belonging to this situation.

[27] The 12 subproblems are simpler than the original classification problem. But there exist some cases that satisfy more than one rule, which would not happen in the original problem.

[28] Generalization ability is the key issue of machine learning algorithms. Interested readers are referred to a survey published by Kushchu in 2002 [11].

[29] "Positive" and "negative" concern the prediction of the rule and "true" and "false" concern the correctness of the prediction.

- **False negative.** The rule predicts that the case does not have disease i, but the case does have it. We use fn to represent the case numbers belonging to this situation.

According to the above categories, $tp+fn$ is the number of cases that have disease i and $fp+tn$ is the number of cases that do not actually have disease i. Then the *sensitivity* (*Se*) of the correct prediction for those who actually have disease i is defined as follows:

$$Se = \frac{tp}{tp+fn} \tag{10.9}$$

And the *specificity* (*Sp*) of the correct prediction for those who do not actually have disease i is defined as follows:

$$Sp = \frac{tn}{tn+fp} \tag{10.10}$$

Sn and *Sp* concern the correctness of the prediction. The larger the better. Bojarczuk *et al.* also defined the third objective, denoted *simplicity* (*Sy*), to limit the size of the syntax tree as follows:

$$Sy = \frac{maxnodes - 0.5numnodes - 0.5}{maxnodes - 1} \tag{10.11}$$

where $maxnodes$ is the maximum node number predefined by the user and $numnodes$ is the node number of the current individual. According to Eq. 10.11, $0.5 \leq Sy \leq 1$; the larger the better.

Then the fitness function of the syntax tree is defined as follows:

$$fitness = Se \times Sp \times Sy \tag{10.12}$$

It is easy to understand that $0 \leq fitness \leq 1$ and that the larger the better.

Bojarczuk *et al.* use 90 cases to train the 12 logic expressions for 12 diseases and then use 48 cases to test the rules. The test results illustrate that the rules evolved by GP has better predicting results than those generated by another rule-induction algorithm, C5.0.

10.4 Summary

Structure optimization or learning is a research field that has attracted a lot of attention. It has strong ties with other hot AI topics, e.g., machine learning, knowledge discovery, and pattern recognition, etc. GP is one of the closest branches of EAs to this field.

The solution process of GP is the same as other EAs. But it has special features with respect to representation, evaluation, and variation. We introduced the basics of GP with an example of curving fitting in Sect. 10.1.2, where tree code is illustrated and its relationship with Lisp's prefix notation programming language are introduced.

Apart from that, two other techniques for mapping between chromosomes and syntax tree are discussed in Sect. 10.2. GEP uses a head-and-tail structure to maintain fixed-length chromosomes and the legality of the chromosomes. Grammatical evolution uses integers and rule labels to find the corresponding expressions, functions, or digits. In this way, grammatical evolution can also generate legal individuals with standard variation operators. Other related methods which be recommended in "Suggested Readings".

Several applications are discussed, including curving fitting, differential function solving, and classification, which is only the tip of the iceberg of the possible GP applications. Other topics will be recommended in "Suggestions for Further Reading."

Due to the space constraints, we do not discuss the following several important issues of GP.

- The details of the data structure of GP for storing, managing, decoding, and evaluating individuals, which is the essence of implementing GP.
- Sometimes, a crossover operator might generate too deep or too large syntax trees, which introduces a too large search space and also limits the comprehensibility of the results. This problem is called *bloat*. A paper published in 2006 by Luke and Panait compared several bloat control methods for GP [12].
- We only introduce the simplest crossover operator and mutation operator. Like the real code situation, there are many other types of variation operators, such as the architecture-altering operator suggested by Koza [13] and transposition operator suggested by Ferreira [2], with different effects on exploration and exploitation for tree code.

We give enough recommendations in "Suggestions for Further Reading" for interested readers on these issues.

After reading this chapter, you should understand the reason for GP's capability of structure optimizing, be familiar with some of the special concepts in GP, e.g., syntax tree, primitive set, fitness cases, etc., and grasp at least one method of two-way mapping between a chromosome and a syntax tree.

Suggestions for Further Reading

Many books on GP have been published since its inception. The most famous four-volume set was published by Koza in 1992, 1994, 1999, and 2003 [1, 13–15]. A more recent textbook on GP, with its JAVA source code, was published by Poli *et al.* in 2008 [16].

Apart from the journals and conferences introduced in Chap. 1, GP still has its own journal, *Genetic Programming and Evolvable Machines*, published by

Springer, and its own conference, the European Conference on Genetic Programming, held every year.

As to Internet resources, Silva proposed a GP toolbox, GPLAB, in the MATLAB® environment,[30] RML Technologies, Inc. sells Discipulus commercial software,[31] and Koza maintain a Web site containing prolific information on GP.[32]

GP for differential equation solving is a fascinating research field. Interested readers are encouraged to read a paper by Sobester *et al.* published in 2008 [17].

For those interested in applying GP to disease diagnosis, one paper published by Zhang and Wong in 2008 might be insightful [18]. The paper on generating classification trees by GP, published by Kuo in 2007, might also be interesting [19]. In 2006, Folino *et al.* discussed the GP-based large-scale data classification problem [20].

Curve fitting, function solving, and knowledge discovery are three applications discussed in this chapter. For more examples, works on combinational logic circuit synthesis by Cheang *et al.* [21], passive filter synthesis by Chang *et al.* [22], autonomous language developing by Hong *et al.* [23], control system evolution for sumo-fighting robots by Sharabi and Sipper [24], and time series forecasting for dynamic environments by Wagner *et al.* [25] are highly recommended.

Apart from GEP and grammatical evolution, in 1999 Folino *et al.* proposed cellular GP [26], Miller and Thomson proposed Cartesian GP in 2000 [27], and in 2006 Hoai *et al.* suggested using a different tree-based representation and local structural modification operators to further improve the search ability of GP [28].

Mating restrictions sometimes play an important role in genetic evolution. In order to introduce mating restrictions into GP, a definition of the distance between two syntax trees is necessary. In 2008 Gustafson and Vanneschi published a paper to deal with this [29].

In 2008 Tay *et al.* implemented GP in a competitive coevolution environment, introduced in Sect. 3.7.1 [30]. Pursuit and evasion strategies are discussed in detail there.

Recall the estimation of distribution algorithm (EDA) discussed in Sect. 3.5.2.6. In 2008 Hasegawa and Iba suggested a new program evolution algorithm employing a Bayesian network for generating new individuals [31].

The must-read papers of this chapter are Chaps. 1–4 of [16] for a general introduction, [2] for GEP, and [10] for an application example.

Exercises and Potential Research Projects

10.1. Generate another offspring of Fig. 10.7.

10.2. Decode the GEP chromosome in Fig. 10.13 into a syntax tree.

---

[30] GPLAB can be downloaded at `http://gplab.sourceforge.net/`.

[31] `http://www.rmltech.com/`

[32] `http://www.genetic-programming.org/`

0	1	2	3	4	5	6	7	8	9	0	1	2	3	4	5	6	7	8	9	0
+	Q	-	/	*b*	*	a	/	Q	*b*	*a*	*a*	*b*	*a*	*a*	*b*	*b*	*a*	*a*	*a*	*b*

Fig. 10.13 The GEP chromosome for Exercise 10.2

10.3. Find a chromosome that can be decoded as $\sin^2(x^2)$ by grammatical evolution using Table 10.2.

References

1. Koza JR (1992) Genetic programming: on the programming of computers by means of natural selection. MIT Press, Cambridge, MA
2. Ferreira C (2001) Gene expression programming a new adaptive algorithm for solving problems. Complex Syst 13(2):87–129
3. Ferreira C (2006) Gene expression programming: mathematical modeling by an artificial intelligence, 2nd edn. Springer, Berlin Heidelberg New York
4. Zhou C, Xiao W, Tirpak T *et al* (2003) Evolving accurate and compact classification rules with gene expression programming. IEEE Trans Evol Comput 7(6):519–531
5. Karakasis V, Stafylopatis A (2008) Efficient evolution of accurate classification rules using a combination of gene expression programming and clonal selection. IEEE Trans Evol Comput 12(6):662–678
6. O'Neill M, Ryan C (2001) Grammatical evolution. IEEE Trans Evol Comput 5(4):349–358
7. O'Neill M, Ryan C (2003) Grammatical evolution: evolutionary automatic programming in an arbitrary language. Springer, Berlin Heidelberg New York
8. Tsoulos I, Lagaris I (2006) Solving differential equations with genetic programming. Genet Programm Evolvable Mach 7(1)
9. Griewank A (1989) On automatic differentiation. In: Iri M, Tanabe K (eds) Mathematical programming: recent developments and applications. Kluwer Academic Publishers, Holland, pp 83–108
10. Bojarczuk C, Lopes H, Freitas A (2000) Genetic programming for knowledge discovery in chest-pain diagnosis. IEEE Eng Med Biol Mag 19(4):38–44
11. Kushchu I (2002) Genetic programming and evolutionary generalization. IEEE Trans Evol Comput 6(5):431–442
12. Luke S, Panait L (2006) A comparison of bloat control methods for genetic programming. Evol Comput 14(3):309–344
13. Koza JR (1994) Genetic programming II: automatic discovery of reusable programs. MIT Press, Cambridge, MA
14. Koza JR, III FHB, Andre D *et al* (1999) Genetic programming III: Darwinian invention and problem solving. Morgan Kaufmann, San Francisco
15. Koza JR, Keane MA, Streeter MJ *et al* (2003) Genetic programming IV: routine human-competitive machine intelligence. Springer, Berlin Heidelberg New York
16. Poli R, Langdon WB, McPhee NF (2008) A field guide to genetic programming. Lulu Enterprises, Raleigh, NC
17. Sobester A, Nair P, Keane A (2008) Genetic programming approaches for solving elliptic partial differential equations. IEEE Trans Evol Comput 12(4):469–478
18. Zhang M, Wong P (2008) Genetic programming for medical classification: a program simplification approach. Genet Programm Evolvable Mach 9(3):229–255
19. Kuo C, Hong T, Chen C (2007) Applying genetic programming technique in classification trees. Soft Comput 11(12):1165–1172
20. Folino G, Pizzuti C, Spezzano G (2006) GP ensembles for large-scale data classification. IEEE Trans Evol Comput 10(5):604–616

21. Cheang SM, Lee KH, Leung KS (2007) Applying genetic parallel programming to synthesize combinational logic circuits. IEEE Trans Evol Comput 11(4):503–520
22. Chang S, Hou H, Su Y (2006) Automated passive filter synthesis using a novel tree representation and genetic programming. IEEE Trans Evol Comput 10(1):93–100
23. Hong J, Lim S, Cho S (2007) Autonomous language development using dialogue-act templates and genetic programming. IEEE Trans Evol Comput 11(2):213–225
24. Sharabi S, Sipper M (2006) GP-sumo: using genetic programming to evolve sumobots. Genet Programm Evolvable Mach 7(3):211–230
25. Wagner N, Michalewicz Z, Khouja M *et al* (2007) Time series forecasting for dynamic environments: the DyFor genetic program model. IEEE Trans Evol Comput 11(4):433–452
26. Folino G, Pizzuti C, Spezzano G (1999) A cellular genetic programming approach to classification. In: Proceedings of the genetic and evolutionary computation conference, pp 1015–1020
27. Miller JF, Thomson P (2000) Cartesian genetic programming. In: Proceedings of the European conference on genetic programming, pp 121–132
28. Hoai NX, McKay R, Essam D (2006) Representation and structural difficulty in genetic programming. IEEE Trans Evol Comput 10(2):157–166
29. Gustafson S, Vanneschi L (2008) Crossover-based tree distance in genetic programming. IEEE Trans Evol Comput 12(4):506–524
30. Tay J, Tng C, Chan C (2008) Environmental effects on the coevolution of pursuit and evasion strategies. Genet Programm Evolvable Mach 9(1):5–37
31. Hasegawa Y, Iba H (2008) A Bayesian network approach to program generation. IEEE Trans Evol Comput 12(6):750–764

Appendix A
Benchmark Problems

All trigonometric functions are calculated with radian.

Benchmark Problems for Chap. 2 and Chap. 3

A.1. Adopted from Spears's problem [1]. The problem is in binary space with ten variables. There is only one peak, $Peak = (0011010101)$. The objective value of a solution *string* is as follows:

$$f(string) = \frac{1}{10}(10 - Hamming(string, Peak)) \quad \text{(A.1)}$$

where $Hamming(\,)$ is the function to calculate two binary strings' Hamming distance. The smaller the objective value is, the better the solution is.

A.2. The sphere model from Eiben and Smith's book [2]. The problem is in real space with n variables. It is a minimum problem whose global optimal value is 0.0 at $\mathbf{x} = \mathbf{0}$. As an exercise, we set $n = 5$. The definition domain for x_i is $[-5, 5]$:

$$f(\mathbf{x}) = \sum_{i=1}^{n} x_i^2 \quad \text{(A.2)}$$

A.3. Ackley's function from Bäck's book [3]. The problem is in real space with n variables. This is a minimum problem whose global optimal value is 0.0 at $\mathbf{x} = \mathbf{0}$. As an exercise, we set $n = 2$. The definition domain for x_i is $[-20, 30]$:

$$f(\mathbf{x}) = -20\exp\left(-0.2\sqrt{\frac{1}{n}\sum_{i=1}^{n} x_i^2}\right) - \exp\left(\frac{1}{n}\sum_{i=1}^{n}\cos(2\pi x_i)\right) + 20 + \mathrm{e} \quad \text{(A.3)}$$

where $\mathrm{e} \approx 2.71828$ is the Euler number.

A.4. From Yao *et al.*'s paper [4]. The problem is in real space with two variables. It is a minimum problem whose global optimal value is -1.0316285. The definition domain for both x_1 and x_2 is $[-5,5]$:

$$f(\mathbf{x}) = 4x_1^2 - 2.1x_1^4 + \frac{1}{3}x_1^6 + x_1x_2 - 4x_2^2 + 4x_2^4 \tag{A.4}$$

A.5. From Lozano *et al.*'s paper [5]. This is a frequency modulation sound parameter identification problem, i.e., to specify six parameters a_1, ω_1, a_2, ω_2, a_3, and ω_3 of the frequency modulation sound model represented by

$$y(t) = a_1 \sin\left(\frac{2\pi}{100}\omega_1 t + a_2 \sin\left(\frac{2\pi}{100}\omega_2 t + a_3 \sin\left(\frac{2\pi}{100}\omega_3 t\right)\right)\right)$$

The objective function is the summation of the square errors between the solution and the real model as follows:

$$f(a_1,\omega_1,a_2,\omega_2,a_3,\omega_3) = \sum_{t=0}^{100} (y(t) - y_0(t))^2 \tag{A.5}$$

where the real model $y_0(t)$ is as follows:

$$y_0(t) = 1.0\sin\left(\frac{2\pi}{100}5.0t - 1.5\sin\left(\frac{2\pi}{100}4.8t + 2.0\sin\left(\frac{2\pi}{100}4.9t\right)\right)\right)$$

This is a minimum problem whose global optimal value is 0. The definition domain for all variables is $[-6.4, 6.35]$.

A.6. Schwefel's Problem from Suganthan *et al.*'s paper [6]. The problem is in real space with two variables. It is a minimum problem whose global optimal value is 0 at $x_1 = 1$ and $x_2 = 3$. The definition domain for both x_1 and x_2 is $[-20,20]$:

$$f(\mathbf{x}) = \max\{|x_1 + 2x_2 - 7|, |2x_1 + x_2 - 5|\} \tag{A.6}$$

Benchmark Problems for Chap. 4

All the benchmark problems for constrained optimization are selected from Koziel and Michalewicz's paper [7] and Runarsson and Yao's paper [8].

A.7. This is a minimum problem with two real variables as follows. The definition domains are $13 \le x_1 \le 100$ and $0 \le x_2 \le 100$:

$$\begin{aligned} &f(\mathbf{x}) = (x_1 - 10)^3 + (x_2 - 20)^3 \\ &\text{s.t.} \quad -(x_1-5)^2 - (x_2-5)^2 + 100 \leqslant 0 \\ &\qquad\; (x_1-6)^2 + (x_2-5)^2 - 82.81 \leqslant 0 \end{aligned} \tag{A.7}$$

The optimum solution is $\mathbf{x}^* = (14.095, 0.84296)$ with $f(\mathbf{x}^*) = -6961.81388$. Both constraints are active.

A.8. This is a minimum problem with five real variables as follows. The definition domains are $78 \le x_1 \le 102$, $33 \le x_2 \le 45$, and $27 \le x_i \le 45$ for $i = 3,4,5$:

$$\begin{aligned} f(\mathbf{x}) &= 5.3578547x_3^2 + 0.8356891x_1x_5 + 37.293239x_1 - 40792.141 \\ \text{s.t.}\quad & 85.334407 + 0.0056858x_2x_5 + 0.0006262x_1x_4 - 0.0022053x_3x_5 - 92 \leqslant 0 \\ & -85.334407 - 0.0056858x_2x_5 - 0.0006262x_1x_4 + 0.0022053x_3x_5 \leqslant 0 \\ & 80.51249 + 0.0071317x_2x_5 + 0.0029955x_1x_2 + 0.0021813x_3^2 - 110 \leqslant 0 \\ & -80.51249 - 0.0071317x_2x_5 - 0.0029955x_1x_2 - 0.0021813x_3^2 + 90 \leqslant 0 \\ & 9.300961 + 0.0047026x_3x_5 + 0.0012547x_1x_3 + 0.0019085x_3x_4 - 25 \leqslant 0 \\ & -9.300961 - 0.0047026x_3x_5 - 0.0012547x_1x_3 - 0.0019085x_3x_4 + 20 \leqslant 0 \end{aligned} \tag{A.8}$$

The optimum solution is $\mathbf{x}^* = (78, 33, 29.995256025682, 45, 36.775812905788)$ with $f(\mathbf{x}^*) = -30665.539$. Two constraints are active.

A.9. This is a minimum problem with four real variables as follows. The definition domains are $0 \le x_i \le 1200$ for $i = 1,2$ and $-0.55 \le x_i \le 0.55$ for $i = 3,4$:

$$\begin{aligned} f(\mathbf{x}) &= 3x_1 + 0.000001x_1^3 + 2x_2 + \frac{0.000002}{3}x_2^3 \\ \text{s.t.}\quad & -x_4 + x_3 - 0.55 \leqslant 0 \\ & -x_3 + x_4 - 0.55 \leqslant 0 \\ & 1000\sin(-x_3 - 0.25) + 1000\sin(-x_4 - 0.25) + 894.8 - x_1 = 0 \\ & 1000\sin(x_3 - 0.25) + 1000\sin(x_3 - x_4 - 0.25) + 894.8 - x_2 = 0 \\ & 1000\sin(x_4 - 0.25) + 1000\sin(x_4 - x_3 - 0.25) + 1294.8 = 0 \end{aligned} \tag{A.9}$$

The best known solution is $\mathbf{x}^* = (679.9453, 1026.067, 0.1188764, -0.3962336)$ with $f(\mathbf{x}^*) = 5126.4981$.

Benchmark Problems for Chap. 5

A.10. Adopted from Spears' problem [1]. The problem is in binary space with ten variables. There are three peaks, $Peak_1 = (0011010101)$, $Peak_2 = (0010010111)$, and $Peak_3 = (1011000100)$. The objective value of a solution *string* is as follows:

$$f(string) = \frac{1}{10} \max_{i=1,2,3} \{10 - Hamming(string, Peak_i)\} \tag{A.10}$$

where $Hamming(\,)$ is the function to calculate two binary strings' Hamming distance. The smaller the objective value is, the better the solution is.

A.11. From Mahfoud's thesis [9]. This is a one-variable maximum problem as follows. The definition domain for the variable is $[0,1]$. There are five peaks at 0.080, 0.247, 0.451, 0.681, and 0.934 with height 1.000 and we are interested in finding them all:

$$f(x) = \sin^6\left(5\pi\left(x^{0.75} - 0.05\right)\right) \tag{A.11}$$

A.12. From Mahfoud's thesis [9]. This is a one-variable maximum problem as follows. The definition domain for the variable is $[0,1]$. There are five peaks at 0.080, 0.247, 0.451, 0.681, and 0.934 with heights 1.000, 0.948, 0.770, 0.503, and 0.250, respectively separately. We are interested in finding them all:

$$f(x) = \mathrm{e}^{-2(\ln 2)\left(\frac{x-0.08}{0.854}\right)^2} \sin^6\left(5\pi\left(x^{0.75} - 0.05\right)\right) \tag{A.12}$$

A.13. This is a two-variable maximum problem as follows. The definition domain for both variable is $[-6,6]$. There are four peaks, which are at $(3.000, 2.000)$, $(3.584, -1.848)$, $(-3.779, -3.283)$, and $(-2.805, 3.131)$ with height 2500.000, and we are interested in finding them all:

$$f(\mathbf{x}) = 2500 - \left(x_1^2 + x_2 - 11\right)^2 - \left(x_1 + x_2^2 - 7\right)^2 \tag{A.13}$$

A.14. Massively deceptive problem from Goldberg's report [10]. It is a maximum problem with 30 binary variables defined as follows. It has 32 global peaks and more than one million local peaks. We are only interested in finding the 32 global peaks at (000000, 000000, 000000, 000000, 000000), (000000, 000000, 000000, 000000, 111111), $\cdots$,(111111, 111111, 111111, 111111, 111111) with height 5.0:

$$f(x_0, \cdots, x_{29}) = \sum_{i=0}^{4} u\left(\sum_{j=0}^{5} x_{6i+j}\right) \tag{A.14}$$

where $\forall x_k \in \{0,1\}, k = 0, \cdots, 29$ and the definition of function $u(s)$ is as follows:

$$u(s) = \begin{cases} 1 & s \in \{0,6\} \\ 0 & s \in \{1,5\} \\ 0.360384 & s \in \{2,4\} \\ 0.640576 & s = 3 \end{cases}$$

Benchmark Problems for Chap. 6

A.15. From Van Veldhuizen's thesis [11]. The problem is in real space with one variable. It has two minimum objectives. The definition domain for x is $[-10^5, 10^5]$:

$$\begin{aligned} f_1(x) &= x^2 \\ f_2(x) &= (x-2)^2 \end{aligned} \tag{A.15}$$

It has a convex PF*.

A.16. Adopted from Van Veldhuizen's thesis [11]. The problem is in real space with n variable. It has two minimum objectives. The definition domain for x_i is $[-5,5]$. As an exercise, we set $n=3$:

$$\begin{aligned} f_1(\mathbf{x}) &= \sum_{i=1}^{n-1}\left(-10\exp\left(-0.2*\sqrt{x_i^2+x_{i+1}^2}\right)\right) \\ f_2(\mathbf{x}) &= \sum_{i=1}^{n}\left(|x_i|^{0.8}+5\sin(x_i)^3\right) \end{aligned} \tag{A.16}$$

It has a disconnected PF*.

A.17. From Zitzler *et al.*'s ZDT set [12]. The problem is in real space with n variables. It has two minimum objectives. The definition domain for x_i is $[0,1]$:

$$\begin{aligned} f_1(\mathbf{x}) &= x_1 \\ f_2(\mathbf{x}) &= g(\mathbf{x})\left(1-\left(\frac{f_1(\mathbf{x})}{g(\mathbf{x})}\right)^2\right) \\ g(\mathbf{x}) &= 1+\frac{9}{n-1}\sum_{i=2}^{n}x_i \end{aligned} \tag{A.17}$$

where $n=30$. Its P* is formed by $g(\mathbf{x})=1$ with a concave PF*.

A.18. From Tanaka *et al.*'s problem [13]. The problem is in real space with two variables. It has two minimum objectives. The definition domain for x_i is $[0,\pi]$:

$$\begin{aligned} & f_1(\mathbf{x}) = x_1 \\ & f_2(\mathbf{x}) = x_2 \\ \text{s.t.} \quad & -x_1^2-x_2^2+1+0.1\cos\left(16\arctan\left(\frac{x_2}{x_1}\right)\right) \le 0 \\ & (x_1-0.5)^2+(x_2-0.5)^2 \le 0.5 \end{aligned} \tag{A.18}$$

It is a constrained MOP with a disconnected PF*.

A.19. Okabe's problem [14]. The problem is in real space with three variables. It has two minimum objectives. The definition domain for x_1 is $[-\pi,\pi]$ and x_2,x_3 is $[-5,5]$:

$$\begin{aligned} f_1 &= x_1 \\ f_2 &= 1-\frac{1}{4\pi^2}(x_1+\pi)^2+|x_2-5\cos(x_1)|^{\frac{1}{3}}+|x_3-5\sin(x_1)|^{\frac{1}{3}} \end{aligned} \tag{A.19}$$

PF* is $f_2=1-\frac{1}{4\pi^2}(f_1+\pi)^2$, $f_1\in[-\pi,\pi]$ and P* is $(x_1,x_2,x_3)=(\xi,5\cos(\xi),5\sin(\xi))$, $\xi\in[-\pi,\pi]$.

Benchmark Problems for Chap. 7

A.20. Solve the 100-item knapsack problem of Eqs. 7.4 and 7.5, where $W = 1000$ and item data are listed in Table A.1:

Table A.1 100-item knapsack problem

Item	p	w	Item	p	w	Item	p	w	Item	p	w
1	75	3	26	32	16	51	54	15	76	72	9
2	80	34	27	37	20	52	69	25	77	44	39
3	88	28	28	76	4	53	23	41	78	72	27
4	70	35	29	76	13	54	43	25	79	95	13
5	8	23	30	55	12	55	14	2	80	60	14
6	23	22	31	30	12	56	94	44	81	30	28
7	6	1	32	1	11	57	35	16	82	69	13
8	59	18	33	89	17	58	6	2	83	35	41
9	12	48	34	14	6	59	25	45	84	71	46
10	67	24	35	76	43	60	90	20	85	92	45
11	88	16	36	36	38	61	91	50	86	64	2
12	72	8	37	66	11	62	86	9	87	97	42
13	20	19	38	34	6	63	16	5	88	65	10
14	51	24	39	92	32	64	38	2	89	61	44
15	58	18	40	70	16	65	35	15	90	89	34
16	31	34	41	18	4	66	5	40	91	50	10
17	20	10	42	83	6	67	80	26	92	95	45
18	67	10	43	23	9	68	65	4	93	48	16
19	38	37	44	10	50	69	66	9	94	73	44
20	44	45	45	23	16	70	78	31	95	80	33
21	61	17	46	11	2	71	11	24	96	10	9
22	19	17	47	79	11	72	74	45	97	62	28
23	17	29	48	12	17	73	89	31	98	73	48
24	36	31	49	80	27	74	98	13	99	21	45
25	76	4	50	24	7	75	25	37	100	44	12

A.21. Solve the 29-city Euclidean TSP.[1] The cities' coordinate data are listed in Table A.2:

[1] This problem is from TSPLIB.

Table A.2 29-city TSP

City	x	y	City	x	y	City	x	y	City	x	y
1	1150	1760	9	790	2260	17	230	590	25	1280	790
2	630	1660	10	710	1310	18	460	860	26	490	2130
3	40	2090	11	840	550	19	1040	950	27	1460	1420
4	750	1100	12	1170	2300	20	590	1390	28	1260	1910
5	750	2030	13	970	1340	21	830	1770	29	360	1980
6	1030	2070	14	510	700	22	490	500			
7	1650	650	15	750	900	23	1840	1240			
8	1490	1630	16	1280	1200	24	1260	1500			

A.22. Solve the four-job, three-machine JSSP from Baker and Trietsch's book [15]. The technological requirements are listed in Table A.3:

Table A.3 Four-job, three-machine JSSP

	p_{ijk}			Machine		
Operation	1	2	3	1	2	3
Job 1	4	3	2	M_1	M_2	M_3
Job 2	1	4	4	M_2	M_1	M_3
Job 3	3	2	3	M_3	M_2	M_1
Job 4	3	3	1	M_2	M_3	M_1

References

1. Spears WM (2004) Evolutionary algorithms: the role of mutation and recombination. Springer, Berlin Heidelberg New York
2. Eiben AE, Smith JE (2003) Introduction to evolutionary computing. Springer, Berlin Heidelberg New York
3. Bäck T (1996) Evolutionary algorithms in theory and practice: evolution strategies, evolutionary programming, genetic algorithms. Oxford University Press, Oxford, UK
4. Yao X, Liu Y, Lin G (1999) Evolutionary programming made faster. IEEE Trans on Evol Comput 3(2):82–102
5. Lozano M, Herrera F, Krasnogor N *et al* (2004) Real-coded memetic algorithms with crossover hill-climbing. Evol Comput 12(3):273–302

6. Suganthan P, Hansen N, Liang J *et al* (2005) Problem definitions and evaluation criteria for the CEC 2005 special session on real parameter optimization. Tech. rep., Nanyang Technological University and IIT Kanpur, Singapore
7. Koziel S, Michalewicz Z (1999) Evolutionary algorithms, homomorphous mappings, and constrained parameter optimization. Evol Comput 7(1):19–44
8. Runarsson T, Yao X (2000) Stochastic ranking for constrained evolutionary optimization. IEEE Trans on Evol Comput 4(3):284–294
9. Mahfoud S (1995) Niching methods for genetic algorithms. Ph.D. thesis, University of Illinois at Urbana-Champaign
10. Goldberg DE, Goldberg DE, Deb K *et al* (1992) Massive multimodality, deception, and genetic algorithms. Tech. rep., Illinois Genetic Algorithms Laboratory, UIUC
11. Veldhuizen V (1999) Multiobjective evolutionary algorithms: classifications, analyses, and new innovations. Ph.D. thesis, Air Force Institute of Technology School of Engineering, Wright-Patterson AFB, OH
12. Zitzler E, Deb K, Thiele L (2000) Comparison of multiobjective evolutionary algorithms: empirical results. Evol Comput 8(2):173–195
13. Tanaka M, Watanabe H, Furukawa Y *et al* (1995) GA-based decision support system for multicriteria optimization. In: Proceedings of the IEEE international conference on systems, man and cybernetics, 2:1556–1561
14. Okabe T, Jin Y, Olhofer M *et al* (2004) On test functions for evolutionary multi-objective optimization. In: Proceedings of the international conference on parallel problem solving from nature, pp 792–802
15. Baker KR, Trietsch D (2009) Principles of sequencing and scheduling. Wiley, New York

Index

S

T

MIX
Papier aus verantwortungsvollen Quellen
Paper from responsible sources
FSC® C105338

If you have any concerns about our products,
you can contact us on
ProductSafety@springernature.com

In case Publisher is established outside the EU,
the EU authorized representative is:
Springer Nature Customer Service Center GmbH
Europaplatz 3, 69115 Heidelberg, Germany

Printed by Libri Plureos GmbH
in Hamburg, Germany